1970

THE PERIODIC SYSTEM OF CHEMICAL ELEMENTS

A History of the First Hundred Years

Dmitri Ivanovitch Mendeleev

THE PERIODIC SYSTEM
OF CHEMICAL ELEMENTS

A History of the First Hundred Years

J. W. van SPRONSEN

Department of Chemistry, University of Utrecht (The Netherlands)

ELSEVIER

Amsterdam - London - New York - 1969

ELSEVIER PUBLISHING COMPANY
335 Jan van Galenstraat
P.O. Box 211, Amsterdam, The Netherlands

ELSEVIER PUBLISHING CO. LTD.
Barking, Essex, England

AMERICAN ELSEVIER PUBLISHING COMPANY, INC.
52 Vanderbilt Avenue
New York, N.Y. 10017

Library of Congress Card Number: 79-88080

WITH 7 TABLES, 139 ILLUSTRATIONS AND 22 PHOTOGRAPHS

STANDARD BOOK NUMBER 444-40776-6

Printed in The Netherlands

Deo optimo maximo

Preface

Although the periodic system of chemical elements has already received much attention, the history of its development has so far been discussed by a limited number of authors and only two books have appeared on the subject during the past thirty years.

The exciting recent history of the periodic system, as well as the international celebration of the centennial anniversary of its discovery to be held in 1969, led me to undertake the work now completed in the form of the present volume.

The approach to the subject has been twofold. In Part I, after a discussion of the concept underlying the approach to the periodic system, the history of classification of matter and the basis on which the system had to be built, a chronological general review is given of precursors, discovery and development of the periodic system. Part II deals with specific problems concerning the discovery of the periodic system itself and various series of elements, as well as—albeit somewhat hesistantly—priority problems.

Wherever possible, the treatment is based on all of the original authors rather than only the most prominent or most frequently cited among them. This holds particularly for the chapters on the prediction of elements and the further development of the periodic system. It is hoped that the inclusion of abundant detail has also contributed to the objectivity of my treatment.

I wish to take this opportunity of acknowledging my indebtedness to all those without whose help this book could not have been completed. The first of these is Professor A. E. van Arkel, who greatly stimulated my interest in chemistry and gave so freely of his time for many valuable discussions. I also wish to thank Professor R. J. Forbes, who as an historian gave the text a critical reading after the first version had been scrutinized by Professor R. Hooykaas.

That this book could be completed on the eve of the commemoration of the hundreth anniversary of the periodic system is also due to the generous co-operation of my employers, Professor J. Smittenberg and Dr. H. A. Cysouw of the Laboratory for Analytical Chemistry of the University of Utrecht. The temporary leave of absence from teaching granted me by the municipal authorities at Alkmaar is also gratefully acknowledged. Technical and administrative assistance was given by Mrs. I. Seeger-Wolf, Miss M. Hollander, Mrs. C. Bulterman, Miss M. Bulterman, Mrs. G. van der Kamp, Mrs. M. M. E. van der Roest and by Messrs. G.J.A. van Eyk, C. Holster, D. Hogewoning, T. Korporaal, G. B. J. Overbeek, G. Selier and F. Spee, and the assisting staff of a number of libraries, in particular the University Library of Leyden.

Finally I wish to express my gratitude to Professor E. C. Kooyman for his contribution to the general frame work, and to Dr. W. Gaade, Dr. A. J. H. Umans and Dr. J. Wolters for their constructive comments on general aspects as well as details of the final manuscript.

J. W. v. S.

Foreword

Of particular importance in the advance of science are: (1) systems of classification and (2) tools. All of the standard sciences have gone through a fact-gathering period. Necessary as such activity is, it fails to lead to a viable science with a broad appeal which attracts enthusiastic new students and investigators. Facts soon reach a point where they become less and less manageable unless an attractive and meaningful system of classification is brought into being. Such a system brings order out of chaos and reveals relationships which tend to remain obscure in a maze of unclassified bits of information.

Equally important is the role of tools in science. It is not difficult to recognize the stimulus given by the introduction of the telescope, thermometer, air pump, microscope, pneumatic trough, voltaic cell, spectroscope, discharge tube, and chromatograph. It is frequently not recognized that tools may be conceptual as well as physical. Mathematical devices such as logarithms, theories such as kinetic molecular, and laws such as the law of definite proportions, may serve as tools in the advance of science just as much as do physical tools.

The Periodic System has fulfilled both of these roles. It has served as a classificatory device but it has contributed much more than mere classification. It has been a conceptual tool which has predicted new elements, predicted unrecognized relationships, served as a corrective device, and fulfilled a unique role as a memory and organization device. The periodic table has contained an innate flexibility which has prevented it from becoming frozen into a rigid structure. It lends itself to a large variety of forms. Although many of these are unique only as schemes representative of the author's originality, certain forms have unique value in bringing out particular relationships.

Although several earlier books and a large number of papers have sought to examine the emergence of periodic classification, all of them fall short of the present volume in comprehensiveness, insight, and excellence. It is appropriate that this book should be published in the centennial year of the contributions of Dmitri Mendeleev and J. Lothar Meyer. It will serve as the definitive study for many decades into the future.

Of particular value in this volume is the careful study of the development of background information which had to be compiled before a valid system of classification could be devised. The book is, as a result, more than a history of periodic classification; it is, to a significant degree, a history of chemistry. There is also evident a recognition that chemistry did not develop in a vacuum, that it recognized and utilized classificatory principles being developed in other sciences, particularly in biology and mineralogy.

The fundamental soundness of the periodic classification is evident in the sequence of crises which it survived. At the time that the periodic concept became evident the chemistry of the rare earth elements was still in a very confused state. As additional rare earth elements were discovered, and the similarities between them became clear, it was found that they were best placed in the same spot in the table. When argon was discovered the periodic system faced a major crisis but this was soon resolved when argon and subsequently discovered rare gases formed a family of elements fitting properly into a zero group.

Problems were also encountered when radioactive elements were discovered. Resolutions of these problems came with the recognition of isotopes. The discovery of isotopes also explained the atomic weight inversions which are found in several spots in the table. The creators of the table were shown to be wise in placing more emphasis on atomic properties than upon a rigid sequence of atomic weights. Of course, the reliability of atomic weights in that day was hardly of an order to inspire confidence, so the practice of disregarding atomic weight sequence on occasion was hardly as bold a step as it might have been in a later period.

It is of interest to note that the periodic table reached its final forms before atomic structure revealed the basis for periodicity. The discovery of sub-atomic particles in no way threw the system of classification into doubt but reinforced the general decisions which had been made. It was only at the end of the table that study of electronic configuration and the creation of transuranium elements brought about a change of arrangement from transition elements to rare earth analogues.

Dr. van Spronsen's book has many values. While serving as a history of the classification of the elements, it also sheds insights into chemical relationships which are not necessarily obvious. Most important, it serves as a case study of the growth of understanding of the elements. The premature attempts at classification, the bare adequacy of atomic weights in 1869, the several approaches toward periodicity, the successful predictions, the capacity of the system to absorb new knowledge, and the flexibility of the system are brought out in such a way as to reveal the nature of scientific investigation.

Aaron J. Ihde
University of Wisconsin
Madison, Wisconsin, U.S.A.

Contents

Part II—Specific Aspects

Chapter 7. Prediction of elements

Chapter 8. Deviation from the order of increase in atomic weight

Chapter 9. The noble gases

Introduction

During the eighteenth century and the first decade of the nineteenth, a large collection of data was assembled on compounds and elements with like properties. This information, which was collected by many scientists, formed the essential basis for a systematization of elements.

There are three distinguishable main stages in the history of the periodic system, namely, that of initiation, that of phenomenological development and that of theoretical development.

It may seem surprising that Goethe should be mentioned as having been one of the principal initiators of the periodic system. As Court Counsellor to Duke Carl August of Saxony-Weimar, this famous philosopher and poet was the consultant on university affairs at Jena. Döbereiner owed his appointment as professor of chemistry at this university to Goethe. The two men became friends and influenced each other. Owing to Goethe's active interest in minerals, of which he possessed a large collection, Döbereiner began to analyze these stones exhaustively and found that the mineral celestine contained the elements forming the calcium triad. This was in 1817. For about half a century the further development remained at low level, although triads were added by Gmelin, Pettenkofer, Gladstone and Cooke; they partly extended these triads to series of four or five elements with analogous properties and a simple numerical relationship between their atomic weights.

A turning point came in 1857, with the discovery by Odling and Dumas of a so-called horizontal relationship between the elements. The actual development of the periodic system seemed to require a catalyst! We think it proper to attribute this catalytic action to Cannizzaro's famous Karlsruhe lecture at the 1860 Congress. He made the distinction between atoms and molecules and defined such concepts as valence; in our opinion this initiated the second stage of the discovery and started the history proper of the periodic system of chemical elements. It may be noted that, once the time was ripe, the periodic system of elements was discovered almost simultaneously in the most leading countries of Europe and in North America.

Nevertheless, formulation of the system was a groping process. For half a century many facts had to be accepted without explanation and a number of difficulties remained, e.g. the arrangement of the rare earth metals, the inversion of the pairs of elements tellurium–iodine and cobalt–nickel, the incorporation of the first noble gases to be discovered, i.e. argon and helium, the discovery of radioactivity and the radioactive disintegration products. Some of these difficulties were rapidly resolved. The atomic theory worked out by Bohr provided the necessary consolidation. This marked the beginning of the third stage.

After 1913 the theoretical background of the periodic system became clear, and the system assumed a new aspect. Even though before this no explanation of the periodicity could be given, the system had been accepted as fundamental to chemistry and its further development. Among the discoverers who adhered to this view were Meyer and Mendeleev. Other firm advocates of the periodic system were Brauner, Crookes, Rydberg, Thomsen, and Werner. However, a few prominent scientists, including Berthelot and Bunsen, either remained sceptical or ignored the system entirely. After 1913 confidence in the periodic system grew, the remaining gaps in the system were filled, a few more elements were discovered in nature and several others were synthetised, e.g. technetium, promethium, and astatine. The periodic system was crowned recently when element 104 was synthetized; in contrast to the 14 immediately preceding elements, this one unequivocally displayed the predicted analogies with hafnium.

Thus, less than two hundred years after Kant's pronouncement (1786) that chemistry, not being amenable to mathematical treatment, could not be included among the sciences, this discipline can and must be treated at least as mathematically as astronomy was in Kant's days. The periodic system rested on a solid experimental basis, and a theoretical basis was achieved after the atomic weight had been replaced by the atomic number that followed, for one thing, from Bohr's atomic model. Now chemistry has grown into a closed system (in the logical sense of the term) and it is the task of physics to prove the postulates used in chemistry, many of which were formulated by the periodic system.

The readers of the following history of the periodic system will, we believe, be astonished at the nonsensical hypotheses, incomprehensible errors, and faulty interpretations made, both before and after its discovery, by many investigators who did valuable work in other fields. We refer here particularly to hypotheses about the complexity of the atom and the existence of a different kind of elements, as well as the interpretation of the spectra of the sun and other stars, all with little or no experimental basis. But other investigators made ingenious predictions of elements that were discovered later, and even predicted the series of rare earth metals, as well as the group of noble gases, from a series of atomic weights (from 1 to over 200) with almost constant differences, because it was soon understood that the periodic system involved more than one form of periodicity. This was an extraordinary achievement, reached from insufficient and even contradictory experimental data. The investigators involved successfully depended on the chemical and physical properties of the elements rather then the seemingly exact criterion of the atomic weights. These great scientists, who prepared the way for, discovered, and developed the periodic system, have justly earned the admiration of posterity.

Early reviews

It seems appropriate to start this history of the periodic system of chemical elements by mentioning Venable [1], whose *The Development of the Periodic Law* appeared in 1896. Although Venable's work was accurate, his approach to the subject was almost entirely descriptive. At the end of the nineteenth century the development of the periodic system, which had then been known for a quarter of a century, was far from complete. Scientists were still engaged in discovering the noble gases and finding the correct way to classify the rare earth metals. The phenomenon of radioactivity had not yet been discovered, and no one expected that the elements missing from the system would ever be synthesised.

When Rudorf [2] published his work on the periodic system of chemical elements in 1904, the situation had already changed considerably. All the noble gases had been discovered. Research on radioactivity had yielded several new elements and even isotopes of known elements. As the significance of the latter for the periodic system was not yet clearly understood, the value of the periodic system was again called in question. Only a small part of Rudorf's book is devoted to the development of the periodic system itself; much of it concerns the periodicity of the many properties of elements, including their physical characteristics.

A history in which the phenomenon of radioactivity and the possible divisibility of the atom received considerable attention was published by Curt Schmidt [3] in 1917, but a study of the periodic system written by Rabinowitsch and Thilo [4] in 1930 refers only to its history. In 1957 Mazurs [5] published a history of the periodic system dealing chiefly with the various forms given to it.[1]

Lastly, there is Kedrov [7], who made a detailed but somewhat biased investigation into the matter of Mendeleev's share in the discovery of the periodic system. We are unable to agree with Kedrov's interpretation of the facts, since he considers Mendeleev to have been the sole discoverer of the periodic system. In the following pages we hope to show that there were six independent discoverers, among whom Meyer and Mendeleev applied the periodic system most consistently to chemistry.

References

1 F. P. VENABLE, *The Development of the Periodic Law*, Easton, 1896.
2 G. RUDORF, *Das periodische System, seine Geschichte und Bedeutung für die chemische Systematik* Hamburg-Leipzig, 1904.
3 C. SCHMIDT, *Das periodische System der chemischen Elementen*, Leipzig, 1917.
4 E. RABINOWITSCH and E. THILO, *Periodisches System, Geschichte und Theorie*, Stuttgart, 1930.
5 E. G. MAZURS, *Types of Graphic Representation of the Periodic System of Chemical Elements*, La Grange, Ill., 1957.
6 S. E. V. LEMUS, *Clasificación periódica de Mendelejew*, Guatemala, 1959.
7 B. M. KEDROV, *Filosofskii Analiz Pervych Troedov D.I. Mendeleeva o Perioditseskom Zakone (1869–1871)*, Moscow, 1959.

[1] The book *Clasificación periódica de Mendelejew*, published in 1959 by the Guatemalan Ministry of Public Education and written by Lemus [6], deals almost entirely with the elements, little attention being devoted to historical aspects.

PART I

GENERAL ASPECTS

Chapter 1

Concepts[1]

1.1. The periodic system of chemical elements

When, at the end of the seventeenth century, the word *Scheidekunst* was introduced into the German language and *scheikunde* into the Dutch language, in both cases to replace the word *Chemie* (chemistry), this deliberate change had a deeper significance than the mere translation of a foreign word. The literal meaning of both *Scheidekunst* and *scheikunde* is: art of separation. The word chemistry itself conveys nothing about the part of the natural sciences it is meant to indicate. It was in all probability derived from the Greek word $\chi\eta\mu\varepsilon\iota\alpha$ ($\chi\nu\mu\alpha$ = casting). Although the term may have been adequate for the earlier aspects of chemistry such as alchemy and applied chemistry, the chemistry of the seventeenth century was less limited. By then, the scientist was well aware that chemical reactions involved separation and decomposition. Even chemical syntheses could be explained as being, in the first instance, reactions of separation. This is the more remarkable in view of the fact that in the seventeenth century nothing was known about the nature of a chemical reaction. Although Boyle had defined the element as the final product of a chemical analysis, he was unable to demonstrate an elementary substance. Lavoisier only succeeded in doing so at the end of the eighteenth century. It was precisely because the chemical reaction was not understood that the new term was adopted. During this period science was dominated by the phlogiston theory, according to which the most spectacular chemical reaction, combustion, releases phlogiston from, for instance, metals, leaving lime (oxide). This view also explains why Boyle, who anticipated the phlogiston theory, could not choose between metal and lime for his element[2].

It is important for our study of the history of the periodic system of chemical elements to realize that in the eighteenth century the purely metaphysical elements (earth, air and fire, or sulphur, mercury and salt) were no longer taken as a basis, but that an empirical search was going on for the simplest chemical substances. A number

---

[1] References to the literature will be found in the individual chapters.

[2] It is perhaps useful to point out at this early stage that Döbereiner, the first discoverer of periodic relationships between elements, did not base his work on the simple substances themselves, but rather on their oxides. It might almost be said that the knowledge and use of elementary substances were not required for the discovery of the periodic system. That this is not true will become clear in the following chapters. Let it merely be noted here that an essential distinction between element and simple substance must be kept in mind.

of components that could not be further decomposed had already been isolated by chemical analysis; more of these substances were found during the following decades and characterized by Lavoisier; they finally numbered a hundred.

These hundred elements did not form an arbitrary collection. As early as the eighteenth century it was known that there were families of elements with analogous properties. In the first half of the nineteenth century it was found that there were certain numerical relationships between the atomic weights (the weight of an atom of an element as referred to the weight of an atom of hydrogen) of the members of these groups of elements. At the beginning of the second half of the last century, these relationships led to the construction of a system encompassing all elements. *This was the periodic system of chemical elements.*

Modern atomic theory has shown that each element is defined by a characteristic number, i.e. the nuclear charge of its atoms or the number of protons. These numbers form an unbroken series. Analogies between the properties of various elements are indicated by simple arithmetic relationships between the characteristic numbers. Consequently, it proved to have been correct to speak of a natural system of elements; this system was indeed the only comprehensive system provided by nature itself, and it was furthermore found to be susceptible of satisfactory interpretation. The atomic weight had to be replaced by the atomic number as the criterion for systematization when more progress had been made in atomic research, but the essence of the periodic system remained unaffected and a few irregularities were even eliminated. This was possible because the results of atomic research demanded a new definition of the element. In a sense, this was a retrograde step because, properly speaking, two new definitions took the place of the old one. It had been found that various kinds of atoms would have to occupy the same place in the periodic system: the number of protons and electrons of these isotopes was the same but their atomic weights differed. Later, this was accounted for on the basis of differences in the number of neutrons in the atomic nucleus. This raised the question of how to define an element: Are two isotopes having the same atomic number different elements? We shall return to this problem later on.

When an adequate number of elements was known and the many simple numerical relationships between them could no longer be denied, even after readjustment of the atomic weights, a periodicity was discovered in the arrangement. This periodicity was based on the atomic weights of the elements. After a given number of elements in the series, the properties were generally repeated. It was now possible to make a distinction between periods and columns or groups of elements, in the form of so-called "horizontal" and "vertical" relationships. In 1913, the atomic weight as a factor of classification had to be replaced, as we know, by the atomic number.

The definition now became: The periodic system[3] of chemical elements is a system in which all the elements are arranged according to increasing atomic weight and

---

[3] We prefer this name to *periodic table*, since *table* does not express the close mutual relationship that exists between the elements in their periodic classification.

elements with analogous properties occur in columns or groups. In contrast to the atomic weight, the nuclear charge of the atom of an element is always a whole number. The Daltonian atom had to be replaced by the Bohrian atom, but the classic chemical techniques could not be used to demonstrate other decomposition products of matter as elementary. The elements remained essentially the same.

The foregoing is a brief characterization of the system of elements that has become the basis of modern chemistry and forms our subject. Before embarking on this story, however, it is necessary to review the history of the various concepts concealed in this definition, firstly, in order to understand why it took relatively so long for this universal chemical system to be discovered. Secondly, we must have a clear grasp of the concepts of system, element, periodicity, analogy, and atomic weight, if we are to understand what could be sought, what was sought, and how these things were sought. Thirdly, only a historical review will explain the nomenclature associated with our subject, which has been called by such names as periodic law, periodic classification, and natural classification of elements.

1.2. The system concept

Classification is as old as science itself, because science was born as soon as man began, knowingly or unconsciously, to arrange natural objects or ideas. He did this, for example, when he assigned names to animals and things. When the Egyptian distinguished between an ibis and another bird, and designated both as birds, he was classifying. The difficulties inherent in classification immediately presented themselves when, for instance, it had to be decided whether a hippopotamus belonged to the fishes because it lives in the water and can swim, or to, say, the four-legged animals, if these were considered to form a separate class. Such difficulties are always associated with systematization and increase when greater differentiation is desired. They can be reduced to the distinction between natural and artificial classification. There is no denying the fact that more than one scientist in the history of biology applied an artificial classification that for many years was very useful for the systematic development of that science. These are historical facts, exemplified by Linnaeus and Cuvier in the 18th century.

But what first seems a mode of classification sanctioned by nature itself, can later prove to have been highly artificial when a more natural classification has been found. It is this comparative "more natural" which requires our attention.

By classifying plants according to the number and form of the pistils, petals, and the stamens of their flowers, i.e., their sexual characteristics, Linnaeus arrived at a much more general system than by classifying according to the shape of their leaves, for example. Both classifications could be called natural, however. When one system is replaced by a more satisfactory system and the new one is called natural because it expresses analogies between more characteristics, the abandoned classification should be considered to have been artificial, and the concept of natural classification proves to be relative. If this is not accepted, the search for the only natural classification must be continued, and every discarded system will appear to have been artificial. A natural system beyond dispute as we understand it, is a classification based on genetic grounds, e.g. in the animal kingdom, a system arranged according to the number of genes. Recently, serum reactions (i.e. chemical reactions) have demonstrated consanguity between different kinds of animals, which points to a genetic relationship. The theory of evolution and the evidence supporting it provided a real basis for this view. If this view is applied to inanimate matter, the idea of a primitive matter (Gr. $\pi\varrho\omega\tau o$ - $\H{\upsilon}\lambda\eta$), so recurrent in natural philosophy, cannot be rejected, at least in principle. In scientific research, the more readily hypotheses lend themselves to experimental or theoretical verification, the more stimulating they are. Admittedly, however, even purely "ad hoc" hypotheses may be useful; they can often only be verified at a much later stage of scientific progress. It is worth noting within this context that Döbereiner's announcement in 1817 of his first triad of elements was made only two years after Prout had advanced his hypothesis on the *materia prima*, in the form of hydrogen.

For a proper understanding of the historical connection between the development of the periodic system of elements and that of the systematization of inanimate matter, i.e. of inorganic substances such as minerals, it should be borne in mind that the two systems classify essentially different entities. The periodic system does not classify simple substances but rather elements, which are the chemical attributes of those substances whose chemical affinity predominates. This explains why, in the nineteenth century, after the periodic system had obtained its basis, a search was made for a mineralogical system that would include the mineral elements.

The history of the periodic system of elements could not start before an accurate idea had been formed of the chemical composition of compounds, and this required not only a unequivocal definition of the concept element but also a demonstration of the elements, practically as well as theoretically. This happened, as we shall show, at the end of the nineteenth century. But the classification of the mineral substances and of the artificially obtained inorganic products before this date, on the one hand, and the numerical relationships found between several of these compounds in the nineteenth century, on the other, undoubtedly influenced the discovery of such a close relationship between elements as exists in the periodic system. As the systematization of inorganic substances was based on Linnaeus' classification methods as well as on Aristotle's first system of natural substances, they deserve some consideration.

Aristotle made a clear distinction between two kinds of substances: the primary ones, i.e. the individuals, the specimens themselves, and the secondary ones, i.e. the species and genera to which these individuals belong. Aristotle himself fully realized that, to classify the species and genera, the properties of like individuals should be regarded as primary. And this is still so. In his *Meteorologia*, Aristotle gave more attention to the chemical substances which, according to him, could all be reduced to the four elementary substances already distinguished by Empedocles, *viz.* water, earth, air, and fire. But these elements must not be identified with the chemical substances earth, water, and air. As the philosophers wished these four elements to include all existing substances, they considered all dry, cold substances i.e., solids lacking water and not liquifiable, to be earths. The liquid substances were kinds of water, and the gases were kinds of air. Substances would therefore be combinations of two or more of these four elements.

Linnaeus created a system according to his division into classes, orders, genera, and species, which served as the basis for subsequent systems of minerals as well as for a classification of chemical substances used by Bergman and Lavoisier and based on chemical composition and binominal nomenclature. It is interesting to note that, according to his own statement, this system was not Linnaeus' most stable contribution to classification.

Whereas Linnaeus, too, often used physical properties, such as behaviour in the presence of fire and acids, as criteria, Haüy achieved a chemical, and therefore more useful, classification by making the chemical composition of the compounds primary.

It was the combination at the end of the eighteenth century of this view of matter with the definition of the element concept, and the determination of the elementary substances by newly developed analytical methods, that ultimately resulted in the discovery of the periodic system of elements. But more was required. So far, we have only considered the qualitative properties of matter. The quantitative properties, whose importance to chemistry was already recognized in the eighteenth century, led to the definition and determination of the equivalent weights of the compounds. These, together with the addition of the atomic weights of the elements in the beginning of the nineteenth century, proved to be the indispensable pillars of the periodic system.

Before going into this more deeply, some attention must be given to the historical development of another concept belonging to the definition of our subject, namely the element.

1.3. The element concept

The earliest philosophers, Indian, Greek and Chinese, were convinced that there were only a small number of simple substances in nature, to which the actually or seemingly complicated substances could be reduced, whether or not they were considered to be primary substances.

The first Greek as well as the Indian philosophers believed that there was only a single substance. Later, this number was increased, but never to more than five. It should be noted that it was not the substance itself but its properties that were regarded as "elemental". This explains why so a small number sufficed, not only in Antiquity but up to the seventeenth century.

Thales of Miletus, the first Greek philosopher to take up the problem of nature, considered water to be the sole element, as did also the earliest Indian philosophers.

Water was bearer of the properties liquidity, mobility, wetness, and coldness. Anaximenes regarded air, under which he apparently also understood wind, breath, and water vapour, as the elementary substance. To Heraclitus of Ephesus all things originated in fire and all things include fire. He ascribed all changes, origin and decay of things ($\pi\alpha\nu\tau\alpha\ \dot{\varrho}\epsilon\iota$), to this sole element, in complete contrast to Parmenides, who refused to accept real change. Empedocles believed matter to be composed not of one of these four elements ($\sigma\tau o\iota\chi\epsilon\iota\alpha$) but of all four. It was Aristotle in particular who extended this doctrine of the four elements and made it fundamental. According to him, earth, water, air, and fire are not absolute components of matter; they can be transformed into one another. The properties, combined in pairs, are the immutable, natural, fundamental qualities of matter: cold and warmth, humidity and dryness. A striking fact is that in ancient Chinese and Indian philosophy the elements do not differ much from these four. Lao Tze assumed four elements: wood, earth, metal, and water, which could be converted into one another. The Buddhist theory of elements closely resembles that of Empedocles, but the properties assigned to these elements are very different, since they are based on the impressions they produce on the respective senses. As in Aristotle's philosophy, a *quinta essentia*, a kind of ether, has been added to them. The Indian philosopher Kanada found three elements to be sufficient: earth, water, and air or heat, i.e. a limited number of components.

It may be generally said that, although the four-element theory had to make way for both the two-element (sulphur and mercury) and the three-element theories (sulphur, mercury and salt) held by the Arab alchemists and Paracelsus respectively, and, although these theories were in turn replaced by the later concept of elements, more importance continued to be attached to the properties than to the substance itself. This holds even though the opposite trend developed after chemists thought they could demonstrate truly elementary fundamental substances.

This departure from the old philosophy of elements arose from Boyle's definition of the element. Boyle had in mind final products of chemical analysis. On the one hand Boyle had disposed, although not finally, of the idea of the transmutation of substance, and on the other he had made it possible to enlarge the number of elementary substances, at least to more than four. This raised questions about the apparent simplicity of the structure of matter. Boyle himself contributed little to the investigation of the elements. He did not mention them by name and made no additions to the number of possibly elementary substances. He could not say whether the metals or the metallic oxides must be regarded as elements. The future phlogistians chose the combustion product as the elementary substance. When, in the course of the eighteenth century, the existence of a far from small number of simple substances seemed possible and Lavoisier demonstrated the 40 elementary substances then known (including some still unanalysed compounds such as soda and potash, which in fact involved undiscovered elements), the foundations of modern chemistry had been laid. Further progress had to await information about the elementary substances in their chemical compounds, again to be provided by Lavoisier, and knowledge of their specific properties in these compounds.

All this was still not enough to permit construction of a unique system of elements. Two things were still missing: a sufficiently large number of known elementary substances and a more clearly expressed quantitative relationship between the equivalent weights of the compounds.

The qualitative connection had been examined by several investigators. The result was a number of series, groups, or families of elements with similar properties, partly chemical, partly physical, but a so-called horizontal connection was still obscure. We shall make this clear in the discussion of the concept of periodicity.

After philosophical considerations of elementary substances and the *materia prima* concept had made way for chemical experimentation (in Boyle's case in theory, in Lavoisier's also in practice), the idea of a small number of primary substances had to be abandoned. This number was therefore determined experimentally. Although research steadily increased the number, it was not expected to be unlimited, for the simple reason that among the very large number of chemical substances discovered, synthetized, and analyzed, a relatively small number of elements nearly always recurred. Attempts have even been made in the present century to deduce the limit along theoretical lines.

The possibility that the elementary substances, demonstrated experimentally, would prove not to be the ultimate simplest substances, was revived by Prout's hypothesis of a *materia prima*, which he saw as hydrogen. This hypothesis continued to attract some scientists even after the discovery of the periodic system of elements. But the influence of positivism on natural sciences prevented acceptance of a hypothesis like this without experimental confirmation. After Dalton's atomic theory had been formulated, an element was described as a collection of atoms of the same kind. Furthermore, each element was characterized by a specific number, i.e. the atomic weight of identical particles. At last, the old controversy of element versus atom had been eliminated.

1.4. Periodicity

As we have already argued, the periodic system of elements could not be discovered until the specific quantitative numerical relationships between elements had been found. Because identification of these relations between elements and compounds was essential to chemistry and to our subject in particular, several sections will be devoted to them. A few remarks will be sufficient here.

In ancient times philosophers were already convinced that in nature the quantitative is as important as the qualitative and that both belong essentially to the natural order. Did not the Scriptures themselves say, "Thou (God) hast ordered all things in measure and number and weight"? Pythagoras was the first Greek philosopher to point to the importance of numerical laws in both mathematics and natural philosophy. He emphatically stated that there are regular relationships in nature, such as the one between the tones of an octave, and tried to compare them with those of other quantities such as, for example, the weights of hammers producing corresponding tones. Pythagoras' influence was so great that Plato did not admit anyone to his school of philosophy unless he knew the elements of mathematics. This is the more remarkable because Plato was not himself a mathematician.

Plato elaborated Pythagoras' ideas in his book *Timaios* and regarded the planetary distances as a series; Aristotle pointed out a relationship between two other series, one in music and one of celestial bodies.

Before any periodicity could be found with respect to the chemical elements, mathematical terms were required for the expression of such substantive quantities. Before the atomic weight had been defined and could be determined experimentally for the respective elements, any system of elements could be only a collection in which these attributes of the primary substances were arranged according to one or more qualitative chemical properties or at most a few numerical characteristics, such as the angles of crystals or density, physical properties which did not convey information about the essence of the elementary substance.

At the present stage of scientific development the only comprehensive and most logical system of inorganic nature is the periodic system. With the fewest possible objects of classification—only the elements—this system provides the utmost possible information. The structure of this system is such that two attributes undergo a periodic change: a numerical attribute, i.e. the atomic weights of the elements, and a qualitative attribute, i.e. the chemical properties.

The numerical quantity changes, though discontinuously, at fairly regular intervals and so forms an unbroken series. The chemical properties change in the same direction very discontinuously, but a repetition of these properties occurs periodically. In this way, groups of elements with analogous properties are formed. In these groups, i.e. vertically, the numerical property also shows a regular pattern.

The periodic system started to provide a natural framework, capable of accommodating new elements and their properties, even though aspects of the system left much to be desired. Thus, the system cannot say anything about nuclear processes. In principle the periodic system is a two-dimensional framework; the third dimension has never been taken very far.

A striking case in point is the important part assigned to the atomic weight. In 1913, it became clear that the atomic number must replace the atomic weight, and the isotope, the atom. However, the principle underlying this law had not yet been completely understood.

The direct historical line began with the finding of numerical and chemical relationships between a small number of elements. Later on, after the general relationship had been discerned, several of these groups proved to be identical to the vertical groups in the periodic system. A few of the groups had already been found long before, partly on qualitative grounds. When a quantitative property was taken into consideration, it was the equivalent weight of the compounds in which these elements occurred. This concept was only defined in 1802 by Fischer, but in the eighteenth century use had been made of its numerical significance for the examination of several acids, bases and salts. We shall consider these simple numerical relationships between the equivalent weights found by Richter, the most important discoverer of stoichiometry, as the basis of the numerical concept in the discovery of the periodic system. He found both arithmetical and geometrical series expressed in the weights of the bases reacting with a constant weight of acid, and conversely.

Strictly speaking, Homberg in 1699 was the first to discover a relationship between the weights of the acids reacting with a constant weight of a basic substance. His interpretation and also his measurements were, however, influenced in some degree by his theory. But a start had been made, and this work was continued in the eighteenth century by Geoffroy, Bergman, Wenzel, Kirwan, Fourcroy, Guyton de Morveau, Higgins, Richter, and Berthollet. Most of these investigators were concerned with the rule of "affinity": the substance the largest weight of which reacts with a certain weight of another substance, has the greatest affinity for that substance. This rule led to a displacement series, for which a quantitative basis was sought. The number of equivalent weights of compounds and the relationships between them finally obtained, was so large that it was difficult to deduce a simple law from them. It is noteworthy that this problem was ultimately solved by theoretical speculation: the definition of the atomic weight of the elements by Dalton after the introduction in 1803 of his atomic theory. But it is also remarkable that Higgins, who had already formulated an atomic theory in 1789 and demonstrated it from the available experimental data, did not come to the point of introducing the atomic weight, even though Kirwan, who was a friend of Higgins, had in 1783 expressed the affinity between acids and bases quantitatively. In 1789 Lavoisier published his *Traité*, which contains his oxygen theory and his definition of the element. The discovery of the periodic system was drawing nearer, but only a little, because much time would have to elapse before a sufficient number of primary chemical substances had become known.

1.5. Atomic theory; atomic weight

The atomic theory has had a long and eventful history. Although this theory influenced the development of chemistry, it lies beyond the scope of this study; many books have already been devoted to the subject. Besides, the atomic concept of matter was not a prerequisite for the discovery of the periodic system of elements.

When Democritus adopted and developed the atomic theory of his teacher Leucippus, he did so to demonstrate indivisible particles of matter. In his view, the motions of these atoms accounted for the changes in matter. But there were two difficulties here. First, the number of atoms assumed by Democritus was not limited. Second, Democritus neglected the weight of these atoms, in spite of the axiom of the conservation of mass, which he accepted. It was only Epicurus who understood the importance of this elementary atomic property. Neither Epicurus nor Democritus, however, related their atomic theory to an existing or new theory of elements.

This explains why little progress was made in applying both the atomic theory and the theory of elements of Empedocles and Aristotle to natural and artificial changes in matter. That ultimately the four-element concept was taken as a better basis to explain these changes, was due to the followers and commentators of Aristotle, whose concepts of the *minima naturalia*, as it were, created *practical* smallest particles. A commentator of great importance in the sixteenth century was Augustus Niphus, who clearly saw the difference between the theories of Democritus and Aristotle. Democritus' atoms are unchangeable; in reactions it is only their relative geometry, i.e. their change of place, which plays a part. According to Aristotle, only an inner change could explain the changed properties after a reaction. Niphus thus had a clear insight into the age-old problem of explaining chemical reactions by an atomic theory. This problem was solved only when the difference between the atom and the ion was recognized.

The revival of the corpuscular theory was started by Francis Bacon, the Renaissance scientist who disagreed with Aristotle. He also made the inductive method primary in the natural sciences. But he did little to bear out his ideas by experimental data.

The development of the atomic theory in the following years not only kept step with that of the element concept, but was also partly grafted on to it. Although the successively assumed elements were not identified with the chemically pure substances bearing the same name, which meant that they could not be used to form an experimental basis for the study of chemistry, an approach was made to the ideal relationship of element (chemically simple substance) to atom (smallest particle of this chemically simple substance).

David van Goirle, who died very young, deserves mention because he formulated his corpuscular theory before Bacon, assuming the two-element concept of Van Helmont as a basis for his atomic theory. Sennert tried to synthetize the four-element theory and Democritus' atomic theory, in order to derive a workable atomic theory capable of explaining the discontinuity of matter. Practical experience forced him to differentiate between two kinds of atoms, viz. elementary atoms and *prima mixta*, which are comparable to what we now call atoms and molecules, respectively. By considering these atoms to be three-dimensional and incorporating Galileo's concept of force, Sennert created the theory of the immutability of elementary particles and thus laid a basis for theoretical physics and chemistry. He had few followers, unlike Gassendi, his elder by twenty years, who, rejecting Aristotle, revived the theories of Epicurus and Democritus and prepared the way for, *inter alia*, Boyle and Newton (atomic mechanics of gases).

The atomic theory of Jungius might be mentioned as an intermediate link; it is not certain that Boyle was acquainted with it. Jungius, too, regarded the elements as the final products of chemical experiment. He was opposed to the *tria prima*, i.e., the three-element theory as applied by Paracelsus, and explained several chemical reactions by means of his atomic theory, including the transmutation of copper sulphate by iron into iron sulphate and copper. It is also interesting to note that he ascribed the various properties of matter to a difference in the spatial arrangement of atoms, as it were anticipating the concept of isomerism. Furthermore, he was one of the first to realize that a chemical laboratory was not complete unless its equipment included an analytical balance.

Boyle could not accept either the theory of Aristotle or that of Democritus. He, too, needed an atomic theory and an element concept suitable for chemistry. According to him, every element has its special kind of atom, though each composed of the same primary substance. The difference between atoms was due to their form, size, and pattern of motion. Boyle called a combination of smallest particles of the same kind *texture* and a combination of smallest particles of different kinds *mixture*. Both result in what we now call molecules of an element and molecules of a compound, respectively. Newton, who formulated the theory of the attraction of macroscopic bodies, did not relate this theory to his atomic theory, although he did reduce the idea of affinity between substances to forces acting between atoms. This might be seen as one of the lost chances to extend the atomic theory. Not until Coulomb applied a formula resembling Newton's law of gravity to the interaction of charged particles did Kossel use this equation in connection with the forces between ions.

Higgins, as an antiphlogistian, was the first to consider actual atoms to be the smallest particles of chemical substances; he then formulated a theory of chemical compounds based on affinity, expressed numerically. His example of the five nitrogen-oxygen compounds shows that, although these five oxides do not completely correspond to those now known, Higgins was in many respects ahead not only of Dalton, but also of Gay-Lussac and Avogadro.

If Higgins' work had become more widely known in 1789, a synthesis of the theory of the smallest particles and the modern element concept might have occurred sooner. The Daltonian atoms are basically different from the atoms of Democritus; according to Dalton, there are just as many different atoms as there are elements. Each kind of atom has its own special properties, such as weight, size, possibilities for bonding, etc. Throughout the nineteenth century Dalton's atomic model was not superseded and scientists had to make do with it. It was hardly modified at all. Classical chemical research was unable to demonstrate the incompleteness of Dalton's hypothesis. Its limitations could only be shown indirectly on the basis of several chemical and physical laws. Although the atomic theory was accepted fairly generally as a working hypothesis, not everyone was convinced of its validity, as shown by the reaction to the discovery of radioactivity.

The particles of matter resulting from this phenomenon were readily regarded as atoms of new elements. After the worst of the confusion had subsided and it had become evident that only a few new elements had been discovered, Rutherford and Bohr developed their atomic model and the phenomenon of isotopy could be defined. This is not to say that modification of Dalton's atomic theory resulted only from the discovery of radioactivity. This discovery in fact crowned the work of the spectroscopists, whose methods provided information about the internal phenomena of the Daltonian atom. We shall return to this point when discussing the influence of the techniques evolved in the nineteenth century on the systematization of elements.

The introduction of atomic weight or equivalent weight of the elements was a very important step. The term atomic weight should not be taken to imply that those who analyzed chemical substances were necessarily convinced by the atomic theory. Most of the research chemists of the nineteenth century were empiricists and were reluctant to accept hypotheses unless they could be readily confirmed by simple experiments. These scientists often preferred to use only the term "equivalent weight". Even at the end of the century there were still scientists who regarded the atomic theory with scepticism, chief among whom was Wilhelm Ostwald. He used the phase as the boundary of matter. He and the other anti-atomists can be considered as followers of Richter who, without employing a theory of the smallest particles, reduced the ratio of weights of chemical substances participating in a reaction, to equivalent weights. Dalton directly related the relative weights in which elements and compounds react with one another to the smallest particles of the substances involved in such reactions. He assumed that hydrogen was the lightest element, and related the atomic weight to it. Berzelius took the atomic weight of oxygen as his standard, but there is no fundamental difference between their approaches.

Because Dalton's assumption about the number of particles involved in a reaction rested on hypothetical grounds and could not be confirmed experimentally until the valence, concept became available, definite values for the atomic and equivalent weights could not be determined during the first half of the nineteenth century. This does not alter the fact that many scientists, of whom we shall mention only Berzelius here, performed many accurate measurements; their atomic weights were often multiples of their actual values[4]. With these values, vertical numerical relationships could be found between the atomic weights, which strengthened the conviction that elements with analogous properties belonged together in groups. But horizontal relationships between elements could not be demonstrated until the real atomic weights had been firmly established.

1.6. The affinity concept

The phenomenon of attraction of opposites interested many philosophers. Ample evidence of this is provided by the frequent mention of the contrast between heaven and earth, day and night, light and dark, land and sea, and male and female in the Bible, in pre- and neo-Socratic and Chinese philosophy. The concept of affinity was applied not only to the macrocosm but also to the microcosm. A unique example is the Yang and Yin principle of Chinese philosophy, which held from ancient times that all things contain two polar forces. Study of the affinity concept shows that, right down

---

[4] Whereas in 1817 only six correct atomic weights were available to Döbereiner, in 1826 there were 30 (Berzelius) and in 1828, 36 (Dumas).

the centuries, men were preoccupied not only with an attraction between dissimilar things, but also with a certain affinity between very similar things. The latter was also stated by Hippocrates and by Empedocles, who was the first to use the term affinity.

Stahl, too, later believed that there must be a certain similarity between substances that combine. In alchemy the classical concepts of love and hatred were used preferentially to explain the occurrence or absence of a reaction. In the phlogiston theory, phlogiston and dephlogisticated air (later hydrogen and oxygen respectively), were diametrically opposed. Newton, as we have seen, ascribed chemical attraction to forces between particles.

In the eighteenth century chemists were on the whole not too happy about the attraction concept employed by Newton. When the main emphasis was put on the concepts of affinity, its relative character became apparent. A good example of this is given by Geoffroy's affinity tables of 1718, which may be compared with displacement series of metals in which the quantity of these substances in a reaction is taken as the measure of affinity. This view was adopted by the leading chemists of the eighteenth century who, like Richter, extended it and then applied it to stoichiometry of acid-base reactions. We are here concerned with macroscopic affinity: the attraction and repulsion of atoms and the possible formation of molecules by these smallest particles became a controversial question only in the nineteenth century. This trend culminated in Berzelius' dualistic theory, stating that a chemical substance is composed of a positive and a negative part. This theory is quite consistent with Dalton's atomic theory, but it is in conflict with both Avogadro's hypothesis and such theories as Gerhardt's unitary theory of the structure of organic molecules. These hypotheses also admit bonding between atoms of like poles.

The latter concepts received a strong impetus as a result of Kekulé's demonstration in 1857 of the quadrivalence of carbon on the basis of the structures of many organic substances. This amplification of the ideas about valence—emerging from organic rather than inorganic chemistry—stimulated the development of horizontal relationships among chemical elements. But even after that, it took five more years before the periodic system was discovered. A solution of the remaining problems in the field of molecular theory was provided by Cannizzaro at the Congress held in Karlsruhe in 1860.

1.7. The analogy concept

The concept of analogy is implicit in the definition of the periodic system. Analogy —sometimes called affinity—is the basis for classification. No classification can be made until a certain relationship has been observed.

Because of the strong analogies between the chemical properties within certain series of inorganic compounds (e.g. potassium halides), some investigators in the 1850s were forced, as it were, to consider certain elements to be each other's "homologues",

sharing certain structures. Homologous series had just been found in organic chemistry, and the molecular weights of compounds showed a relationship similar to that between the atomic weights of the elements. However, the members of these organic
series were compounds, the molecules of which were composed of atoms of well-
known elements, among them carbon and hydrogen. Recognition of these analogies
was useful as long as their hypothetical character was kept in mind. From the analogy
with organic compounds, Dumas came to observe a horizontal relationship between
elements. Three of the discoverers of the periodic system, Hinrichs, De Chancourtois,
and Lothar Meyer, were willing at least to consider the divisibility of the Daltonian
atom. We shall refer to this in our discussion of Prout's hypothesis. Although the
genetic relationships between elements could not be demonstrated, several investigators nonetheless based a system on it. Those of Crookes, Preyer and Müller are
worth mentioning. The last even tried to show an analogy between the affinity of
elements and that of living creatures.

1.8. Properties

Generally the chemist is primarily interested in the chemical properties of matter,
but he does not ignore its physical properties. The essential distinction between the
physical and chemical properties of a substance is that the former concern the substance
itself, whereas the latter are expressed only in reactions with other substances. It
was the chemical properties that determined which series of analogous elements were
given a vertical arrangement in the periodic system. The periodic system is therefore a
system of elements—or rather of atoms—and not one of simple substances.

The elements occur in three physical states and polymorphism is frequent, which is
why a universal system based on physical properties is an impossibility. However, the
atomic weight—provided it be not regarded as a chemical attribute—is one physical
property that proved to be a very useful tool for systematization. This property, which
could be expressed numerically, was found to show a pattern of relationships among
elements and was taken as the basis for the periodic system, even though the series of
atomic weights turned out not to form a series of characteristic numbers. The latter
fact explains the long delay in the discovery of the periodic system. In the absence of a
characteristic numerical series it was impossible to know which atomic weights, and
therefore which elements, were missing. That the atomic weights had discrete values
became increasingly clear, however, as their determination became more accurate.
After the discovery of the periodic system, other physical properties, e.g. the atomic
spectra of the elements, became pertinent factors.

In considering the relationships between the properties of substances with respect to
classification, it must be borne in mind that only substances showing an observable
analogy in properties can be classified. But these analogies must not be so close that
species become impossible to distinguish. Although philosophers thought there was a

law of continuity in nature, exemplified by such characteristics as the velocity of atoms and molecules in gases, temperature, pressure, etc., showing an extremely wide range of variation in degree, matter itself is divided discontinuously. Atoms, once they were known, were thought to be the elementary particles of which matter is composed. Later, electrons, neutrons and protons were discovered as elementary particles, followed later by the positron and more and more kinds of mesons and hyperons having different masses.

Whether or not matter is really discontinuous, chemistry is, for the present, adequately served by a system based on both the Bohrian atom and the chemical properties of about 100 elements. It was directly after the discovery of a pre-eminently discontinuous property, valence, with its concrete values, that the periodic system was created. It had become clear that the phenomenon of analogy is most strongly expressed by elements with the same valence. The horizontally related elements also show clear analogies, for example magnesium and aluminium, potassium and thallium, barium and lead, and particularly the rare earth metals. The existence of very concrete values for the atomic weights, most of which are not whole numbers, made it difficult to find a correct formula to express a law relating the atomic weights to the properties of the elements, either vertically or horizontally. Mendeleev in fact made the condition that such a formula must exclude places *between* the elements in the periodic system.

1.9. Influences of the development of analytical techniques

Lavoisier knew only about 40 elements, several of which had not yet been isolated. When Döbereiner set up his first triad in 1817, the constituent elements calcium, strontium and barium had just been separated. The discovery of galvanic electricity and its application to the electrolysis of substances, until then considered indivisible, led to the discovery of more elements. In 1817 Döbereiner counted 48 elementary substances. In the next fifty years 10 new elements were separated either by electrolysis or classical analysis. More elements were discovered after spectral analysis had become possible. The first two—caesium and rubidium—had just been discovered by this method when, in 1862, De Chancourtois established the periodic system. He had 62 elements to work with. His colleagues knew of only one additional element, viz. indium, discovered in 1863.

The perfection of analytical methods also influenced the determination of atomic weights. Increasingly accurate determinations were made by such scientists as Berzelius, Stas, Dumas, and Marignac. That the periodic system was comparatively so late in coming was, therefore, due not so much to technical imperfections in atomic weight determinations as to the already mentioned fact that the theory of chemical bonding, based on Avogadro's hypothesis, was not unanimously accepted. Many atomic weights consequently still retained the character of equivalent weights. Faraday's discovery in 1834 of electrochemical equivalents did not help to solve this problem.

1.10. Influence of laws, hypotheses and theories

The periodic system was the product of chemistry in all its aspects, and all chemical laws influenced its discovery, e.g. the law of multiple proportions and the law of gas volumes, as well as the fundamental law of conservation of mass and Avogadro's hypothesis. The law governing the relationships between the specific heat and the atomic weight of the elements, formulated by Dulong and Petit, and Mitscherlich's law of isomorphism, exerted considerable influence. These laws were discovered in 1819, just after the first triad had been set up and had provided a basis for conclusions as to the order of magnitude of the atomic weights.

Dulong and Petit's law was further developed by Neumann, who extended it to the molecular weights of chemical compounds and who was able to correct some doubtful atomic weights by applying the new law. But Neumann's law was incapable of solving the atomic weight problem, the solution of which in any event depended on more exact figures for the specific heats. These were obtained by Regnault in 1856.

Mitscherlich's law or rule of isomorphism contributed to the determination of the correct atomic weights of what were later called the transition metals. It is noteworthy that Mendeleev preluded his chemical studies by a discussion of this rule: the thesis he prepared for his doctorate in 1855 dealt with this subject, and his magister thesis of 1856 with the specific volume of gases. Later, he often referred to this rule of isomorphism in dubious cases. Unfortunately, this rule was not always properly applied. Because of the isomorphism of potassium tellurate and potassium osmiate, Retgers believed that he could incorporate the element tellurium, which has a greater atomic weight than iodine, in the eighth group, i.e. the transition group, thus confusing the classification.

Resulting, as it did, in the theory of valence, the unitary theory of organic bonding had more influence than any other.

1.11. Prout's hypothesis

Prout's hypothesis, which stated that all elements are built up of hydrogen atoms, forms a special case because it was never confirmed, or at any rate not in the form given by its author. We have seen how early speculation began about a primary substance, a *materia prima*, (πρώτη ὕλη, ἀρχή). For the pre-Socratic thinkers Thales of Miletus, Anaximenes, and Heraclitus, this primary substance, i.e. water, air, or fire, respectively, was also the only existing element. Before that, earth had been regarded as the primary substance by Babylonian and Greek philosophers. Many later philosophers also thought that both animate and inanimate matter had originated from a single substance.

Prout stated in 1816 that the primary substance was hydrogen. He knew both the chemical elements and Dalton's atomic theory, and concluded from them that the atoms of the elements might have been built up of the hydrogen atom, but provided no argumentation supporting his hypothesis, which was even contradicted by contemporary experimental data. Such later data did not greatly impress Prout and some of his followers. Indeed, Dalton, who was the first to determine the atomic weights he had defined, was not a first-class experimentalist. Consequently, no great significance need be attached to the fact that his atomic weights deviated from a whole number, i.e. a multiple of the atomic weight of hydrogen, which equals 1. But in 1816 the atomic weights determined by Thomson and by Wollaston were also known and these, too, deviated from whole numbers. That Prout had expected many objections is evident from the fact that he published his hypothesis anonymously.

In those days, most scientists were empiricists, and in the second half of the eighteenth century many, together with Comte, became positivists. Prout's hypothesis therefore elicited less response than the author had hoped, especially as, soon after its publication, Berzelius succeeded in making more accurate determinations of atomic weights. Berzelius was completely uninterested in a quest for whole atomic weights to satisfy Prout's hypothesis and simple mutual numerical relationships. Yet some investigators, even among the predecessors of the periodic system and the discoverers themselves, did not, in principle, automatically reject Prout's hypothesis, for instance Leopold Gmelin, Pettenkofer, Dumas, De Chancourtois.

Dumas, who determined many atomic weights himself, was also a follower of Prout, although he modified the hypothesis somewhat by assuming the primary particle to be, not a whole hydrogen atom, but half or a quarter of it. He was convinced that the atomic weight of some elements was not a whole number. One of the most convincing exceptions was chlorine, the atomic weight of which was exactly between 35 and 36. Prout's hypothesis was defended for a long time because comparatively many atomic weights were found to be almost precisely a whole number. Furthermore, particularly in the period from 1850 to 1860, very many of the simple numerical relationships found between the atomic weights of elements with analogous properties also suggested that the atomic weights should be whole numbers.

Not until isotopy had been discovered and many elements had been found to consist of a mixture of isotopes in which one kind is preponderant, did it become clear that the assumption of a primary particle need not be rejected. After the discovery of protons and neutrons, these could be regarded as elementary particles. The view that the primeval atom was successively built up of these particles remained a hypothesis that became less convincing as the number of newly-discovered elementary particles, such as mesons and hyperons, increased.

At the end of the last century, interest in Prout's hypothesis revived when Crookes attempted in 1887 to deduce a genetic system of elements. His suggestion was that, in prehistorical times, the atoms of the elements could have originated from the cooling of a primary substance he called "protyle". To explain the fact that the atomic weights deviated from whole numbers, Crookes postulated so-called meta-elements. A few years later, Preyer followed up this idea and gave concrete pedigrees for the elements on solely hypothetical grounds. In 1900 Lockyer, the discoverer of solar helium, formulated a theory of the evolution of elements based on their spectra, which he related to the evolution of the stars.

Several scientists even went so far as to advance what they called a hypothesis of condensation. They required a hypothesis on the origin of the elements from primary substance to explain such phenomena as the irregularities in the progression of atomic weights (Te–I, Co–Ni, Ar–K). If Prout's hypothesis had a positive influence on the discovery of the periodic system of elements, it was because some scientists sought for better atomic weight determinations to prove the hypothesis. The results of these determinations made Prout's hypothesis improbable, but provided more accurate atomic weights to form a basis for the periodic system. At the same time, however, many of the simple numerical relationships found between the atomic weights of the elements with analogous properties, which should have supported Prout's hypothesis, were not consistent with it. After the discovery of the periodic system, much less attention was paid to Prout's hypothesis; the system could be set up without its aid. The empirically derived system could be used quite adequately without reference to a primary substance and a genetic relationship.

Chapter 2

Classification of inanimate matter

2.1. Introduction

╷In the works of Aristotle, we already find advanced classifications[1], e.g. in the field of arranging minerals, especially because of his related theory concerning the origin of stones and ores. Aristotle also originated the concept of species, which has been used so universally. It is therefore interesting to note how minerals were classified in the 18th and 19th centuries, when many of them were discovered [2, 3, 4]. The Arab scholar Geber (Dsjebir, Jabir) (721 or 722–803 or later) applied logic in his research. He subdivided the minerals into (1) spirits, substances that are evaporated completely by fire; (2) metals fusible and malleable; and (3) minerals, fusible or infusible, broken or pulverized by hammering. The contents of these classes are as follows: class 1 contains species of sulphur, arsenic (arsenic sulphides), mercury (sometimes mentioned with the metals), sal-ammoniac[2], and camphor; class 2 comprises lead, tin, gold, silver, copper and iron; and class 3 is subdivided into minerals containing (a) some spirit, though having the form of a solid substance, e.g. malachite, turquoise, and mica; (b) a small amount of spirit: shells, pearls, vitriols; and (c) hardly any spirit: onyx, sand, and weathered vitriols.

Other Arab scholars were also interested in classifying various forms of matter. Their distinctions for classes and groups were of course based mainly on physical properties and not on chemical composition. Fundamentally, in his *Book on the Secrets of the Secrets*, Al Razi [5] (860–925) followed Geber. His three classes are animal matter, vegetable matter, and earth matter. This last class contains six groups: (1) fourteen spirits including mercury, sulphur, and sal-ammoniac, (2) the seven metals, (3) stones, (4) five vitriols, (5) boraxes, and (6) salts. In his *Book of Medicines*, Avicenna (980–1037) classified the minerals into (1) stones, (2) fusible matter, being mercury and the other metals, (3) sulphurs (sulphur and some sulphides), and (4) salts[3].

The classification by Agricola (Georg Bauer) (1494–1555) shows the Arabian influence very clearly. This mining engineer classified the minerals as follows: (1) earths, (2) stones, (2a) rocky stones, (2b) precious stones, (3) *solid liquids* (salts), (4) metals, and (5) compounds such as pyrites and lead glance.

---

[1] Forbes [1] also described Chinese and other mineral classifications.
[2] Geber, who mentioned this substance for the first time, called it *nuschādir*.
[3] Used, of course, in the old sense of soluble mineral substances; to Avicenna, these were common salt, alum, vitriols, and sal-ammoniac.

References p. 40

Becher [6] (1635–1682), who broadened the phlogiston theory, tried to apply a chemical classification and divided the unworked minerals, together with the prepared chemicals, such as nitric acid and sulphuric acid, into nine groups. The metals gold, silver, copper, iron, tin, lead and mercury are included in the second group.

With the passage of time, more substances were discovered and more and more specification was required for their classification. Of the systems set up in past centuries, the classification of plants, to which Linnaeus had made the most important contribution, stands supreme. We should also mention the arrangement for zoology, with which the name of Cuvier is inseparably connected. The classification of minerals was originally placed on the same morphological basis. Mineralogy belonged to the natural history sciences, but in the nineteenth century this branch of science was drawn into the domain of chemistry, which cleared the way for a classification of minerals based on their chemical composition.

We must now distinguish between natural and artificial classification, even though it is sometimes difficult to demarcate them. In biology, systematization was long purely morphological and could therefore lead to a natural classification of animals and plants. In the kingdom of inanimate matter, however, the morphological classification of minerals did not lead to a natural system; systematization remained artificial until the chemical properties of substances were considered. This way of thinking in turn led to the ultimate discovery of the periodic system of elements as a natural classification.

The qualities of the elements were compared with one another just as in the former philosophy of elements. Now, however, these qualities had acquired a material basis in the atomic theory, which furthermore introduced a quite new conception of element.

2.2. Discovery of a periodic classification of chemical elements

The history of the discovery of the periodic system spanned only a little more than seven years, but was preceded by a long preliminary phase. This phase falls into two periods. The second period opened in 1817 with the discovery of the triads of elements by Döbereiner. The first period represents the history of systematization or classification in chemistry, which continued after 1817, but it also concerns the history of the numerical relationships between chemical compounds, especially salts, acids and bases, and later between elements. These two historical developments took a generally parallel course, but a few points of contact are observable.

2.3. Classification by Boyle and systematization in the eighteenth century

Boyle [7, 8] already realized that it was best to classify the chemical substances according to their chemical properties, and also their composition, rather than according to such physical properties as viscosity and volatility, as most of his predecessors and contemporaries had done. To arrive at a classification, Boyle made use of the classic terms class, genus and species, but often as relative concepts. He generally classed metal as a genus or gender and, for instance, silver and gold as species, but when he considered the metals with respect to the *mixed bodies*, i.e. the compounds, Boyle called them species relative to the so-called *fossilia*. His classes or categories are formed by acids, bases and neutral substances, each of which is a family of substances with analogous properties. With this relative classification Boyle in a certain sense followed Aristotle, who applied his classification primarily to the animal kingdom[4].

Linnaeus' classification of inanimate matter is principally artificial and derived from external appearance, i.e. the nature of species. He took the crystalline form as his starting-point. Buffon [10], as a naturalist, sought a natural classification of minerals. His system already included salts, acids and bases as substances classified according to their chemical properties, i.e. qualities. Chemistry itself seems to be susceptible of more than one form of systematization. A consideration of Linnaeus' work and influence is imperative in this connection [11]. Not only the story of systematization, but also that of nomenclature begins with the work of this scientist.

In 1753 Carolus Linnaeus (1707–1778) first published [12] his system of plants as part of the tenth edition of his famous *Systema Naturae* [13], which appeared in 1758–59. The system became an influential example for the description and classification of the vegetable world. Linnaeus divided the plants into classes, which in turn were subdivided into orders, these again into families (genera) and these into kinds (species). Thus every plant acquired a double name: its family (generic) name and the corresponding specific name. Linnaeus' studies of the mineral kingdom enabled him [14], as early as 1756, to produce a systematic classification that differs greatly from the one he published later [15]. The main divisions differ little from the classic pattern. The three classes of his first division are: (1) petrae, (2) minerae and (3) fossilia. Each of these is subdivided into three orders. The chemical elements are placed in the second and third orders of the second (= mineral) class, i.e. the sulphura and mercuralia, respectively. The second order consists of four families. The third family, that of the pyrites, includes sulphur and some sulphides, seven kinds of all. The fourth family contains arsenic and some arsenides. The classification in this order was made according to colour and smell, in the third order according to the behaviour of the substances vis-a-vis fire and their solubility in, for instance, nitric acid and aqua regia.

---

[4] Aristotle [9] made a distinction between primary substances (πρώτη οὐσία), which were individuals, specimens, and secondary substances (δεύτερα οὐσία), which were the kinds to which the individuals belong; for example, the kind (species) of man, but also the family to which this kind belongs, i.e., the kind (genus) of animals. Thus, man and animal are secondary substances, but a specific animal is a primary substance. Since by γένος Aristotle sometimes meant the kind and sometimes the class, the translator expressed the distinction by using the word species for the former and class or group for the latter. Therefore, the classes of birds and fishes include certain species of these. Some species can be subdivided into groups; e.g., that of man and lion into a group of wild and a group of tame individuals. For the subdivision of the γένος Aristotle used the concept εἶδος, which in view of the above, must be translated as species or as group (genus).

Here, too, we are concerned with an artificial system. The metals deserve special attention. These elements, with their compounds, are the species of the ten families of the third order. In his subsequent systems Linnaeus incorporated the metals together with their compounds into a separate class, the fourth. Linnaeus was a true product of

Carolus Linnaeus in Lapponian dress
[Copyright Rijksmuseum voor de Geschiedenis der Natuurwetenschappen, Leiden]

his age, a time in which the phlogiston theory still held sway. This is evident from the third class, that of the phlogista. The seven families of this class consist exclusively of the modifications of carbon, such as charcoal and soot, and the kinds of sulphur and sulphurous substances.

It is clear that this system does not yet show any of the characteristics of a natural classification and was therefore doomed to failure. But the first great step had been taken. The scientist who could be expected to compose a better classification and to provide a nomenclature of chemical substances was Torbern Bergman (1735-1784), Linnaeus' countryman, student and future colleague. Bergman's special interest was chemistry. He chose a binominal basis for his Latin nomenclature [16]. Bergman rightly considered nomenclature and classification to be equally important. His improved mineral system entitled *Meditationes de Systemate Fossilium Naturale* [17], which appeared in 1784, the year of his death, bears distinct traces of the influence of his great countryman, but its arrangement of acids, bases and salts is more effective. For Bergman was aware of the fact that salts are built up of acids and bases, and made the base the genus and the salt itself the species; a self-explanatory example is nitrosum argentatum. The result is a natural classification such as Buffon [18] also favoured, but Buffon's nomenclature reversed the classes and orders. He included the metals in the fifth order, which consists of four classes arranged according to both the physical and chemical properties of elements. In his chemical studies Buffon was influenced [19] by his friend Guyton de Morveau. Bergman died not long after Guyton de Morveau had approached him about a new chemical nomenclature. After Bergman's death, Guyton de Morveau applied to Lavoisier, with the result that they, together with Berthollet and Fourcroy, published the *Méthode de Nomenclature chimique* in 1787.

Torbern Bergman   *Jeremia Benjamin Richter*

More or less related to Bergman's work is that of René Just Haüy (1743–1822). This mineralogist did not concern himself with *chimie pure*, but, in studying and arranging the minerals, he adopted their chemical composition as his guide. As a consequence, he removed mineralogy from the sciences comprising natural history, which on the one hand cleared the way for a mineral classification based on properties and composition, to the advantage of chemistry, but on the other hand limited systematization, because the contemporary standard of chemical analyses was not yet high enough to provide an adequate amount of data.

Of Haüy's many publications, we mention the one written in 1801 and entitled *Traité de Minéralogie [20, 21]*, bringing us into the nineteenth century. Lavoisier's point of view greatly influenced Haüy, who had begun his investigations in 1783. When summing up the chemical compounds, Lavoisier [22] took the reaction between acids and bases as his starting-point. As a result of this innovation he had need of a nomenclature, which he found in a logical terminology for the reaction products—1152 salts derived from 24 bases and 48 acids—which in turn gave him a certain system of compounds [23]. His work may therefore be called a consistent extension of the studies of Bergman and Linnaeus, but with the incorporation of the results of his own oxygen theory.

Thus, salts are always composed of two substances, as the names given by Haüy indicate, for instance, *sulphate de baryte*. Acids, oxides, etc., which also have binary names, e.g., *acide sulfurique, l'oxide de plomb*, consist of two substances. After the approved manner of the time, Lavoisier classified the bases (carbonates) according to their affinity with the corresponding acids. Affinity was understood to mean the weight needed for neutralization. The order, therefore, became roughly: barytes, potash, soda, lime, magnesia, and so on (see further *2.5*).

He classified the elements—or, to be more exact, the simple substances left after analysis—into four groups. This division is based on the metallic or non-metallic nature of substances, the oxidizability and the formation of acids playing the dominant role [24].

2.4. Systems of the nineteenth century

To deal with this subject we must return to Haüy [25]. Like Linnaeus, he divided the mineral kingdom into four classes, to only the last class of which both scientists assigned the same kind of substances, viz., metals. As a mineralogist, however, Haüy made more interesting subdivisions of the other classes than Linnaeus. This is immediately clear from the titles he gave his classes and from the contents of the species and genera. Here, for a start, is a chart illustrating Haüy's system of 4 classes, 8 orders, 29 genera, and 141 species.

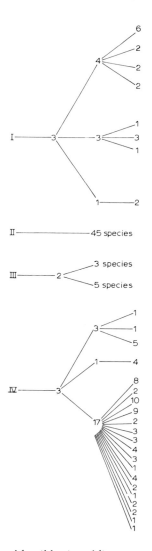

The first class includes the *substances acidifères composées d'un acide uni à une terre ou à un alcali, et quelquefois à l'un et à l'autre*. The orders are successively the alkalis, the earths, ammonia, and the salts of the light metals. The first order contains the product of an acid and an alkali or an earth; one of its genera is lime, which again contains the species of calcium carbonate, calcium phosphate, etc. The teaching of Lavoisier is immediately recognizable here, even in the nomenclature.

As can be seen, Haüy used genus as a collective noun indicating the individuals grouped as a species. In the example mentioned above, that collective noun is lime; the individuals here are a number of compounds of very similar chemical composition, form and other properties, collected under the name of their species, e.g., *chaux carbonatée*. The second class comprises the *substances terreuses, dans la composition desquelles il n'entre que des terres, unies quelquefois avec un alcali* and has only one subdivision into species, e.g., the minerals of quartz, spinel, etc. The third class holds the *substances combustibles non métalliques*, e.g., sulphur, diamond, anthracite, bitumen, while the fourth class contains the *substances métalliques*, by which Haüy apparently meant the heavier metals with their salts. As the classification of metals is the most interesting aspect of his system in terms of our study of the systematization of elements, we shall discuss it in some detail.

This fourth class was divided into three orders for the following reasons. The first order contains the metals *non oxydables immédiatement, si ce n'est à feu très violent, et réductibles immédiatement*, which are only Pt, Au, and Ag. The second order includes the metals *oxydables et réductibles immédiatement* with only Hg as a genus, while the third order comprises the metals *oxydables, mais non réductibles immédiatement*, namely Pb, Ni, Cu, Fe, Sn, Zn, Bi, Co, As, Mn, Sb, U, Mo, Ti, W, Te and Cr. Most of the metals of this fourth class were also included[5] by Lavoisier [26] ,who placed them in one of the four groups to which he assigned the *substances simples*, namely in the group of the simple substances that are *métalliques, oxydables et acidifiables*. In the genera of the metals, the species in many cases comprise the pure metal and its compounds [27].

[5] It should be noted that Lavoisier also classified light and heat *(caloric)* as elements, the latter as a consequence of the phlogiston theory. He incorporated the elements chlorine, fluorine and boron, not yet isolated, into his classification, namely with the non-metals.

In 1807 Thomson [28] divided the chemical substances into two classes of salts, which he in turn subdivided into orders, genera and species. He had previously classified the elements as metals, which he subdivided according to specific weight, hardness and melting-point, and non-metals, arranged according to density [29].

Black [30] divided chemical substances into five classes: salts, earths, combustible substances, metallic substances and waters. The class of metallic substances, for instance, comprised 15 genera, all of them elementary metals.

One of the systems [31, 32] of Berzelius, namely that of 1815, included only those simple substances that in nature are found in their native state. They are placed in the three orders that form a subdivision of the single class of inorganic substances. The first order, comprising the non-metals, includes six families; the family of sulphur has among its species sulphur itself and the family of carbonium, diamond. The second order—with the electronegative metals—is also subdivided into six families and comprises arsenic and antimony. The electropositive metals in the third order are split up into two subdivisions. Only the first subdivision contains metals occurring in nature in the free state, namely Pt, Au, Hg, Pd, Ag, Bi, Pb, Cu and Fe.

The discovery of isomorphism [33] forced Berzelius in 1824 to modify his system. Exactly as in Rammelsberg's 1847 system [34], the compounds were now arranged according to the acid radical and not the element itself. The notion of species had thus become, chemically speaking, somewhat weaker. Many double compounds were included, and the morphological properties received more emphasis, which in this case represents a backward step in the direction of an artificial classification.

2.5. Numerical relationships [35]

The second trend in the historical development, that towards the numerical relationships, began in the last year of the seventeenth century, when Wilhelm Homberg (1652–1715), a German born in the Dutch East Indies, found that a constant weight of salts was formed when he used various acids to neutralize a certain weight of potassium carbonate [36]. Although Homberg was a theoretician who adapted his experiments to his hypotheses, he nevertheless established the foundations of stoichiometry in 1699. He established that the weight of an acid required to neutralize a certain weight of different alkalis is a measure of the *passive force* of these alkalis. A notable fact is that in Homberg's work an anticipation can be seen of the OH-ions of Arrhenius' acid-base theory. Homberg attributed the basic character of the alkalis to *des particules ignées*.

In 1718 Etienne François Geoffroy (1672–1731) published his *Table des différents Rapports observés en Chimie entre différentes Substances* [37], which already included a displacement series of several metals. By *Rapports* Geoffroy meant the affinity of a given substance for another. Thus, the affinity for hydrochloric acid decreases in the series of elements: tin, antimony, copper, silver, mercury. This view was propagated by authors as didactic as Macquer [38].

We must also include Bergman in this discussion. Besides his systematic work, he also considered the idea of affinity in connection with numerical relationships, as shown by his publication *De Attractionibus electivis*, which appeared in 1775.

Very important work was done by Carl Friedrich Wenzel [39] (1740–1793). Quantitative experiments of the utmost precision revealed that a reaction between acids and bases involves very definite ratios, and that the rate of dissolution of metals in acids is proportionate to the concentration. He is consequently considered to be the precursor of Berthollet, who formulated the theory of mass action. Wenzel's book *Lehre von der Verwandtschaft der Körper* had been published in 1777, only a few years before the publication of Lavoisier's *Traité*, after which Jeremia Benjamin Richter (1762–1807) made his research results known to the world.

Following his dissertation entitled *De Usu Matheseos in Chemia*[6], in which he applied mathematics to chemistry as early as 1789, he published between 1792 and 1794 the triple *Anfangsgründe der Stöchiometry oder Messkunst chymischer Elemente* [41] in which he formulated his stoichiometric laws. Around the same time his *Ueber die neuern Gegenstände der Chymie* (in eleven volumes) also appeared. In this book he first reported the remarkable results of his attempts to find numerical relationships. He had discerned two kinds of series, viz., arithmetical series representing the relations between the weights of bases that neutralize a certain weight of acid, and geometrical series representing the weights of the acids that neutralize a certain weight of base. An example of the first series is the ratio of the weights of bases of aluminium, magnesium, calcium, beryllium, strontium and barium:

$$a : (a + b) : (a + 3b) : (a + 6b) : (a + 9b) : (a + 19b).$$

The geometrical series $c : cd^3 : cd^5 : cd^7$ applied to acids, i.e., hydrofluoric acid, hydrochloric acid, sulphuric acid and nitric acid. Ernst Gottfried Fischer [42] worked from these relationships, to which he made important additions in 1803 (see also p. 43).

Thus, we see the number entering chemistry little by little[7]. To Richter, this appeared to be the equivalent weight. Some years later, partly on the basis of stoichiometric laws, Dalton created his atomic theory in which the atomic weight functions as number. In the same period Proust formulated his law of the constant composition of compounds, in which the ratio of the weights of the elements reacting together is expressed in numbers [43, 44]. In further efforts during the nineteenth century to systematize chemical substances and elements, numbers could therefore be employed. In fact, it

[6] This work may be considered a reply to a pronouncement by Immanuël Kant, who had taught him mathematics at the University of Königsberg. Kant had refused to class chemistry with the natural sciences because it was not susceptible of mathematical treatment [40], and characterized chemistry as a systematic art.

[7] The great importance of numbers was already understood in ancient times. Aristotle said: "Things are numbers." The Pythagorean school attributed a dominant significance to numbers and proportions (see p. 13). We have seen that a similar judgment can be found in the Bible: in the *Book of Wisdom* (which Protestants consider apocryphal) we read in Ch. 11, verse 20: "......Thou [God] hast ordered all things in measure and number and weight", a line which Richter adopted as his motto.

would have been impossible to devise a periodic system before this time. Now that mathematics had entered chemistry, mineralogy, being closely related to this discipline, ceased to be a part of natural history.

To complete our review of scientists who investigated the mutual relationships between chemical compounds, we must certainly mention Richard Kirwan (1750–1812), who developed the theory of affinity according to numerical rules. In 1783 he gave some striking examples [44], a few of which are given below in modern notation. They represent the weights of a basic substance that must be added to neutralize 100 grams of a certain acid (in a solution of a given strength):

	K_2CO_3	Na_2CO_3	CaO	NH_3	MgO	Al_2O_3
H_2SO_4	215	165	110	90	80	75
HNO_3	215	165	96	87	75	65
HCl	215	158	89	89	71	55

The affinity of a base for an acid is, according to Kirwan, proportional to these weights[8].

Berzelius who, as we have already seen, took a leading part in the systematization of chemical substances, can undoubtedly be called the greatest inorganic chemist of the first half of the nineteenth century. It is therefore all the more surprising that he himself never attempted to find a closer relationship between simple substances, and never showed any interest in the investigations of those who did. This was perhaps due to the fact that Berzelius saw so much resemblance between the mineralogical systems and those of botany and zoology, to which mathematical relations could not be applied. The relations between numerical constants of simple substances found by his contemporaries must have seemed irrelevant to him. Before considering these new systems, however, we must briefly examine the purely arithmetical aspect of the problem.

2.6. Significance of numbers

Whewell [45] made a distinction between "necessary" truth and "contingent" or "experimental" truth. The former is true in itself and arises from the idea; the latter is inherent in a thing. An example of a necessary truth is a mathematical equality; as an instance of an experimental truth we may take, suitable to our context, a triad such as the one observed first by Döbereiner: the atomic weight of strontium (i.e. one of its

---

[8] Curiously enough, these affinities run parallel to the base strengths. The above series are no more than ratios of the equivalent weights. Kirwan himself did not mention this point, although he was a friend of William Higgins, whose work anticipated Dalton's atomic theory. These equivalent weights run parallel to the atomic weight of the metals, which in turn parallel the ionization potentials of the atoms on which the base strengths largely depend [35].

properties) is roughly the average of those of calcium and barium[9]. We can represent this relationship by symbols as follows: $(Ca + Ba) : 2 \sim Sr$. The equation $(Ca + Ba) : 2 = Sr$ approaches the truth (see *4.2*). A necessary truth, inseparably bound up with this relation, as will be shown below, is found in the equality $(20 + 56) : 2 = 38$. The use of these numbers is justified by the fact that this equality was later found to be an experimental truth. After the development of the periodic system, these numbers acquired the following meaning: the element strontium has the order number 38, which indicates a location equally distant from calcium (number 20) and barium (number 56). The explanation of this was provided by nuclear physical experiments: a strontium atom has 38 protons in its nucleus and the other elements also have as many protons as their order numbers indicate.

There would seem to be little reason why this experimental relationship should not be accepted as a necessary truth as well, seeing that $(20 + 56) : 2$ is always 38, whether these numbers refer to a place in the system or to the number of protons. But we need to be very cautious here. Has it not been found, for example, that the mass of the two protons in the nucleus of an atom of helium is not equal to the sum of the masses of the two protons that form this element in a nuclear synthesis? We should also allow for the possibility that the law of the conservation of matter may prove to be invalid. Evidently, therefore, an essential distinction should be made between the number itself, as a constant, and the number belonging to a physical or chemical constant, even though the latter would seem to be additive; or, to be more precise, the extent to which the abstract system in which $1 + 1 = 2$ is applicable to concrete cases must always be verified. Although this difficulty was not felt at the beginning of the nineteenth century, others did then exist. The equivalent weight and the atomic weight excluded relationships of whole numbers. This phenomenon was later explained by the discovery of isotopy.

One might be inclined to conclude that the search for a system expressing periodicity forced every investigator towards an experimental system based on the triads of Döbereiner and the elementary relationships found by Gmelin, Pettenkofer, Dumas and others. And this actually was so in the case of most of those who discovered the periodic system; but the mineralogist Béguyer de Chancourtois, the first to do so, held mathematics (in his terms mathematics being simple numerical relationships represented by graphs) to be primary (see *5.2*). A system of elements[10] cannot be based on simple numerical relationship[11], because the result would be too artificial. Nevertheless, a natural system makes it possible to take such numerical constants as the equivalent weight, atomic weight and molecular weight, as well as the affinity expressed in numerical values, by way of a guide, however inexact they may be mathematically.

---

[9] It is a striking fact that Döbereiner was so impressed by this relationship that he thought strontium might possibly be seen as a mixture of calcium and barium (see *4.2*).

[10] Unless certain isotopes are classified instead of elements.

[11] This holds only if the *number* of protons (or electrons) in the atoms is taken into account and no attention is paid to their mutual bonds.

Some constants, such as the highest valence of the elements and the lengths of the periods in the periodic system, can be expressed—keeping in mind the actual possibilities—in simple arithmetical relationships[12].

At the beginning of the nineteenth century, therefore, we see two main tendencies. Some investigators, among them Haüy and Berzelius, proposed to classify the inorganic compounds, and a few applied themselves to the organic compounds as well. These scientists—partly because they were mineralogists—started with the raw materials found in nature. It is hardly surprising that the systems they produced show considerable kinship with those evolved in botany and zoology, sciences that also systematize individuals found in nature. We have already seen that scientists who classified chemical compounds, among whom Mohs and Naumann, also used their composition and thus strengthened the resulting systems. Naumann used both the common chemical properties and the external resemblances of minerals resulting from the molecular form. Some of these scientists attempted to collaborate in classifying simple substances. On the one hand, this was easier to do than classifying compounds, the number of simple substances being, of course, smaller than the number of compounds, and their properties therefore less divergent. On the other hand, the properties of simple chemical substances do not form a basis for a classification, as will be shown in 3.7. Thus it turned out to be necessary to classify elements, i.e., the simple substances as they occur in their compounds, in which form their analogous properties could be more easily recognized. The deliberate use of numerical data simplified this problem.

The ultimate result was a system including all the elements and based on the periodicity of their properties. This could not have been accomplished before the numerical constants discussed above, and particularly the atomic weight, had been unequivocally defined and determined. As we shall see in the next chapter, this point could not be reached until after 1860.

2.7. The periodic system as a natural classification[12]

Now that we have followed the two historical phases, one concerning the classification of compounds and simple substances, the other numerical relationships between compounds, up to the beginning of the nineteenth century, we must try to ascertain whether these phases merged at any point. The story of the discovery of the periodic system shows us, however, that these two phases generally followed parallel courses. As we have seen, Berzelius, in spite of being well informed about Richter's results, denied, or considered accidental, any numerical relationship between simple substances, perhaps because of a tendency to avoid any conflict with his electrochemical theory. Just before his death, Berzelius classified the elements according to electrochemical

---

[12] See also 3.6.

properties [46, 47], obviously not having succumbed to Ampère's system based on colour and volatility, i.e. purely physical properties, no matter how attractive it seemed [48].

In fact, only the work of Leopold Gmelin can be considered to represent a synthesis of the two historical lines. In 1825 Gmelin published two versions of a system of chemically pure substances [49, 50]. The second and most interesting version has 9 classes, comprising water, oxygen, fluorine, chlorine, selenium, sulphur, carbon, the metals and the organic compounds. Gmelin deliberately chose the acid radicals as his main systemic factor (Fourcroy [51] did the same). Just two years later, even before Döbereiner had published his further investigation of the triads, Gmelin discovered numerical relationships connected with the *inner essence* of the elements, to which he later added others (see *4.3*).

If we wish to consider the periodic system of elements mainly as a classification—the other classifications already mentioned have been discussed exhaustively by Whewell [52, 53]—we must first examine its status. In the preceding sections we could speak of substances that had been incorporated into a system. This is not so where the periodic system is concerned. The scientists of that time, and above all Mendeleev, understood that simple substances were difficult to classify. This was due in part to the phenomenon of polymorphism, which compelled Haüy [54, 20] to consider individuals as different species, due to their differences in form, but mainly to the dissimilar physical properties of these simple substances. The group of halogens forms a striking instance of this. The chemical properties have many more points in common, however, and were therefore considered primary in the search for relationships. These chemical properties even came to be seen apart from the simple substance itself. But the essential factor continued to be the properties these substances presented when combined with another simple substance or radical, in other words what would later be called the ions, whether in a solid state or in solution. In this bound state the simple substance was called an element. And these elements can be incorporated into a periodic system. From this point of view it is striking that Döbereiner formed his first triad with the oxides of the three alkaline earth metals, viz., calcium, strontium and barium.

The periodic system cannot be equated with a natural-historical classification, because individuals (substances) are not classified in the periodic system. To draw a parallel between the two, the elements must be taken as abstract species, each species representing only one element. The groups of elements would then be the genera and the whole system would form the order of elements, a subdivision of the class of the chemical substances. The result tends to be a theoretical chemico-physical classification based on numerical relationships, though not simple ones where the atomic weights are concerned. The quantitative properties of the elements play a decisive part here, in contrast to a natural-historical classification in which only qualitative properties may be used.

In this context, Whewell's criticism [55] of Berzelius' mineral system becomes understandable. This philosopher objected to the fact that Berzelius considered both sulphur and the metal sulphides to be species of the same genus; the compounds of sulphur should be compared with analogous compounds of other elements. On the other hand, Whewell thought that sulphur, selenium and phosphorus belong together. Although his criticism in effect opened new perspectives for scientists searching for a system of elements, it was not justified. Whewell did not discuss the point further, even though in 1858, the year in which his study appeared, the foundation of this system was completed. Berzelius, Haüy, and others who had constructed a system of chemical substances, classified minerals, i.e. substances occurring in nature in a free state. Selenium and, of course, phosphorus do not belong to this category. Of the non-metals Berzelius included only sulphur, two modifications of carbon, i.e. diamond and anthracite, and arsenic[13]. In Haüy's system we find tellurium as well[14]. Linnaeus [15] assigned the species of plumbago[15], carbo (carbon), and soot to the class of *phlogista* in the genus of graphite, in which only simple substances occur, while one of the six species of the genus sulphur is pure sulphur (nativum). He, too, classed arsenic among the metals. The other metals classified by these three scientists have already been mentioned in *2.4*.

If Whewell himself, however, had been convinced that a chemical system was possible, as he claimed, he would not have treated chemistry separately in his book—as an analytical science—and he would have sought more analogies with mineralogy. Actually, Whewell never went further than a division into metals and non-metals [56].

To discuss the concept of natural classification, we must examine certain problems in more detail. We must know which of the concepts axiom, hypothesis, law, definition, or thesis, we have to associate with natural classification. After the consideration in *2.6*, it will be clear that the periodic system, which encompasses the triads named in that section as well as other triads, quartads, etc., is a collection of experimental truths. We need not consider the purely mathematical ideas of axiom and thesis, and we can also eliminate the hypothesis if we are not restricted to simple purely mathematical relationships. (Does nature ever satisfy exact simple mathematical relationships?) It remains to choose between definition and law, or to relate both concepts to the periodic system of elements. In fact, these two concepts are not mutually exclusive. What we are concerned with is an experimental truth (contingent truth) containing a *law*: a natural law or periodic law; a natural law inferred from experimental facts, from which other experimentally verifiable facts can be deduced. The English were therefore right to choose the term "periodic law". De Chancourtois, Mendeleev, and Hinrichs were also right to designate their systems "natural systems". Years before his discovery of the periodic system or classification, in 1857, Odling sought a natural classification on the basis of the existing natural families (see *4.12*). For that matter, the natural systems in botany and zoology set up by Linnaeus and Cuvier, respectively, were generally known and were even used in secondary schools. In a manner of speaking, children

---

[13] These investigators, however, considered arsenic to be a metal.
[14] Tellurium was considered by Haüy to be a metal.
[15] A combination of carbon and lead.

were brought up on them. The arithmetical relationships do not exclude use of the concept *natural*; on the contrary, they make it possible to give a clear definition of the system in which the properties of the elements[16], their atomic weight (later atomic number) and periodicity are the dominant factors.

Mendeleev stated that this natural law recognizes no exceptions, and he therefore did not hesitate—in contrast to Meyer who was more cautious—to make bold alterations, for instance in atomic weights, and to formulate predictions. That he in particular found the tellurium–iodine and other inversions indigestible, need hardly be said.

It may be mentioned here that the law of constant proportions of simple substances in compounds provides us with a good example for comparison. Haüy formulated this law, on the basis of data collected by the chemical analysis of minerals, long before Proust had stated his law but after Romé de L'Isle had assumed the same law on the strength of the crystal theory *[21]*, developed by himself. Later, however, Haüy used the law to define the species *[57]*. Whether this law should really be considered a hypothesis, is a point we shall not go into. To do so would require consideration of the controversy between Proust and Berthollet *[21, 58]*.

Another law propounded at about this time, Gay-Lussac's law of combining volumes of gases, might be a better model for comparison, because it is entirely experimental in character.

The periodic system is an experimental law to the extent that it is not based on atomic theory, which does not mean that the atomic weight may not serve as one of its main pillars. For even without the acceptance of concrete atoms, this quantity can be defined as well as determined experimentally.

2.8. *Conclusion*

Logical systems cannot be achieved unless the chemical properties and the composition of substances are taken as a basis. This first became apparent in the eighteenth century. To obtain the most compact chemical system possible, however, elements had to be defined as the simplest materials not susceptible to further decomposition. This was done at the end of the eighteenth century, but in the course of that century it became clear, on practical grounds, that some relationships between chemical substances could be resolved into displacement series (affinities), especially those of metals, which were immediately placed on a quantitative basis according to the weights of reacting substances. This development ran parallel to the systematization of minerals. Not until the composition of salts was understood, and elements, as elementary parts of salts,

---

[16] One difficulty here is the place to be given to the noble gas of helium. If the elements are classified according to the structure of the electron orbits of the atoms, viz., s, p, d and f, helium with its s^2-structure would fall into the series of the alkaline earths, to which it does not belong on the basis of its physical and chemical properties.

References p. 40

were mutually compared on the basis of numerical relationships between equivalent and atomic weights, was the foundation laid for a periodic classification of elements. The further development of the systematization of minerals exercised no influence on the prehistory of the periodic system of elements. Although some scientists might account for the composition of elements in metaphysical terms, no further division of the Daltonian atoms could be demonstrated, however, and the relationships between the properties of elements that ultimately provided the key to the periodic system had to remain purely empirical.

In the end it became apparent that this method of classification was not artificial; every element received a place in the "natural" order. The many empirically established properties of elements finally became part of a general law, the periodic law, as a "natural" classification of elements.

References

1 R. J. FORBES, *Studies in Ancient Technology*, Leyden 1963, Vol. VII p. 79 ff.
2 H. M. LEICESTER, *The Historical Background of Chemistry*, London–New York 1956.
3 R. HOOYKAAS, *Chem. Weekblad*, 33 (1936) 599.
4 P. KRAUS, *Jābir ibn Hayyán, Contribution à l'Histoire des Idées scientifiques dans l'Islam*, Vol. II; *Jābir et la Science grècque*, Cairo 1942, p. 18 ff.
5 J. RUSKA, *Al-Rázi's Buch Geheimnis der Geheimnisse*, Berlin 1937, p. 84.
6 J. J. BECHER, *Tripus Hermeticus Fatidicus, Pandens Oracula Chymica*, Frankfort 1689.
7 R. BOYLE, *The Origin of Forms and Qualities*, Oxford 1666, in *Works of Boyle*, ed. by Thomas Birch, London 1744, Vol. II pp. 469, 484.
8 M. BOAS, *Robert Boyle and the Seventeenth Century Chemistry*, Cambridge 1958, p. 146 ff.
9 See e.g. ARISTOTLE, *Parts of Animals*, The Loeb Classical Library, transl. by A. L. Peck, London–Cambridge 1937, p. 79 ff.; J. BRUN, *Aristote et le Lycée*, Paris 1961, pp. 33–34, 80 ff.
10 LE COMTE DE BUFFON, *Histoire naturelle des Minéraux*, Paris 1785, Vol. III p. 611 ff; new ed., Dordrecht 1798, p. 225 ff.
11 J. W. VAN SPRONSEN, *Chem. Weekblad*, 60 (1964) 244.
12 C. LINNAEUS, *Species Plantarum, exhibentes plantas rite cognitas, ad genera relatas, cum differentii specificis nominibus trivalibus, synonimis selectis, locis natalibus secundum Systema Sexuale digestas*, 2 vols., Holmiae (Stockholm) 1753; *ibid.*, facs. ed., London 1959.
13 C. LINNAEUS, *Systema Naturae*, 10th ed., Holmiae (Stockholm) 1758–59, Vol. II.
14 C. LINNAEUS, *Systema Naturae*, Editio multo auctior & emendatior, Leyden 1756.
15 C. À LINNÉ, *Systema Naturae*, 13th ed., Leyden 1793, Vol. III.
16 J. W. VAN SPRONSEN, *Chem. Weekblad*, 60 (1964) 75.
17 T. BERGMAN, *Nova Acta Reg. Soc. Scient. Upsaliensis*, 4 (1784) 63.
18 LE COMTE DE BUFFON, *Histoire naturelle des Minéraux*, Paris 1785, Vol. III p. 611 ff.; Paris 1783, Vol. II p. 160.
19 M. P. CROSLAND, *Historical Studies in the Language of Chemistry*, London–Melbourne–Toronto 1962, p. 162.
20 R. J. HAÜY, *Traité de Minéralogie*, 5 vols., Paris 1801.
21 R. HOOYKAAS, *Archives internationales d'histoire des sciences*, 18/19 (1952) 45.
22 A. L. LAVOISIER, *Traité élémentaire de Chimie*, 2nd ed., Paris 1793, Vol. I p. 180 ff.
23 loc. cit., p. XIX ff.
24 loc. cit., p. 192 ff.
25 R. J. HAÜY, *op. cit.*, Vol. V.
26 A. L. LAVOISIER, *op. cit.*, p. 192.
27 R. HOOYKAAS, *Centaurus*, 5 (1958) 307.
28 TH. THOMSON, *System of Chemistry*, 3rd ed., Edinburgh 1807; *Système de Chimie*, Paris 1809, Vol. IV.

29 TH. THOMSON, *System of Chemistry*, Edinburgh 1802.

30 J. BLACK, *Lectures in the Elements of Chemistry*, 2 vols; London–Edinburgh 1803, Vol. I p. 420; *Chymia*, 6 (1960) 271; see also J. R. PARTINGTON, *A History of Chemistry*, London–New York 1962, Vol. III, p. 130 ff.

31 J. J. BERZELIUS, *Schweigg. Journ.*, 15 (1815) 301, 419.

32 *Berzelius' neues chemisches Mineralsystem*, ed. by C. F. Rammelsberg, Nuremberg 1847, p. 78.

33 *loc. cit.*, p. 183.

34 *loc. cit.*, p. 225.

35 J. W. VAN SPRONSEN, *Chem. Weekblad*, 58 (1962) 655.

36 J. W. VAN SPRONSEN, *Chem. Weekblad*, 59 (1963) 535.

37 E. F. GEOFFROY, *Histoire de l'Académie royale des Sciences*, (1718) p. 202.

38 P. J. MACQUER, *Eléments de Chymie théorique*, Paris 1749.

39 C. F. WENZEL, *Lehre von der Verwandtschaft der Körper*, Dresden 1777.

40 J. W. VAN SPRONSEN, *Chem. Weekblad*, 60 (1964) 157.

41 J. B. RICHTER, *Anfangsgründe der Stöchiometrie oder Messkunst chymischer Elemente*, 3 vols., Breslau (Wroclaw) und Hirschberg (Jelenia Góra) 1792/93.

42 E. G. FISCHER, *Claude Louis Berthollet über die Gesetze der Verwandschaft*, Berlin 1802, p. 229 ff.

43 W. WHEWELL, *History of Scientific Ideas*, 3rd ed., London 1858, Vol. II p. 25.

44 R. KIRWAN, *Phil. Trans. Roy. Soc. (London)*, 37 (1783) 15.

45 W. WHEWELL, *op. cit.*, Vol. I p. 57 ff.

46 J. J. BERZELIUS, *Jahresber. Fortschr. Chem.*, (1846) 219.

47 *Berzelius' neues chemisches Mineralsystem*, *op. cit.*, pp. 78, 213 ff.

48 *loc. cit.*, p. 203.

49 L. GMELIN, *Z. Mineralogie*, (1825, I) 322, 418, 490.

50 L. GMELIN, *Z. Mineralogie*, (1825, II) 33, 97.

51 A. F. FOURCROY, *Philosophie chimique ou Vérités fondamentales de la Chimie moderne*, 3rd ed., Paris 1806, p. 205.

52 W. WHEWELL, *History of the Inductive Sciences*, 3 vols., London 1847.

53 W. WHEWELL, *History of Scientific Ideas*, 2 vols., 3rd ed., London 1858.

54 R. J. HAÜY, *Tableau comparatif des Résultats de la Cristallographie et de l'Analyse chimique, relativement à la classification des minéraux*, Paris 1809, p. XXI.

55 W. WHEWELL, *History of Scientific Ideas*, *op. cit.*, Vol. II p. 7.

56 W. WHEWELL, *History of the Inductive Sciences*, *op. cit.*, Vol. II p. 197.

57 R. J. HAÜY, *Tableau comparatif*, *op. cit.*, p. VII.

58 J. W. VAN SPRONSEN, *Chem. Weekblad*, 59 (1963) 437.

Chapter 3

Period of maturation

3.1. Introduction

It seems surprising that it took so long to achieve a molecular theory, because the principal laws of chemistry and the fundamentals of the atomic and molecular theories, set forth by Avogadro and Ampère, were already at hand during the first decade of the nineteenth century:

1789[1] The law of conservation of matter (Lavoisier)
1806[2] The law of definite proportions (Proust)
1808   The law of multiple proportions (Dalton)
1808   The law of combining volumes of gases (Gay-Lussac)
1808   Dalton's atomic theory
1811   Avogadro's hypothesis.

One of the reasons for the general reluctance to combine Dalton's atomic theory with the hypothesis of Avogadro stems from the fundamental difference between the two scientists concerning the concept of chemical bonds; whereas Dalton assumed there are bonds only between dissimilar atoms (in contrast to Democritus *[3]*), Avogadro believed bonds also existed between similar atoms.

A complete review of the development of the concept of atomic weight would take us beyond the scope of our subject. It is necessary, however, to record the atomic or equivalent weights which were available to those attempting to systematize the numerical relationships found between elements.

It is evident that a comprehensive periodic system of elements could not be constructed until most of the chemical elements were known and their atomic weights had been unequivocally determined. This moment came after September, 1860, the year in which the memorable Karlsruhe Congress—one of the most important congresses in the history of chemistry—was convened to solve "once and for all" the problems presented by the atomic and molecular concepts and the partly allied concept of the atomic weight. It need hardly be said that the conceptions of atom and molecule were

---

[1] These dates, indicating the years of publication (not those of the formulation), show that these laws and theories became available to everyone doing scientific work. For the original publication, see references *[1]* and *[2]*.

[2] 1808 is often taken as the year in which this law received recognition; see also *3.2*.

not conclusively resolved. Nor, indeed, did all the members present at this meeting[3], including the leading authorities at that time [5, 6], fall into line, despite a concrete solution[4] suggested by Stanislas Cannizzaro. Cannizzaro's values for the atomic weights, based mainly on the views of Avogadro and Gerhardt, were not made public until two years after the Congress, in the *Jahresbericht über die Fortschritte der Chemie*. It was thanks to this Congress that in the decade following 1860 several scientists were able to discover the periodic system of elements and that De Chancourtois was actually able to do so as early as 1862 by the use of atomic weights that were correct or almost correct in all but a few cases (see *3.5*).

The fact that the first discoverer of the periodic system did not consider Avogadro's hypothesis to be an essential prerequisite would seem to refute our argument that the periodic system of elements could not emerge until that hypothesis had received general acceptance. De Chancourtois seems not to have noticed, however, that he started from atomic weights determined on the strength of this hypothesis.

3.2. Development of the concept of atomic weight

Quantitative experiments on the formation of salts, carried out between 1792 and 1794, led Richter to introduce the concept of equivalent weight (*Verbindungsgewicht*) (see *2.4*). For he found that in the case of neutralization of a certain weight of a base (or acid) it required the weights a, b, or c in grams of acids (or bases); the ratio a : b : c was at all times constant for every base (or acid). This ratio of weights must, according to Richter, be considered along with the ratio of the equivalent weights of the acids (or bases). In 1802 the values of the equivalent weights themselves were collected in a table by Fischer, the author of the German translation of Berthollet's *Essai d'une Statique chimique* [7], even before Richter [8, 9] had published his Table. Fischer took the equivalent weight of sulphuric acid, i.e. 1000, as his basis.

As a result of the method employed by Richter to determine the equivalent weight, this quantity was first defined for compounds. The first list giving equivalent weights of elements was drawn up by Dalton.

An authority like Berthollet, however, was by no means convinced that compounds react with one another in constant proportions. He waged a fierce battle with Proust who finally, in 1806, formulated his definitive law concerning the constant composition of compounds. The controversy arose mainly because Berthollet's studies mostly concerned equilibrium reactions, whereas Proust only reluctantly admitted that two elements can often form more than one compound with each other [10].

A few years earlier, in 1803, Dalton had developed his atomic theory (as is to be seen

---

[3] Although Kekulé had taken the initiative in calling the first international chemical congress at Karlsruhe, in 1861 he was not convinced of the necessity of taking Avogadro's hypothesis as the starting point in the determination of atomic weights and vapour densities. In fairness to Kekulé and, for instance, also to Wurtz, it should be observed that they later played a prominent part in establishing the new atomic weights [4].

[4] Distributed in pamphlet form.

References p. 61

from his note-book) and put forward his law of multiple proportions. Yet Dalton's theories did not immediately gain general acceptance. One of the reasons for this delay was their rejection by certain prominent scientists.

Dalton must also be credited with having been the first to determine atomic weights. Because he worked with some mistaken conceptions and also with atomic weights

John Dalton. Bronze statue in front of the John Dalton College, Manchester

calculated on the basis H = 1 (which involved considerable inaccuracy), his values differ by more than a little from those accepted to-day. The number of elements (actually only gases) whose atomic weights were determined by Dalton, was small. In principle, however, it had become possible to determine the atomic weights of other elements. But the greatest difficulty was the persistent lack of the concept now called valence [11]. Until this concept had been formulated and applied, it remained impossible to differentiate between atomic and equivalent weights.

Dalton avoided the complication presented by multiple compounds of two elements, in a somewhat simplistic way, supported by his heat theory. He assumed the most simple formulae, successively, AB, AB_2, A_2B, etc., for these compounds. Thus, he assigned to ethylene the formula CH, and to marsh gas the formula CH_2. The formula for water became HO. It is understandable that with such a general assumption many atomic weights deviated from the correct ones by a factor of 2 or another simple number. Dalton's first list of atomic weights[5] [12, 13], containing those of five ele-

[5] Initially, Dalton did not use the word *atomic weight*. In 1803 he headed his list "relative weight of the ultimate particles of gaseous and other bodies". In 1808 he used the word *elements*; in 1810 the concepts of *simple substance* and *element* seemed to him to be synonymous.

ments, dates from 1803 (Table 1). In this year Dalton read a paper before the Manchester Literary and Philosophical Society that was not published until two years later. This is one reason why in 1805 he had not yet published the atomic weights of some metals listed in his 1804 note-books[6]. Doubt as to the correctness might also have led him to delay publication.

Amadeo Avogadro

Another list known to us (Table 1) *[15]* dates[6] from 1808. It is interesting to see that this list does not include a number of atomic weights, such as those of arsenic, antimony, bismuth, tin, and manganese, although these atomic weights occur in his 1806 and 1807 notebooks. We find these atomic weights first among those of some other elements in the list[7] dating from 1810 *[14,16]*. Dalton probably did not consider these

[6] For the several tentative lists recorded by Dalton in his notebooks, see the detailed article by Hooykaas *[14]*. In the caption to Fig. 7 on page 327 of this publication the word *elements* is to be understood as *matter*.

[7] In Dalton's *New System of Chemical Philosophy*, published in 1808 and 1810, we find several lists of atomic weights. One of these *[16]*, bearing the heading *Weights of ultimate particles*, contains the atomic weights of metals. Here Dalton gave an atomic weight of 21 for sodium and 35 for potassium, which latter element he sometimes wrote with one s. In other lists, headed *Relative weight of chemical elements or ultimate particles [15]* and *Weight of simple elements [17]*, explaining diagrams of atoms, we find the atomic weight of *potash* given as 42 and *soda* as 28. The oxides of both these metals are evidently meant here. Although Dalton himself pointed this out *[18, 19]*, he apparently did not think it necessary to insert these oxides in a list of *compound elements*. In the first part of his book he nevertheless discussed in detail the distinction between simple body and compound and chose as an example the oxides of several metals, which had only shortly before been found to be compound bodies.

TABLE 1

ATOMIC WEIGHTS DETERMINED OR USED BY DIFFERENT SCIENTISTS

	Dalton[1] 1803 [12]	Dalton 1808 [15]	Dalton 1810 [16]	Thomson 1810 [22]	Wollaston 1814 [23]	Berzelius 1815 [21]	Berzeliu 1827 [32]
H	1	1	1	1	1[3]	1.06	1
Li (1817)	——	——	——	——	——	——	20.5
Be (1828)*	——	——	——	——	——	10.9[5]	53
B (1808)*	——	——	——	—	—	11.7	21.7
C	4.3	5	5.4	—	5.7	12.0	12.2
N	4.2	5	5	—	13.3	12.6	14.2
O	5.5	7	7	—	7.6	16	16
F (1869)*	——	——	——	——	——	9.6	18.7
Na (1808)*	——	21	21	23.3	22.0	93.3	46.5
Mg (1808)*	——	13[2]	10[2]	17.6	11.1[4]	50	25.3
Al (1825)*	——	——	——	——	——	55	27.5
Si (1823)*	——	——	——	——	——	49	44
P	7.2	9	9	—	13.2	27	31
S	14.4	13	13	—	15.7	32	32
Cl (1807/10)*	——	——	——	—	33.4	70[6]	35.5
K (1808)*	——	35	35	38	37.2	156	78
Ca (1808)*	——	16[2]	17[2]	21.8	19.3	82	41
Ti (1825)*	——	——	40?	——	——	28.8	62.2
V (1831)	——	——	——	——	——	——	——
Cr	—	—	—	—	—	113	56
Mn	—	—	40?	—	—	113	57
Fe	—	38	50	—	26.1	111	54
Co	—	—	55?	—	—	117	59
Ni	—	—	25? 50?	—	—	117	59
Cu	—	56	56	—	30.3	128	63
Zn	—	56	56	—	31.1	128	64.5
As	—	—	42?	—	—	134	75.3
Se (1818)	——	——	40	——	——	——	79.1
Br (1826)*	——	——	——	——	——	——	——
Rb (1861)	——	——	——	——	——	——	——
Sr (1808)*	——	39[2]	39[2]	37.6	40.1[4]	178[7]	87.5
Y	—	—	—	—	—	141	64.5
Zr (1824)*	——	——	——	——	——	——	67
Nb	—	—	—	—	—	—	—
Mo	—	—	—	—	—	96	96
Di (1841)**	——	——	——	——	——	——	——
Ru (1844)	——	——	——	——	——	——	——
Rh (1803)	——	—	—	—	—	238	120
Pd (1803)	——	—	—	—	—	227	114
Ag	—	100	100?	—	102.3	430	216
Cd (1817)	——	——	——	——	——	——	112
Sn	—	—	50	—	—	235	118
Sb	—	—	40	—	—	256	128
Te	—	—	—	—	—	128	125
I (1811)	——	——	——	——	—	—	125.5
Cs (1860)	——	——	——	——	——	——	——

(Footnotes, see pp. 48 and

Berzelius 1828 [58]	Dumas 1828 [29]	Berzelius 1835 [59]	Gerhardt 1843 [31]	Berzelius 1845 [60]	Cannizzaro 1860	De Chancourtois 1862	1969
1	1	1	1	1	1	1	1
20.5	20.4	12.9	7	13	7	7	7
53.1	53.1	53	—	—	—	9	9
21.7	10.9	21.8	11	22	11	11	11
12.2	6[9]	12.3	12	12	12	12	12
14.2	14.2	14.2	14	14	14	14	14
16	16	16	16	16	16	16	16
18.7	18.6	18.7	19	18.7	19	19	19
46.6	46.6	46	23	46	23	23	23
25.4	25.2	25.4	12	25	24	24	24
27.4	27.4	27.4	13.5	27.4	27	27	27
44.5	14.8	44	—	44.5	28	28	28
31.4	31.4	31.4	31	31	31	31	31
32.2	32	32.2	32	32	32	32	32
35.5	35.4	35.5	35.5	35.5	35.5	35	35
78.5	78	78.5	39	78	39	39	39
41.0	40	41	20	40	40	40	40
48.7	62.2	49	45	48	50	48	48
—	— —	137	—	137	69	137 à 140	51
56.4	56.4	56	26	53	53.5	53	52
57.0	57	55	27.5	55	55	55	55
54.4	54.2	54	28	56	56	56	56
59.1	59	59	29.5	59	59	60	59
59.2	59.2	59	29.5	59	59	59	59
63.4	63.3	63	31.5	63	63	63	63
64.6	64.4	65	33	65	65	65	65
75.3	75.2	75	75	75	75	75	75
79.3	79.2	79	79.5	79.5	79.5	80	79
79	74.6	78	80	80	80	79	80
— —	— —	— —	— —	— —	— —	87	85
87.7	87.6	88	44	87.5	87.5	88	88
64.4	64.4	64	—	—	—	64 and 100	89
67.4	89.6	67	—	67	90	67	91
—	—	—	—	—	—	—	93
95.8	95.6	96	48	95	96	96	96
—	— —	— —	—	—	—	99	
— —	— —	— —	— —	197	197	205	102
104.4	120	104	— —	104	104	104	103
106.7	112.8	107	—	106	106	107	107
216.6	216	217	108	216	108	108	108
111.7	112	112	56	112	112	111	112
117.8	117.6	118	59	118	118	115	119
129.2	129	129	122	129	122	121	122
129.2	64.4	128	129	128.5	129	128	128
126	125.3	127	127	127	127	127	127
—	— —	— —	— —	— —	— —	124	133

(continued)

TABLE 1 *(continued)*

	Dalton[1] 1803 [12]	Dalton 1808 [15]	Dalton 1810 [16]	Thomson 1810 [22]	Wollaston 1814 [23]	Berzelius 1815 [21]	Berzelius 1827 [32]
Ba (1808)*	— —	61[2]	61[2]	63	65.8[3]	273	137
La (1839)	— —	— —	— —	— —	— —	— —	— —
Ce (1803)	—	—	45?	—	—	184	92
Ta	—	—	—	—	—	—	—
W	—	—	56?	—	—	387	189
Os (1804)	— —	—	—	—	—	—	—
Ir (1804)	—	—	—	—	—	—	—
Pt	—	100	100?	—	—	193	195
Au	—	140	140?	—	—	397	199
Hg	—	167	167?	—	95.1	405	202
Tl (1861)	— —	— —	— —	— —	— —	— —	— —
Pb	—	95	95	—	98.1	416	207
Bi	—	—	68?	—	—	283	213
Th (1828)	— —	— —	— —	— —	— —	— —	— —
U	—	—	60?	—	—	503[8]	434

   * Element had been separated.
  ** Proved not to be an element.
— — Element had not yet been discovered or separated.
  —   No atomic weight had been determined, although the element was known.
[1]   Only Dalton's published atomic weight tables are included here (see also p. 43 ff.).
[2]   Calculated by the present author with use of the atomic weight of the oxide analogous to that of Dalton *[15]* who did the same for K and Na, i.e. by subtracting 7.
[3]   Converted by the present author into H=1. Wollaston had taken O=10 as the standard and thus obtained H=1.32.

atomic weights sufficiently definite to publish them in 1808, and in 1810 we in fact see that some of the elements carry different values for the atomic weight (Table 1, list 3). Dalton was unable to accept the volume law published by Gay-Lussac in 1808, because he could not explain it theoretically. Dalton held tenaciously to his diffusion theory, according to which the molecules (which he called atoms) of reacting gases could not be of the same size and therefore could not occupy the same volume. Gay-Lussac's law states that these volumes are equal to, or differ by, a simple whole number. Dalton refused to accept the rounding-off of measurements, i.e. of the values found for the volumes. As Hooykaas *[14]* had pointed out, this is all the more surprising because Dalton very frequently rounded off his own measurements to comply with his theoretical views. Dalton maintained his own theory too firmly to accept inconvenient results of others' research. Inaccurate determinations of atomic weights and densities, for some of which he was responsible, also led him to reject Avogadro's hypothesis, which was based on Gay-Lussac's law. Even after more exact atomic

melin	Berzelius	Dumas	Berzelius	Gerhardt	Berzelius	Canniz-zaro	De Chan-courtois	1969
827	1828	1828	1835	1843	1845	1860	1862	
[0]	[58]	[29]	[59]	[31]	[60]			
58.6	137.3	137.2	137	68.5	137	137	136	137
-—	—	——	——	—	—	—	91	139
46	92.1	92	92	—	—	—	92	140
34	184.9	184.5	185	—	—	—	184	181
96	189.6	189.2	190	92	190	184	185	184
-	199.4	184.5	199	—	199	199	208	190
-	197.6	184.5	198	98.5	197	198	197	193
48	197.6	194.4	197	98.5	197	197	199	195
56	199.2	198.8	199	—	197	197	200	197
01	202.9	101.3	203	100	200	200	204	201
-—	——	——	——	——	——	204	103	204
04	207.5	207.2	207	103.5	207	207	207	207
71	213.2	213	142	210	213	210	209	209
-—	——	——	120	—	119	—	119	232
7	434.6	433.8	434	60	119	120	120	238

[4] Converted by the present author into metal; XO was taken as the formula for the oxide.

[5] The original mentions 683.3 instead of 68.33 (O=100).

[6] Called muriaticum by Berzelius, the present element chlorine being regarded as a compound MO_3. Atomic weight converted by considering chlorine as an element by the present author.

[7] The original contains a misprint, viz. 1418.14. This must be 1118.14 (with O=100 as the standard).

[8] Berzelius considered UO_3 to be a metal.

[9] A list in which Dumas considered some atomic weights as whole numbers mentions 12 as the atomic weight of carbon.

weights had become available, little attention was given to Avogadro's hypothesis[8] [5, 20]. Dalton and many others continued to ignore it[9].

At this time the concept of equivalence also came into use in England, but it was confused with the atom concept[10]. Mainly as a result of this confusion, it was only after 1840 that the problem of the atomic weights could advance a step towards its final solution. Even Berzelius did not realize the importance of Avogadro's hypothesis. This scientist, who became so famous for his many and accurate atomic weight determinations, based his work only on Gay-Lussac's law. This eliminated Dalton's arbitrary assumptions. Berzelius made use of the correct formulas H_2O, NH_3, etc., but he made no distinction between atoms and molecules in gases such as hydrogen and

[8] Until 1825 no attention was paid to Avogadro's work by any scientist.

[9] Not until 1833 did Berzelius employ Avogadro's hypothesis. Kopp did not even mention Avogadro's name in his Geschichte der Chemie, published in 1844.

[10] Kekulé believed that the concept of atomicity (valence) which he introduced was inconceivable without any atomic theory.

oxygen. Later, Berzelius also enlisted other aids to determine atomic weights, viz., the doctrine of isomorphism and the law of Dulong and Petit, both introduced in 1819.

Berzelius made his first list of atomic weights [21] (Table 1), based only on Gay-Lussac's law, in 1815, after the appearance of Dalton's [15] and Thomson's [22] lists in 1810 and Wollaston's list [23] in 1814. Thomson estimated only about six equivalent weights, some of which he determined more exactly than Dalton. Wollaston's 1814 list contains far more elements, some with better atomic weights. As he took AO for the formula of the oxides, the values of the univalent metals have the correct order of magnitude, but those of the bivalent metals are too small by a factor of 2. In Berzelius' table drawn up only one year later, we see many atomic weights—which he called *Gewichte des Körperelements*—with values surprisingly close to the present ones. There are also differences, some values being two or four times too high. He achieved still better results in 1827, after he had understood the implications of the research done by Dulong and Petit [24] on the specific heats of the elements and the work of Mitscherlich [25] on isomorphism (Table 1). The divergence from the atomic weights accepted to-day can logically be accounted for by the confusion caused by the distinction between atomic and equivalent weight. For example, when Berzelius quotes an atomic weight of 17.74 (H = 1) for chlorine, he assigns to the same element an equivalent weight of 35.5, thus two times higher, by which he meant to convey the fact that, in forming a compound, chlorine always enters into that reaction with paired atoms. Berzelius took for an equivalent of chlorine an amount equal to Cl_2; e.g., the formula for hydrochloric acid is $HCl = H_2Cl_2$. Here, H has an equivalent weight of $2 \times 0.5 = 1$ (recalculated for H = 1 used at present). For the molecular weight of hydrochloric acid, the correct value is indeed $1 + 35.5 = 36.5$. Another example is given by the formula for calcium chloride, which was written $CaCl$. It is now clear that $CaCl_2$ is meant by this, and so it is not difficult to calculate the molecular weight, i.e., the sum of the equivalent weight of calcium, which is equal to its atomic weight, and the equivalent weight of chlorine (20+35.5). As we can see, this result is equal to half the correct value. This is therefore due to Berzelius' atomic weight standard: $H = 1$, i.e. H = 0.5.

The fact that Berzelius initially disagreed with Avogadro's view, caused him to assume, for instance, unequal numbers of particles in equal volumes of the gases hydrogen chloride and hydrogen: the ratio of the number of particles is, according to him, 1 : 2.

In 1826 Dumas [26, 27] at first accepted Avogadro's doctrine in his attempts to determine the formulas of the inorganic hydrogen compounds, which led to a fierce conflict with Berzelius. However, after Dumas' determinations of vapour density at higher temperatures gave quite different results from what he had expected (retrospectively explained by the fact that he worked with dissociating gases [28]), he abandoned Avogadro's hypothesis in 1832.

In 1828 Dumas [29] published two lists, one with ratios he had found in many cases by taking AO for the lowest oxide of an element, and the other with atomic weights he had deduced from the first list by applying the laws of Dulong and Petit, Gay-Lussac

Stanislao Cannizzaro *Jöns Jacob Berzelius*

and Mitscherlich. The latter list shows a general agreement with Berzelius' 1826 list. Dumas' weights for the alkali metals are also too high by a factor of 2. Many of the atomic weights are the same as the modern ones, however. But Gmelin's list [30] published in 1827 (Table 1) includes the exact values for the alkali metals. With HO (i.e. O = 8) as the formula of water and MO for the metal oxides, however, most of the elements have values, twice too small. In this respect we may call Gmelin the predecessor of Gerhardt, but for some elements his values diverged widely from Gerhardt's.

Many famous scientists, including Gay-Lussac, Liebig, and Gmelin, took part in the controversy about the molecular theory. Around 1840 the opposition to Berzelius reached a climax. In 1843 Gerhardt [31] changed some of Berzelius' atomic weights; as a result the atomic weights of the alkali metals became correct but those of the bivalent metals, for instance, became too small by a factor of two (Table 1); on the analogy of H_2O, he accepted A_2O as the formula for the metal oxides where Berzelius took AO.

As already mentioned, it was Cannizzaro who, after 1858, restored order out of chaos. He was the first to make the distinction between atom and molecule comprehensible. Many of Berzelius' atomic weights Cannizzaro found to be correct; only his values for the alkali metals, as well as those of silver, boron, silicon and zirconium proved to be twice too high. Gerhardt had determined these atomic weights correctly. It should indeed be mentioned that several atomic weights had been determined more

Wollaston's slide-rule of equivalent weights (upper part)
(University Museum, Utrecht)

accurately by Gerhardt than by Berzelius, although with values that were half the actual weights.

3.3. Parallels between the chronology of atomic weight determinations and the discovery of the relationships between elements

In 1817, Döbereiner first formed a triad. Table 1 shows that this scientist could have made use of Berzelius' 1815 atomic weights if he had known of them. Berzelius had determined only eight atomic weights correctly and three approximately. Of his other determinations, most were half what they should be. This does not alter the fact that triads could be formed with elements carrying these incorrect atomic weights.

A case in point is the Ca–Sr–Ba triad. Döbereiner could also draw on Wollaston's 1814 determinations and on those made by Dalton in 1808 and 1810 and by Thomson in 1810. We should point out however (see 4.2) that Döbereiner used an approximation of Wollaston's values for the equivalent weights of calcium, strontium and barium. His triad is based on the equivalent weights of the oxides of these three elements.

After re-calculating the atomic weights with $O = 16$ as the standard (Döbereiner used $O = 7.5$), a comparison of Döbereiner's triad with the present one shows agreement to a large extent, notably:

$$\text{Döbereiner's triad:} \quad SrO = \frac{CaO + BaO}{2} = 107 = \frac{59 + 155}{2};$$

$$\text{The present triad:} \quad = 104.75 = \frac{56 + 153.5}{2}.$$

In 1829 it was Döbereiner himself who pointed out that the atomic weights had been more correctly determined by Berzelius [32] two years earlier. Making use of these values, Döbereiner was enabled to add some new triads and to point out some other relationships.

When in 1850, thirty-three years after Döbereiner's first publication and one year after his death, Pettenkofer referred to this subject again—the first to do so—the situation in the field of the atomic weights had changed. Pettenkofer used Berzelius' 1845 list. As we have seen, most of Berzelius' atomic weights agreed with the present ones.

3.4. Development of the valence concept

The origin and further development of the valence (or valency) concept has already been described by several authors [33]. For this subject, as for the preceding one, it will therefore suffice to discuss briefly the influence of this development on the systematization of elements. When the preparatory work on the classification of elements was proceeding during the 1850s, only tentative attempts were made to deduce a specific relationship between atoms, molecules and radicals from the course taken by chemical reactions, especially those between organic substances.

Frankland, in 1852, formulated the concept of *combining power* for the elements of the N-group, i.e. expressed in the numbers 3 or 5. In 1857 Kekulé introduced an analogous concept, i.e. an atomicity of 1 for H, Cl, Br, K, etc., of 2 for O, S, etc., of 3 for N, P, As, etc. This concept, also known as *Sättigungscapacität, Werthigkeit* and *Valenz*, became important to systematization when the quadrivalence of carbon was established in the same year by Kekulé [34]. A few months later [35] Couper published his constitution formulae, which also represented an important contribution to the concept of valence, even though Couper used a valence of 2 as well as 4 for carbon. From that moment, the interdependence of groups of analogous elements was taken into account and the development of the concept of valence kept pace with that of the systematization of elements, despite the fact that most of the chemists working on valence did not see the broader connection between the two. Although Odling saw it as early as 1857, Williamson [36], who had long worked on the problem of combination in organic compounds, in 1864 still classified elements according to the number of atoms that an element contributed to a compound, taking as his basis *atomicities*, i.e. the valences of the elements. In this year, publications dealing with the valences of the elements appeared in rapid succession, and Kekulé, Wurtz, Erlenmeyer and Butlerow stated their opinions.

That Odling, in 1857, recognized the relationship between groups of elements and was able to construct a periodic system as early as 1864, was partly due to his own contribution in the field of valence. In 1855, as an extension of Gerhardt's theory of types as well as Williamson's work and probably independent of Frankland, he developed ideas similar to those of the latter [37].

Dumas' discovery, in 1857, of the regularities in the differences between the atomic weights of elements belonging to the same group, which he compared with the homologous series of organic substances, can also be attributed to the contributions he had made to the theory of organic compounds. In 1839 Dumas had already advanced his "theory of types" to replace his substitution theory. It was mainly Gerhardt and Laurent who cleared the way for the construction and application of the correct molecular formulae. These two scientists were interested in the individual atoms and molecules, but they also suspected that a principle of classification could be applied to these elementary particles. Gerhardt [38] stated this point in his textbook (1853) with the words: "Classer, c'est formuler des analogies". His premature death prevented Gerhardt from developing his ideas further. A year later, Laurent [39] commented in connection with the theory of types: "Mais la question est loin d'être aussi simple: elle est entourée de mille difficultés, et établir une théorie des types, c'est établir une classification chimique basée sur le nombre, la nature, les fonctions et l'arrangement tant des atomes simples que des atomes composés."

3.5. Use of atomic weights by the founders of the periodic system

We do not intend to identify the atomic weights available to each scientist and their use. We need only point out again that Béguyer de Chancourtois' publication of the first periodic system of elements in 1862, as well as the influence of the 1860 Karlsruhe Congress, brought some order into the confused field of atomic weights. The time was therefore certainly ripe for a system based on the numerical value of the atomic weights. Although nothing like all the elements were known yet, the only consequence was the inevitable incompleteness of the systems of De Chancourtois and the other contemporary scientists.

Among those who did much to prepare the way for this mineralogist and geologist, we mentioned Odling and Dumas first, because it was they who, in 1857 and 1858, saw some connection between the groups of analogous elements until then regarded as independent entities (see also *4.11, 4.12, 4.13* and *4.16*). As De Chancourtois mentioned only the name of Dumas, there is no way of knowing whether he was aware of Odling's work.

De Chancourtois was not the first investigator to see a more general connection, however. Adolph Friedrich Ludwig Strecker [40] (1822–1871) stated as early as 1859: "Es ist wohl kaum anzunehmen, dasz alle hervorgehebenen Beziehungen zwischen den Atomgewichten (oder Aequivalenten) in chemischen Verhältnissen einander ähnlichen Elemente blosz zufällig sind. Die Auffindung der in diesen Zahlen durchblickenden *gesetzlichen* Beziehungen müssen wir jedoch der Zukunft überlassen...". Strecker had long had a strong interest in this subject [41] and therefore attended the Chemical Congress at Karlsruhe. After holding a professorship at Christiania for ten years, he became a professor at the University of Tübingen from 1860 to 1870, where he was one of Lothar Meyer's predecessors. To what extent he influenced Meyer is unknown.

As early as 1852, Faraday [42] spoke in enthusiastic terms of Dumas' first work on the relationship between the atomic weights and his view of the transmutability of elements (see *4.6*). Faraday concluded that there might very well be a law governing the elements, for "...when we come to examine the combining powers of the three (chlorine, bromine, iodine), as indicated by their respective equivalents or atomic weights, the same mutual relation will be rendered evident. This circumstance has been made the basis of some beautiful speculations by M. Dumas—speculations which have scarcely yet assumed the consistence of a theory, and which are only at the present time to be ranged amongst the poetic day-dreams of a philosopher—to be regarded as some of the poetic illuminations of the mental horizon, which possibly may be the harbinger of a new law".

Study of De Chancourtois' system, which contained nearly all the then known elements (i.e. 57 of the 63), discloses that he could not classify lead and barium, despite the fact that they were well known. The other elements missing from his system are vanadium, niobium, erbium and terbium. Because of uncertainty about its valence and atomic weight, yttrium occurs in two places (with the atomic weight of 64). On further consideration, the atomic weights of thallium, thorium and uranium are found to have been taken at about half their actual value and that of ruthenium at twice the actual value, while the atomic weights of yttrium, lanthanum, osmium and caesium also deviate, in another way, from the correct values.

Newlands, in contrast to De Chancourtois, was not immediately convinced of the exactness of the atomic weights given by Cannizzaro, and in his first system he used those of Gerhardt, which explains why Newlands arrived at two differences between the atomic weights of analogous elements, namely 8 and 16. Odling, on the contrary, worked from the start with Cannizzaro's atomic weights, though not all the values in his system are correct. This pertains chiefly to the transition metals.

Some elements, however, were classified differently by the investigators, because they did not accept the same atomic weight values. To give an impression of this situation, Table 2 makes it possible to compare values taken for the atomic weights of these elements by five of the discoverers of the periodic system, with the values given by Cannizzaro and those accepted today. A glance at this table shows that most of those used in 1860 were correct. The correct atomic weights of osmium and iridium only became known after 1870, while for vanadium and uranium the exact atomic weights were used by Mendeleev in 1869 and 1870, respectively. Accordingly, Mendeleev classified uranium below tungsten.

He was also the first to classify vanadium, niobium and tantalum correctly, after the correct atomic weights had been given by De Chancourtois. Indium, discovered in 1863, was classified for the first time by Mendeleev, though wrongly on the basis of half the value of its true atomic weight. This element, together with thallium, was soon classified by both Meyer and Mendeleev in the boron group. The latter element had already been properly treated by Odling in 1868 (Fig. 28, p. 115). Odling regarded both zirconium and thorium as homologues of titanium. Mendeleev at first assigned thorium only half its true atomic weight.

For a long time no one knew what to do with yttrium, erbium, lanthanum and cerium, although Mendeleev had taken the correct atomic weights for yttrium and cerium in 1870. Didymium, also classified by Mendeleev, was later found not to be a pure element. Yttrium was correctly classified by Mendeleev between strontium and zirconium, but lanthanum was not recognized as a homologue of yttrium. It should be noted that two elements of this group (scandium and actinium) had not yet been discovered.

The difficulties caused by the transition elements (Group VIII) can easily be understood, because on the one hand these elements resemble each other closely and on the other their atomic weights, taken three by three, show little difference (see *11.3*).

Difficulties in the classification of beryllium will be discussed in another section (*12.6*).

TABLE 2

SOME ATOMIC WEIGHTS USED BY THE DISCOVERERS OF THE PERIODIC SYSTEM OF ELEMENTS

Underlining indicates that these weights were properly determined for the first time.

	Cannizzaro 1860	De Chancourtois 1862	Odling 1864	Newlands 1864	Hinrichs 1867	Odling 1868	Meyer 1868	Mendeleev 1869	Meyer 1870	Mendeleev 1871	1967
Ru	104	205	–	–	–	104	104	104	103.5	104	102
Rh	104	104	–	–	104.4	104	104	104	104	104	103
Pd	107	107	106.5	106.5	106.6	106.5	106	107	106	106	107
Os	199	208	–	199	–	199	199	199	199	195 ?	190
Ir	198	197	–	–	198	197	197	198	197	197	193
Pt	197.5	199	197	197	198	197	197	197	197	198 ?	195
Au	197	200	196.5	196	197	196.5	197	197 ?	196	199 ?	197
La	–	91	–	–	–	–	–	94	–	180 ?	139
Ce	–	92	–	–	–	92	–	92	–	140 ?	140
Ni	59	59	–	–	58	59	59	59	59	59	58
Co	59	60	–	–	60	59	59	59	59	59	59
Er	–	–	–	–	–	–	–	56	–	178 ?	167
Di*)	–	99	–	–	–	–	–	95	–	138 ?	–
Yt	–	64 and 100	–	–	–	–	–	60	–	88 ?	89
Nb	–	–	–	–	–	–	–	94	–	94	93
In 1863	–	–	–	–	71	–	–	76	113	113	115
Tl	204	103	203	203	204	203	204 ?	204	202.7	204	204
Th	–	119	231	–	231	231.5	–	118 ?	–	231	232
U	120	120	–	–	–	–	–	116	–	240	238
Zr	90	67	89.5	–	89.6	89.5	90	90	90	90	91
V	69	–	138	137	137.2	137	137	51	51	51	51
Ta	–	184	–	–	137.6	138	137.6	182	182	182	181

*) proved to be a mixture.

3.6. Investigators in the United States of America

During their early investigations, Odling, Newlands, Hinrichs, Meyer and Mendeleev had no contact with one another and did not know of De Chancourtois' work, in spite of Kopp and Will's references (see *16.7*) to their publications in the *Jahresbericht*.

The work of the predecessors was somewhat better known, in Europe as well as in the U.S.A., the more so because Dumas, Cooke and Carey Lea published in American periodicals. The first and the last of these authors were consequently mentioned in 1866 by Hinrichs. That not all the American scientists immediately learned of the results achieved in Europe, is shown by a paper by Gibbes[11] *[43]*. As late as 1875, he constructed a system in which only 41 elements were classified according to increasing atomic weight. The other elements were placed outside the system. Another system by Gibbes showed the form of De Chancourtois' *Vis tellurique*, but also included only 41 elements. These *Synoptical Tables of the Elements* by Gibbes are based on the research of Dumas, Gladstone, Cooke and Odling. Venable *[44]* later came to the conclusion that Gibbes knew nothing of Meyer's and Mendeleev's systems because of the roughness of his sketch, which he certainly would not have used for teaching had he been aware of the existence of better material.

All the same, we should not regard Gibbes as an independent American discoverer of a periodic system. Not only was Hinrichs before him, but Gibbes furthermore appears to have known of Odling's 1868 publication; the system set up by Gibbes is based entirely on Odling's results. According to Taylor *[45]*, the fact that the contemporary work on the periodic system was not known to Gibbes is to be ascribed to the difficulties of the period in which he worked. The war between France and Germany and his own financial troubles must have hindered his contact with Europe just at the time when the periodic system was discovered. Gibbes read only *The Chemical News*, which did not publish even a brief summary of Mendeleev's work until 1875.

Taylor commented that Gibbes' purpose was not so much to draw up a general classification according to atomic weight as to provide illustrative teaching material. This does not, in our opinion, alter the fact that Gibbes should have composed the best possible system, also for use in his lectures.

3.7. Distinction element vs. simple substance [46]

The distinction between the concepts of *element* and of *simple substance* is immediately apparent from a study of Mendeleev's share in the discovery of the periodic system. Mendeleev *[47, 48]* chose this distinction as the starting-point for his study of the relationship between elements. It should be emphasized that he distinguished be-

---

[11] Gibbes' treatise was not published until 11 years after it had been submitted. This was due to the fact that in 1875 the Elliot Society had to suspend its activities and thus could not print Gibbes' paper in its periodical, which did not appear again until 1886.

tween "simple substance", i.e. metal or non-metal as such, and "element", by which he meant one of the components of a compound consisting of a number of different "elements". His periodic system was constructed in the form of an arrangement of "elements" rather than of "simple substances". Thus, there is not much resemblance between the halogens themselves, but they reveal close analogies when considered as "elements" e.g. in the series of potassium halides.

Both Mendeleev and Odling clearly understood the distinction, although both sometimes used expressions by which the distinction became blurred [49]. The assignment of the atomic weight only to the element and not to the simple substance, as Mendeleev thought proper, is disputable, however. Understandably, not all investigators took this distinction sufficiently into account. It is often assumed that the modern definition of element made its appearance with Lavoisier [50], who defined the element as a substance whose separation into other substances had not yet been accomplished. On the basis of this definition, scientists could at least search for what we now call elements and were later able to classify these elements. But Lavoisier could not know which were the true simple substances. Consequently, it was difficult for him to make a practical distinction between *simple substance* and *element*. This distinction must not be lost sight of, however. Historians such as Paneth [51] and Hooykaas [52] have studied this point closely and have also stressed its importance.

As we have already seen (2.1), it was the qualities of the elements on which a natural classification, i.e., the periodic system, was based. Not only Mendeleev but also Paneth and Urbain accepted this criterion as a matter of principle [52]. The distinction drawn by Janet [53, 54] between *éléments chimiques simples* and *éléments chimiques complexes* lies in another field, and arose from the discovery of the isotopes (see also 13.2 ff). Dalton [15, 18] used the terms *element* and *simple substance* (as well as *ultimate particle*) interchangeably, which shows that he did not consider the distinction between the two concepts to be imperative (cf. footnote 5, p. 44 and 2.7). Although only a few scientists pointed explicitly to the difference between *simple substance* and *element*, some of them apparently based their conclusions on this distinction. Döbereiner, who used neither the term *simple substance* nor the term *element*[12], first demonstrated his discovery of triads on the basis of compounds of elements, e.g., the oxides of calcium, strontium and barium. It is therefore quite clear that the analogy of properties of the *elements* was primary. Pettenkofer, in 1850, used the term *simple body* interchangeably with the term *element* (and in 1858 only the term *element*), but he too considered, although he mentioned it only incidentally, a natural series to consist of the compounds of certain elements, e.g., the carbonates of iron, calcium and strontium, which are isomorphous. Gmelin, too, used the two terms in the same sense. Gladstone only used the word *element*. Béguyer de Chancourtois used the term *élément* interchangeably with *corps simple*. Odling, as far back as 1857, made an exhaustive study of the analogies between the elements in their compounds, and became convinced that it was the

---

[12] In another connection—also in 1817—Döbereiner did make use of the *element* concept to indicate *simple substances*, however.

element, not the simple substance, that possessed the properties requiring study (see also *4.12*). Odling stressed the necessity for finding the correct analogies, although he used the terms "*element*" and "*elementary body*" interchangeably.

Meyer first examined the differences and the similarities between the properties of the elements more closely in 1870. Although he made neither a nominal nor an essential distinction between simple substance and element, he did not ignore the difference. He remarked, for instance, that titanium and zirconium of the fourth group are related to silicon of the same group by isomorphism, which referred to the compounds of these simple substances and consequently to the elements. But when Meyer indicated the atomic volume and the specific heat of elements, and the malleability of metals as properties constituting a periodic function of the atomic weight, he was referring to the simple substances. His view of the possible complexity of atoms may also have prevented him from making a strict distinction between element and simple substance. This also seems to hold for Hinrichs, because of his highly specific concept of the atom (see *5.6*); but he, too, clearly understood the distinction between the element in its natural state, and the element as part of a compound. He said in his *Atomechanik*: "Die strenge Sonderung der Elemente und ihrer Verbindungen wäre unzweckmässig."

3.8. Conclusion

As the result of a whole complex of factors, the period leading to the discovery of a periodic system of elements spanned close to half a century. After one generation of chemists had laid the necessary foundations, it was reserved to a new generation to make the real discovery of the periodic system. In the foregoing we have attempted to show that this discovery would have been impossible before 1860.

In those days communications on chemical matters were so restricted that the final discovery of the periodic system was made independently in Europe by five scientists and simultaneously in North America by Hinrichs.

We do not mean to imply, however, that after 1870 all chemists in Europe and America were acquainted with the system of elements or were convinced of its significance. To cite only a single example, the American Hodges [55] was still searching for triads in 1875. What he actually wanted to find was groups of six elements differing in atomic weight by a certain number, but not more than three or four elements could be arranged in this way. The other elements, in his opinion, were still undiscovered. Hodges clung to the figure six because his beryllium group had a sixth member, thorium, although this group still lacked three elements. Carey Lea [56], even though he had played a part in the preparatory stage (*4.15*), in 1895 still had serious objections to the system which by then was well established (*6.23*).

With our present knowledge, it is obvious that the essence of the periodic classification was not, and could not be, grasped by anyone at that time. In later years, however, the system had to be used for purely practical reasons, much as its users would have preferred to understand its theoretical background! Mendeleev *[57]* expressed this very clearly in 1891, when he wrote:

"To explain and express the periodic law is to explain and express the cause of the law of multiple proportions, of the difference of the elements, and the variation of their atomicity, and at the same time to understand what mass and gravitation are. In my opinion this is now premature. But just as without knowing the cause of gravitation, it is possible to make use of the law of gravity, so for the aims of chemistry it is possible to take advantage of the laws discovered by chemistry without being able to explain their causes."

References

1 J. W. VAN SPRONSEN, *Chem. Weekblad*, 53 (1957) 129.

2 J. W. VAN SPRONSEN, *J. Chem. Educ.*, 36 (1959) 565.

3 See e.g. G. DE SANTILLANA, *The Origins of Scientific Thought*, New York 1961, pp. 148, 149.

4 J. GILLIS, *Verhandel. Koninkl. Vlaam. Acad. Wetenschap Belg., Klasse Wetenschap*, No. 62, p. 69 (1959).

5 J. W. VAN SPRONSEN, *Chem. Weekblad*, 58 (1962) 484.

6 CLARA DE MILT, *Chymia*, 1 (1948) 153.

7 E. G. FISCHER, *Claude Louis Berthollet über die Gesetze der Verwandschaft*, Berlin 1802, p. 229 ff.

8 J. B. RICHTER, *Chemisches Handwörterbuch nach den neuesten Entdeckungen entworfen von D. Dav. Ludw. Bourguet*, 6 vols., Berlin 1802–1805, Vol. III p. 164.

9 See also J. R. PARTINGTON, *A History of Chemistry*, London–New York 1962, Vol. III, p. 674 ff.

10 J. W. VAN SPRONSEN, *Chem. Weekblad*, 59 (1963) 437.

11 See e.g. A. C. W. ROODVOETS, *Het Ontstaan van het Begrip Valentie*, Thesis, Leyden 1934.

12 J. DALTON, Lecture on 21 October 1803, publ. in *Mem. Lit. Phil. Soc. Manchester* [2], 1 (1805) 287.

13 See also *Ostwald's Klassiker No. 3, Die Grundlagen der Atomtheorie, Abhandlungen von J. Dalton und W. H. Wollaston (1803–1808)*, Leipzig 1889, p. 13; *Alembic Club Reprints No. 2, Foundations of the Atomic Theory by J. Dalton, W. H. Wollaston, T. Thomson (1802–1808)*, Edinburgh 1961, p. 26.

14 R. HOOYKAAS, *Chem. Weekblad*, 44 (1948) 229, 321, 339, 407.

15 J. DALTON, *A New System of Chemical Philosophy*, London 1808, Part I p. 219; see also J. DALTON, *Ein neues System des chemischen Theiles der Naturwissenschaften* (transl. by F. Wolff) Berlin 1812, Vol. I, p. 246; *Ostwald's Klassiker No. 3, op. cit.*, p. 18; *Alembic Club Reprints No. 2, op. cit.*, p. 33.

16 J. DALTON, *A new System of Chemical Philosophy*, Manchester–London 1810, Part II, p. 248.

17 J. DALTON, *op. cit.*, Part II, p. 546.

18 J. DALTON, *op. cit.*, Part I, p. 220.

19 J. DALTON, *op. cit.*, Part II, p. 260, 262.

20 J. W. VAN SPRONSEN, *Chem. Weekblad*, 60 (1964) 10.

21 J. J. BERZELIUS, *J. Chem. Physik (Schweigger)*, 15 (1815) 277; see also W. A. KAHLBAUM, *Monographien aus der Geschichte der Chemie*, Vol. III, *Berzelius' Werden und Wachsen*, von M. G. Söderbaum, Leipzig 1899, p. 222 ff.

22 TH. THOMSON, *System of Chemistry*, 4th ed., Edinburgh 1810, Vol. V.

23 W. H. WOLLASTON, *Ann. Chim.*, 30 (1814) 138.

24 P. L. DULONG and A. T. PETIT, *Ann. Chim. Phys.* [2], 10 (1819) 395; *Journ. de physique, de chimie et d'histoire naturelle*, 89 (1819) 81.

25 E. MITSCHERLICH, *Abhandlungen Akad. Berlin*, 1818–19, p. 427.

26 J. B. DUMAS, *Ann. Chim. Phys.* [2], 33 (1826) 337; transl. in *Ann. Phys. Chem.*, 9 (1827) 293, 416.

27 J. B. DUMAS, *Traité de Chimie*, Paris 1828, Vol. I, p. LXXV.

28 J. B. DUMAS, *Ann. Chim.*, 49 (1832) 210.

29 J. B. DUMAS, *op. cit. ref. 27*, Vol. I, p. XXXII and L.

30 L. GMELIN, *Handbuch der theoretischen Chemie*, Frankfurt a/Main 1827, Vol. I, Part I, pp. 34, 35, 36.

31 C. F. GERHARDT, *Ann. Chim.*, 7 (1843) 129; 8 (1843) 238.

32 J. J. BERZELIUS, *Lehrbuch der Chemie*, 2nd ed., Dresden 1827, Vol. III, Part I p. 112 ff.

33 See e.g. A. C. W. ROODVOETS, *op. cit.*; T. R. A. BEUKEMA, *De Ontwikkeling van het BegripValentie*, Thesis, Leyden 1935.

34 See e.g. J. W. VAN SPRONSEN, *Chem. Weekblad*, 61 (1965) 35.

35 See e.g. J. W. VAN SPRONSEN, *Chem. Weekblad*, 59 (1963) 665.

36 A. W. WILLIAMSON, *Proc. Roy. Inst. Gr. Brit.*, 4 (1864) 274; *Z. Chem. Pharm.*, 7 (1864) 697.

37 W. ODLING, *J. Chem. Soc.*, 7 (1855) 1.

38 C. F. GERHARDT, *Traité de Chimie organique*, Paris 1853, Vol. I, p. 121; *Lehrbuch der organischen Chemie*, Leipzig 1854, Vol. I, p. 138.

39 A. LAURENT, *Méthode de Chimie*, Paris 1854, p. 358.

40 A. STRECKER, A., *Theorien und Experimente zur Bestimmung der Atomgewichte der Elemente*, Braunschweig 1859, p. 146.

41 See e.g. A. STRECKER, *Ann.*, 68 (1848) 47.

42 M. FARADAY, *The Subject Matter of a Course of Six Lectures on the Non-metallic Elements*, London, delivered 1852, printed 1853, p. 158 ff.

43 L. R. GIBBES, *Proc. Elliot Society*, 2 (1875) 77, publ. in 1886.

44 F. VENABLE, *The Development of the Periodic Law*, Easton 1896.

45 W. H. TAYLOR, *J. Chem. Educ.*, 18 (1941) 403.

46 See also R. HOOYKAAS, *Chem. Weekblad*, 43 (1947) 526.

47 D. MENDELEJEFF, *Ann.*, Suppl. VIII (1871) 133.

48 See also D. MENDELEJEFF, *Principles of Chemistry*, London 1891, Vol. I, p. 22; Vol. II, p. 21; D. MENDELEJEFF, *Grundlagen der Chemie*, St. Petersburg 1891, pp. 27, 688.

49 D. MENDELEJEFF, *op. cit.*, Eng. ed., Vol. I, p. 207; German ed., p. 232.

50 See e.g. F. M. JAEGER, *Elementen voorheen en thans*, Groningen–The Hague 1948, p. 152.

51 F. A. PANETH, *Über die erkenntnistheoretische Stellung des chemischen Elementenbegriffs*, Halle/Saale 1931, p. 16 ff., reprinted from *Schriften der Königsberger Gelehrten Gesellschaft, Naturwissenschaftliche Klasse*, 8 (1931) 116 ff.; see also H. DINGLE and G. R. MARTIN, *Chemistry and Beyond, a Selection from the Writings of the late Professor F. A. Paneth*, New York–London–Sydney 1964, p. 53; *Scientia*, 58 (1935) 219, 272.

52 R. HOOYKAAS, *Het Begrip Element*, Thesis, Utrecht 1933, p. 215 ff.

53 CH. JANET, *Essais de Classification hélicoïdale des Eléments chimiques*, Beauvais 1928, p. 10.

54 CH. JANET, *La Classification hélicoïdale des Eléments chimiques*, Beauvais 1928, p. 9.

55 D. C. HODGES, *Am. J. Sci.* [3], 10 (1875) 277.

56 M. CAREY LEA, *Am. J. Sci.* [3], 49 (1895) 357.

57 D. MENDELEJEFF, *op. cit.* ref. *48*, Eng. ed., Vol. II, p. 21; German ed., p. 688.

58 J. J. BERZELIUS, *Ann. Phys. Chem.*, 14 (1828) 566.

59 J. J. BERZELIUS, *Lehrbuch der Chemie*, 3rd ed., Dresden–Leipzig 1835, Vol. V.

60 J. J. BERZELIUS, *Lehrbuch der Chemie*, 5th ed., Dresden–Leipzig 1845, Vol. III, p. 1237 ff.

Chapter 4

Precursors of the periodic system (1817–1862)

4.1. Introduction

The first observation of a relationship between the atomic weights (equivalent weights) of elements (initially, compounds of elements) with analogous properties was made in 1817 by Döbereiner. The time for such a discovery was far from ripe. This is evident from the fact that Döbereiner himself had to wait for more than twelve years before he was able to follow up his own early discovery. Actually, except for a few studies done by Gmelin, it was not until 1850, more than 30 years later, that serious research was undertaken in this field.

Some of these early workers, who dealt with the simple arithmetical relationships between atomic weights, considered the divisibility of atoms a subject for argument. It is a remarkable fact that a relatively large number of these investigators, including such scientists as Pettenkofer, Dumas, Gladstone, Cooke and Mercer, saw a correspondence with the series of homologues in organic chemistry that Gerhardt had discovered a few years before. As a result, the relationships between the atomic weights were subjected to closer examination. Carey Lea assumed even negative atomic weights, while Kremers thought the atomic weight depended on temperature. The most striking findings, which concerned mutual relationships between the groups and which led directly to the discovery of the periodic system, were made in 1857 and 1858 by Odling as well as by Dumas, who also took an active part in the development of the theory of organic radicals, first suggested in 1843 by Gerhardt.

4.2. Döbereiner's triads

In 1817, one year after Ampère [1] had published his detailed articles on a natural classification in science, Wurzer reported that Johann Wolfgang Döbereiner [2] (1780–1849), professor of chemistry at Jena, stated the equivalent weight of strontium oxide (50) to be the average of those of calcium oxide (27.5) and barium oxide (72.5):

$$SrO^1 = \frac{CaO + BaO}{2} = \frac{27.5 + 72.5}{2} = 50$$

[1] We have included the equations here as a matter of interest; Döbereiner himself used no formulas. He called the oxides *Strontia*, *Kalk* (lime) and *Baria*.

As his basis he took H $= 1$ and O $= 7.5$, the values on which Wollaston's atomic weight table (Table 1) had been constructed[2]. Döbereiner believed that there was a similar relationship between the specific weights of the sulphates of these elements:

$$SrSO_4{}^3 = \frac{CaSO_4 + BaSO_4}{2} = \frac{2.9 + 4.4}{2} = 3.65$$

He considered this relationship to be very remarkable and believed that the strontium sulphate mineral celestine contained equal parts of anhydrate and heavy spar. He therefore evidently saw no reason not to consider the element strontium, which had been separated in 1808, as a mixture of calcium and barium. The research on celestine was inspired by Goethe, who was advised in chemical matters by Döbereiner and who had a special interest in minerals [3]. Döbereiner also discussed the concept of stoichiometry with Goethe. In the same year Döbereiner [4] also noticed that elements having many chemical and physical properties in common, for example Fe, Co and Ni, do not differ greatly in equivalent weight.

These few observations must be considered as the first attempt to introduce the atomic weight[4] as a parameter in the relationships between elements. Consequently, Döbereiner must be credited with preparing the way for the discovery of a system in which all elements can be classified and in which the atomic weight—much later the atomic number—is the essential factor (see 3.3).

A study of Döbereiner's textbook on chemistry reveals how he arrived at his discovery [5]. In the introduction to the first part, which appeared in 1811 (§ 2), the author already expressed his faith in the systematic approach to chemistry: "Die Grundlage der Chemie ist *Erfahrung*, wozu wir durch *Beobachtungen* und *Versuche* gelangen. Aus der Erfahrung leitet der Chemiker durch richtige Vernunftschlüsse eine Theorie her, welche alle einzelne Tatsachen zu einem wissenschaftlichen Ganzen, oder zu einem Systeme verknüpft. Ein chemisches System ist daher der Inbegriff aller chemischen Erfahrungen."

Although there is no reference, either in this introduction or in the rest of the book, to a system of elements, it is evident that Döbereiner was on the right track, the more so because the development of his argument shows that he adhered to the doctrine of *Verwandtschaft* and was in fact continuing the work of the eighteenth-century investigators in this field. When applied to the alkaline earth metals, the *Verwandtschaftsfolge* of the base of these elements reacting with a constant weight of a certain acid—i.e. the weights of the bases—gives the sequence Ba, Sr, Ca, Mg, Be, thus from heavy to light.

Two years later, Döbereiner began to construct a list of the equivalent weights of elements and compounds. This work was completed in 1815 and was published a year

---

[2] The values used by Döbereiner and those in Wollaston's list are not precisely the same. With Wollaston's values the equivalent weights would be: CaO $= 19.3 + 7.6 = 26.9$; SrO $= 40.1 + 7.6 = 47.7$ and BaO $= 65.8 + 7.6 = 73.4$.

[3] Döbereiner used simply the names, i.e. *Schwefelsauren Strontia*, *Schwefelsauren Kalk* and *Schwerspath*.

[4] For the problematical difference between atomic weight and equivalent weight, see 3.2 ff.

later *[6]*. The list facilitated the estimation of weight relationships between reacting compounds. From the resulting numerical series, a table of equivalent weights can be drawn up which, on closer examination, shows agreement with the one Wollaston composed in 1814, but Döbereiner's list is more complete. It even contains two more elements (I and Ir) than the list Berzelius published in 1815 (see *3.2* and *3.3*).

Johann Wolfgang Döbereiner
Statue in front of the Chemical University Laboratory, Jena

In 1816 Döbereiner proposed an analytical method for separating calcium and magnesium (by means of ammonium carbonate and ammonium chloride), from which his familiarity with this material became clearly apparent; also that his discovery of a triad was the result of very thorough chemical research. Döbereiner sought a system into which he could fit his triads, the elements with analogous properties and roughly the same equivalent weight (e.g. Fe, Co, Ni, Cr, Mn), and the phenomenon that the rarest elements (e.g. Sn, Hg, Au, Ag, I) have the largest equivalent weights. In this he did not succeed. He found only a single horizontal relationship and a single vertical one of the future periodic system.

References p. 96

In 1829, twelve years after his first publication of the triads, this scientist puhblsied his second and last article *[7]* on this subject. It is odd that in his books *[8, 9]* published in 1826 he made no reference whatever to the triad he had already found in 1817. Berzelius' improved determinations of equivalent weights confirmed a surmise which Döbereiner had not previously committed to paper, viz., that bromine should have an atomic weight equal to the average of those of chlorine and iodine:

$$Br = \frac{Cl + I}{2} = \frac{35.470 + 126.470}{2} = 80.470 \; ^5.$$

The value determined by Berzelius for the atomic weight of bromine, 78.383, lies close to this and could, in Döbereiner's opinion, tend towards the calculated weight.

It is most interesting that Döbereiner set up this triad only one year after the discovery of bromine as an element. All three elements of this triad occur in the sea-salts and belong together just like the alkaline earth metals.

Using more accurately determined equivalent weights, he now restated the triad he had constructed twelve years before for three of the alkaline earths:

$$SrO \; ^6 = \frac{CaO + BaO}{2} = \frac{56.96 + 153.10}{2} = 105.03 \text{ (converted at O} = 16).$$

The value for the equivalent weight of SrO, determined by Berzelius at 103.57, corresponds well with this. Comparing the above values with those used by Döbereiner in 1817, we see distinct progress in the mode of equivalent weight determination. At the same time, a doubling of the equivalent weight values appears to have been accepted. After conversion into the equivalent weights of the metals, these values are found to compare very closely with the present ones:

Berzelius' values		Present values
Ca	41	40
Sr	87.5	88
Ba	137	137

Furthermore, Döbereiner found[7] for:

$$\dot{N}a = \frac{\dot{L}i + \dot{K}}{2} = \frac{195.310 + 589.916}{2} = 392.613$$

---

[5] A printer's error may have altered this value; it should be 80.970.

[6] Döbereiner here still followed Berzelius' nomenclature for these oxides: Cȧ, Bȧ and Sṙ. He also put O = 100 in contrast to the case of the halogen triad.

[7] With our formulas, these notations would be: NaO, LiO, and KO. This gives a doubled value for the atomic weight of the metals.

(converted at $O = 16$ and with metals instead of the oxides[8]: $Na = \dfrac{Li + K}{2} =$
$\dfrac{15.25 + 78.39}{2} = 46.82$. Berzelius gave the atomic weight of Na as 390.897; for
$O = 16$, it was 46.55).

It should be noted that Döbereiner was obliged to use for lithium the equivalent weight given by Gmelin in 1827, i.e., 7.6. Although he did not indicate that he had done so, he took two equivalents of lithium in the oxide but still used the Li formula. This was done to obtain agreement with the equivalent weight values of sodium and potassium, also given by Döbereiner, which were too large by a factor of 2. With Berzelius' value, i.e., 20.5, no triad would have been found. All this shows how strongly Döbereiner believed in his law of triads.

In the group of phosphorus and arsenic, the third member was missing. Curiously enough, this scientist did not recognize antimony as such. It is true that the triad does not altogether apply here, but he could have attributed this to an erroneous determination of the atomic weight. This would have given him the following triad:

$$As = \frac{P + Sb}{2}; \qquad 75 \sim \frac{31 + 128}{2} = 79.5.$$

Another triad found by Döbereiner is:

$$Se = \frac{S + Te}{2} = \frac{32.239 + 129.243}{2} = 80.741 \text{ (Berzelius' value was 79.263).}$$

In his opinion fluorine could not belong to the group Cl–Br–I, because $Cl \neq \dfrac{F + Br}{2}$. This element would therefore have to be considered as belonging to another group of halogens. He believed that the two missing members of this group would be discovered if the law of triads was valid. It is clear that Döbereiner refused to include the elements in triads except on reliable grounds. When he found that

$$N = \frac{C + O}{2} = \frac{12.256 + 16.026}{2} = 14.141 \text{ (Berzelius' value was 14.138),}$$

he attributed this result to chance, because these elements show no resemblance to each other. For several other pairs of elements he did not find a third member, notably B and Si, Al and Be, Yt and Ce. This proves that at that time boron was still considered to be a homologue of silicon, and aluminium a homologue of beryllium, also as regards valence; this view later caused widespread disagreement (see *12.6*). Döbereiner did not relate magnesium to the alkaline earths. He continued to see it as a separate element. This case is comparable to the exceptional position of fluorine. Other triads

---

[8] With the correct formulas for the oxides, M_2O, the values would be halved.

References p. 96

found by him, are the following[9]:

$$\frac{Fe_2O_3 + Mn_2O_3}{2} = Cr_2O_3 = \frac{979.426 + 1011.574}{2} = 995.500 \ (O = 100)$$

$$(\text{converted at } O = 16: \frac{156.4 + 161.8}{2} = 159.1)$$

$$\frac{FeO + CoO}{2} = MnO = \frac{439.213 + 468.991}{2} = 454.102 \ (O = 100) \text{ or}$$

$$\frac{70.2 + 72.9}{2} = 71.6 \ (O = 16)$$

$$\frac{NiO + ZnO}{2} = CuO = \frac{469.675 + 503.226}{2} = 486.450 \ (O = 100) \text{ or}$$

$$\frac{75.2 + 80.4}{2} = 77.8 \ (O = 16).$$

For $O = 16$, the correct values of the averages are 160.6, 72.9 and 79.3, respectively. Thus, real differences were demonstrated, and Döbereiner himself was aware of them. He noted that the specific weight of $Cu \neq \dfrac{Ni + Zn}{2}$, and thought that the six iso-morphous oxides mentioned should be arranged differently and that a revision of the atomic weights and specific weights would be useful. The sequence of the elements of the first two triads should not be Fe–Cr–Mn and Fe–Mn–Co, but rather Cr–Mn–Fe and Mn–Fe–Co. Döbereiner further indicated the following relationships:

$$\frac{Pt + Os}{2} = Ir = \frac{1233.260 + 1244.210}{2} = 1238.735 \ (O = 100) \text{ or}$$

$$\frac{197.3 + 199.1}{2} = 198.2 \ (O = 16).$$

Berzelius found 1233.260 (197.3) for the atomic weight of iridium. Finally, Döbereiner found the relationships for the atomic weights:

$$\frac{Pd + Pluranium}{2} = Rh \text{ and further:}$$

$$\text{specific weight Pb} = \frac{\text{specific weight Ag} + \text{specific weight Hg}}{2}.$$

In 1828, Sann separated pluranium[10] from raw platinum. This later appeared to be highly impure ruthenium. A pure ruthenium was not produced until 1844 by Klaus. The sequence was later found to be Ru–Rh–Pd instead of Pd–Rh–Ru.

---

[9] Döbereiner's notation for Fe_2O_3 is $\overset{\cdots}{Fe}$, and that for the two other oxides is similar to this.

[10] In a footnote, Poggendorff, the editor of the periodical, pointed out to Döbereiner that there was some uncertainty about Sann's discovery.

4.3. Contributions of Gmelin; 1827, 1843 and 1852

Leopold Gmelin (1788–1853), professor of chemistry and medicine at Heidelberg, was, after Döbereiner, the only inorganic chemist of high repute to study the numerical relations between the atomic weights of the elements in this period. From the argumentation of the third edition of his *Handbuch der theoretischen Chemie [10]*, which appeared in 1827, it is not apparent whether he was aware of the relationships established by Döbereiner in 1817, but later publications (e.g. in 1843) show that Gmelin became very interested in Döbereiner's work. He did more, however, in this field, than simply keep abreast of his contemporaries' work and mention their results in the various editions of his *Handbuch [10, 11, 12]*. Gmelin also took an active part in the demonstration of relationships. This deserves special mention because it is generally believed that, after Döbereiner's two publications, no important additions were made in this field until the eighteen fifties.

In 1827 Gmelin proposed two triads, the first of which had been formed by Döbereiner as early as 1817: $\dfrac{Ca + Ba}{2} = Sr$ and $\dfrac{Li + K}{2} = Na$.

As we have already seen, Döbereiner had included magnesium among the alkaline earths, but could not make it part of a triad. Gmelin now found the following relationship:

$$\frac{Mg + Ba}{4} = \frac{12 + 68.6}{4} = 20.15 = Ca\ (= 20.5).$$

Gmelin thought Prout's hypothesis might be a natural law. He assumed that these and other remarkable relationships were unquestionably related to the "innerste Wesen der Stoffe". Like Döbereiner, he did not consider these relationships to be accidental. He also observed that the atomic weights of other elements sometimes form series too. An example[11] is:

$$Al : Gl : Yt : Ce = 1 : 2 : 3\tfrac{1}{2} : 5.$$

Furthermore, he pointed out a number of elements that, taken in groups, have almost the same atomic weight or a twofold, fourfold or eightfold value of that weight:

Mo = Pt = W = 96; Ta = 184 ($\sim 2 \times 96$)
Cr, Mn, Co, Fe, Ni = 28–29; Cd = Pd = 56 (2×28); U = 217 ($\sim 8 \times 28$)
Ti = Te = Zn = Cu = 31–32; Sb = Au = 64–66 ($\sim 2 \times 32$) etc.

The neglect of the properties of the elements is quite evident here.

In 1843 Gmelin *[11]* was able to give an extension of Döbereiner's triads: "Es gibt Gruppen von Elementen, welche ähnliche physikalische und chemische Verhältnisse zeigen. Ob eine solche Gruppe gerade aus 3 Elemente bestehe, wie Döbereiner will,

---

[11] Gl = glucinium = beryllium.

References p. 96

Cr   28.1   Mn   27.6   Fe   27.2
Co   29.6   Ni   29.6
Zn   32.2   Cu. 31.8
Pt   98.7   Ir   98.7   Os   99.6

O   8   S   16   Se   40   Te   64   Sb 129 $= 1:2:5:8:16$
F   18.7   Cl   35.4   Br   78.4   J   126 $= 2:4:9:14$
Mg   12.7   Ca   20.5   Sr   44   Ba 68.6 $= 3:5:11:17$
Si   14.8   Zr   22.4   Th   59.6 $\sim 2:3:8$
Ti   24.5   Mo   48   W   95   Ta 185 $\sim 1:2:4:8$
Cr   28.1   V   68.6 $\sim 2:5$
Mn   27.6   U   217 $\sim 1:8$

$$\frac{Cl\ 35.4\ +\ J\ 126}{2} = 80.7 = Br$$

$$\frac{Li\ 6.4\ +\ \acute{K}\ 39.2}{2} = 22.8 \sim Na$$

$$\frac{Ca\ 20.5\ +\ Ba\ 68.6}{2} = 44.55 \doteq Sr$$

$$La \sim 3 \times Mg$$

$$P\ 31.4\ +\ As\ 75.2 = 106.6 = W$$

$$Pd\ 53.4 \sim \tfrac{1}{2}Ag\ 108.1$$

$$\frac{Hg\ 101.4\ +\ Ag\ 108.1}{2} \sim Pb = 103.8$$

Fig. 1   Atomic weight relations formulated by Gmelin in 1843 and 1852.

welche die Elemente nach der Trias gruppiert, bleibe dahingestellt." As in 1827, Gmelin clearly distinguished three possibilities. There are elements having analogous properties, such as chromium, manganese and iron, whose atomic weights are about equal. Secondly, an atomic weight can be a multiple of that of another element. The examples given by this author in 1843 and in 1852 in the fifth edition of his *Handbuch* *[12]* are to be found in Fig. 1, together with triads serving as an example of the third possibility, namely the existence of mean atomic weights, i.e., triads. This in fact represents a large group of relationships in which both atomic weight and analogy were taken into consideration.

   In 1843 Gmelin also tried to find a relationship existing between all elements. This, however, meant demoting the atomic weight. Fig. 2 shows one of his attempts. The elements oxygen, nitrogen and hydrogen, for which he apparently could find no homologues, form a basis for his classification. In the 1849 English edition of the *Handbuch* *[13]* Watts[12], the translator, added some elements, including erbium, yttrium, terbium and didymium, as well as niobium and pelopium. The existence of

---

[12] Note that Watts here made a contribution to the known relationships between the elements, although in his own textbook, *A Dictionary of Chemistry*, Odling had dealt with this point (see 5.5).

```
        O                                N                              H
    F  Cl  Br  J                                                 L  Na  K
       S  Se  Te                                            Mg  Ca  Sr  Ba
          P  As  Sb                                         G  Y  Ce  La
              C  B  Si                                     Zr  Th  Al
                 Ti  Ta  W                              Sn  Cd  Zn
                    Mo  V  Cr        U  Mn  Co  Ni  Fe
                       Bi  Pb  Ag  Hg  Cu
                          Os  Ir  R  Pt  Pd  Au
```

Fig. 2   System of Gmelin (1843).

```
        O                                N                              H
        8¹⁾                              14                             1

  F    Cl   Br   J                                              L    Na   K
  19   35   80   127                                            6    23   39

       S    Se   Te                                       Mg   Ca   Sr   Ba
       16   39   64                                      ₁12   20   44   68

            P    As   Sb               G³⁾  Er   Y   Tr⁴⁾  Ce   Di   La
            31   75   129              5    -    -    -    47   50   47

                 C    B    Si                            Zr   Th   Al
                 6    11   21                            22   60   14

            Ti   Ta   Nb   Pe²⁾  W               Sn   Cd   Zn
            25   184  -    -     92              58   60   53

                 Mo   V    Cr         U    Mn   Co   Ni   Fe
                 46   69   27         60?  28   29   30   28

                      Bi   Pb   Ag   Hg   Cu
                      208  104  108  100  32

                      Os   Ru   Ir   Rh   Pt   .Pd   Au
                      100  52   99   52   99   53    197
```

¹⁾ atomic weights from Liebig's Jahresbericht 1851
²⁾ pelopium
³⁾ glucium (= beryllium)
⁴⁾ terbium

Fig. 3   Arrangement of the elements by analogous properties according to Gladstone (1853), based on the classification of Gmelin.

this last element has never been demonstrated. It is a remarkable fact that, although these elements were mentioned, they were not included in the system published in the next German edition [12], which Watts helped to prepare. This was the fifth edition, which appeared in 1852, one year after Gmelin's retirement and a year before his death. Gladstone based the relationships he found on the system given in the English edition of Gmelin's textbook (Fig. 3) (see also 4.8).

Leopold Gmelin   *Max von Pettenkofer*

4.4. Interlude

Little attention was paid to Döbereiner's work, as can be seen from the fact that
only one publication on the subject appeared before 1850. Berzelius apparently did not
consider the triads worth mentioning in either his *Jahresbericht* or his textbook. He did
notice *[14]* that the ratio of the atomic weights of oxygen, sulphur, selenium and
tellurium took the form of simple whole numbers, viz., 1 : 2 : 5 : 8, but he preferred
to postpone drawing any conclusions until the scientific developments had reached a
more advanced stage.

We cannot help wondering whether Gmelin was the only scientist who attempted to
find relationships between the elements during this long period (from 1829 to 1850).
He was not, but neither the atomic weight nor the equivalent weight was considered
to be a determining factor. Kopp *[15]* was also searching for relationships between
elements. He observed that the specific volumes of homologous elements, such as Cl,
Br and I; W, Mo and Cr; Mn, Ni and Co, are roughly equal or differ by a factor of 2, as
for Ag and Au; K and Na. He also found that the isomorphous elements show relationships in their specific weight, atomic weight and atomic volume *[16]*. Persoz
wanted to demonstrate that the specific weights of the elements are proportional to
the atomic weights or to multiples of them *[17]*, but this conclusion took him too far.

4.5. Pettenkofer; 1850

In 1850, Liebig's former pupil Max von Pettenkofer (1818–1901), the extraordinary professor of medical chemistry at the University of Munich and founder of the science of experimental hygiene, opened a new era during which one or more publications on the present subject appeared almost every year. He gave a lecture on the relationship between the equivalent weights of the elements, which was published [18]. Pettenkofer considered that the results of his investigations supported Prout's hypothesis. The hypothetical equivalent weights used by Pettenkofer are whole numbers, which express the relationships very clearly. He drew attention to Gmelin's conclusion about these relationships. Although Pettenkofer was one of the very first scientists to extend

element	eq. w. hyp.	eq. w. det	difference
Li	7	6.51	0.49
Na	23	22.97	0.03
K	39	39.11	0.11
Mg	12	12.07	0.07
Ca	20	20.00	0.00
Sr	44	43.92	0.08
Ba	68	68.54	0.54
Cr	26	26.00-26.30	0.00
Mo	46	46	0.00
W	66	?	?
O	8	8	0.00
S	16	16	0.00
Se	40	39.62	0.38
Te	64	64.14	0.14
C	6	6	0.00
N	14	14	0.00
Hg	100	100	0.00
Ag	108	108	0.00
C	6	6	0.00
B	11	11	0.00
Si	21	21.34	0.34

Fig. 4   Classification of Pettenkofer (1850).

the study of the relationship between analogous elements, his statements show that he did not see the connection between his results and those of Döbereiner. Pettenkofer put it down to mere chance that the equivalent weight of Br is the mean of those of Cl and I and also that the equivalent weight of Sr is the mean of those of Ca and Ba, because, he said, F, Cl and Br, as well as Mg, Ca and Sr, are also analogous elements and the relationships mentioned above do not hold for them.

From this statement it could be inferred that Pettenkofer did not consider Döbereiner's or Gmelin's work, showing a similar relationship between the last-mentioned elements, as a basis on which to continue work. This inference would be mistaken, however. He too created triads and larger groups of analogous elements (Fig. 4), but his approach was different. He came to the conclusion that there is a constant difference between the equivalent weight of, for example, Li, Na and K. He failed to realize that this conclusion is in fact another way of saying that Na has the mean equivalent weight of Li and K. It is conceivable that Pettenkofer chose this approach merely as a means of extending the relationships by taking not only a constant difference between equivalent weights into account, but a multiple of them as well. This freed him from limitation to triads. The four elements Mg, Ca, Sr and Ba, the last three of which form one of Döbereiner's triads, he regarded as elements whose atomic weights differ by eight units or by a multiple thereof.

For the alkali metals, the alkaline earths, and the sulphur and chromium groups, this unit of difference is also 8. For the group C, B and Si, and for the halogens, it is 5 [13]. In what Pettenkofer called the *natural group*, comprising N - 14, P - 32, As - 75 and Sb - 129, the differences consist of multiples of 5 and 8. He also noticed that the difference between the atomic weights of C and N, Mg and Ca, and Hg and Ag, is likewise 8. Pettenkofer attributed the fact that the frequently occurring difference of 8 is equal to the equivalent weight of oxygen to coincidence, because it is by mere chance that, in the homologous series methyl - 15, ethyl - 29, butyl - 57, and amyl - 71, the difference of 14 or a multiple thereof is equal to the equivalent weight of nitrogen. Because of the parallel between the natural group of simple substances and the natural group of organic radicals, he was nevertheless more or less compelled to assume that the elements were further divisible, the more so as the radicals ammonium and cyanogen, for instance, have properties analogous to those of elements. This scientist was the first to see a numerical connection between inorganic and organic substances, in which he was soon to be followed by Dumas.

If this theory of constant differences or multiples thereof is valid, said Pettenkofer, it should be possible to calculate the equivalent weights of elements for which these are difficult to determine. He thus predicted atomic weights of elements whose properties must be known (see also *7.2*). In 1858 Pettenkofer, meanwhile appointed ordinary professor, extended his studies with reference to similar investigations by Dumas (see *4.8*).

4.6. Dumas; 1851

In 1851 Jean Baptiste André Dumas *[19, 20]* (1800–1884), the great French chemist, and Minister of Agriculture and Commerce, expounded his views in a lecture to the British Association for the Advancement of Science. From the contents of this lecture it is evident that he was not aware of Pettenkofer's work. Although as early as 1828 he had already classified the non-metals into three groups *[21]* (the fluorine, oxygen and nitrogen groups) and although he knew that the elements of these groups could com-

---

[13] This value gives only an approximation of the difference.

bine with $\frac{1}{2}$, 1 and $1\frac{1}{2}$ hydrogen atoms, respectively, per $\frac{1}{2}$ atom of the elements, in 1851 he noticed various triads, e.g.,

S	Se	Te		Cl	Br	I
16	40	64		35	80	125

$$\left(=\frac{S + Te}{2}\right) \qquad\qquad \left(=\frac{Cl + I}{2}\right)$$

Moreover, only by the names of their components, he noted the triads lithia–soda–potassa, and lime–strontia–baryta. He wondered whether the middle element might not be a compound of the two extremes. Reviving the alchemists' views on the trans-mutation of metals, Dumas suggested that the elements of a triad can merge into one another, and recommended that research should be undertaken with the aim of discovering the mechanism of the transmutation of molecules. In support of his view he cited the combined occurrence in nature of the three elements of a triad or two of the three elements of a triad, e.g., Cl, Br and I; Co and Ni; Fe and Mn; S and Se. Like Pettenkofer, Dumas called attention to the resemblance to organic radicals. We see that Dumas' opinions were similar to those of his predecessors. He had little to add to Döbereiner's statements about the triads, even though he was bolder in his concept of the middle term of a triad, but was no more prepared than Pettenkofer to regard this term as a simple substance. Where Döbereiner saw this element as a mixture of the other two, Dumas assumed it to be a compound of them.

4.7. Kremers; 1852 and 1856

Peter Kremers [22] (b. 1827) was a private chemist at Bonn and later at Cologne, who performed many experiments with salt solutions.

In his quest for relationships between analogous elements, he began to examine numerical connections between elements having little in common. It surprised him to find that some elements differ in atomic weight by 8 units, e.g.,

O	S	Ti	P	Se
8	16	24.12	32	39.62

and that the following triads could also be identified:

$$Mg = 12.07 = \frac{O + S}{2}, \quad Ca = 20 = \frac{S + Ti}{2}, \quad Fe = 28 = \frac{Ti + P}{2}.$$

Like most of his contemporaries, Kremers formed concepts about the composition of substances. He noticed that division of the atomic weights of the metals and metalloids by 4 left an odd number in the case of the former and an even number in that of the latter. This led him to surmise that 4 was the atomic weight of a basic element; therefore, multiplication of this atomic weight by an odd number would give that of a metalloid. The explanation of the affinity of a metal for a non-metal would then be reduced to this relationship of even and odd atomic weights.

Four years later, Kremers [23] thought he had found the reason for the divergence of the triad's middle terms from the mean atomic weight of the two extremes (Fig. 5). He made the astonishing assertion that such deviations must be dependent on temperature; at a given temperature these deviations would disappear. Kremers arrived at this concept after learning that the solubility of KBr, for example, is equal to that of a mixture of equimolecular weights of KCl and KI, only at one particular temperature.

h	m	$\dfrac{h-m}{h}$
$\dfrac{K^{*)} + Li}{2}$	Na	-0.007
$\dfrac{Ba + Ca}{2}$	Sr	+0.010
$\dfrac{Ag + Hg}{2}$	Pb	+0.003
$\dfrac{J + Cl}{2}$	Br	+0.016
$\dfrac{S + Se}{2}$	Cr	+0.038
$\dfrac{S + Te}{2}$	Se	+0.017
$\dfrac{Cr + Va}{2}$	Mo	+0.035
$\dfrac{P + Sb}{2}$	As	+0.009

*)Atomic weights from Liebig and Kopp's Jahresbericht 1854; only that of antimony (120.3) is from Schneider

Fig. 5 Differences between calculated and determined atomic weights of the elements of some triads of Kremers.

With this concept Kremers cannot be said to have done much to advance the development of science. He also gave some consideration to known triads, the only new contribution, that of Ag–Pb–Hg, being based on analogies, in respect to analytical chemistry. Kremers did not refer to his concept again (5.9) until after the time of the discovery of the periodic system.

4.8. Gladstone; 1853

John Hall Gladstone (1827–1902), reader at St. Thomas' Hospital, London, too, was interested in triads, as shown by his participation in the discussions after Dumas' lecture in 1851 (see *4.6*). In 1853, while arranging elements according to increasing atomic weight *[24]*, he made several observations, none of which were very striking. He came to the conclusion that many elements have atomic weights around 28, 52 or 99 (Cr, Mn, Fe, Co, Ni; Ru, Rh, Pd; Pt, Ir, Os). Between 80 and 99 there is only one element, tungsten. Gladstone thought these irregularities were so striking that he asked a mathematician to make a probability calculation. His arrangement of elements in the order of analogy (Fig. 3), however, also interested him. As we have seen, this arrangement, which follows the English edition of Gmelin's 1849 textbook *[13]*, seems at first sight to be an extensive system, but on closer examination it is evident that the elements (almost *all* known elements) had not yet been arranged in the order of increasing atomic weight. A large number of them, however, had been compared.

Gladstone himself did not always find the analogy striking. For example, tin shows little resemblance to cadmium and zinc, and the same holds for mercury and copper; fluorine differs considerably from the other halogens, and magnesium from the other alkaline earth metals.

His main contribution was the classification of several kinds of analogies. He defined three categories:

(1) *Elements having almost the same atomic weights*

There is one large group of metals with strongly analogous properties:

Cr	Mn	Fe	Co	Ni
26.7	27.6	28	29.5	29.6

and two small groups:

Pd	Rh	Ru	and	Pt	Ir	Os
53.3	52.2	52.2		98.7	99	99.6

Gladstone considered whether to add Hg (100) to the latter. Ce (47), La (47) and Di (50) occur in the same mineral. He wondered whether the atomic weights of the elements of these small groups would prove to have just the same value for each of these groups. He, unlike Döbereiner, did not try to form triads with these elements.

(2) *Elements having atomic weights that are multiples of each other*

The elements of the Pt-group have about double the atomic weights of the elements of the Pd-group; other examples being B (10.9) and Si (21.3); O (8) and S (16).

He also noticed a series with the difference of 11.5:

$2 \times 11.5 =$	23	Ti $=$	25
$4 \times 11.5 =$	46	Mo $=$	46
$5 \times 11.5 =$	57.5	Sn $=$	58
$6 \times 11.5 =$	69	Y $=$	68.6
$8 \times 11.5 =$	92	W $=$	92
$16 \times 11.5 =$	184	Ta $=$	184.

References p. 96

This series is not very useful, in contrast to the following group. Both were, however, already well known.

(3) *Triads*

Li	–	Na – K		Ca – Sr	–	Ba		Cl	–	Br – I		S	–	Se	–	Te
6.5		23	39.5	20		43.9	68.5	35.5		80	127	16		39.5		64.2

The resemblance between the elements of the first category was compared by Gladstone to the phenomenon of allotropy, that of the second to polymerism, and that of the third to the homologous series found in organic chemistry, thus following Pettenkofer and Dumas. The radical series consists of hydrogen and the alkyl groups:

hydrogen	methyl	ethyl	......
1	15	29	......
a	$a + x$	$a + 2x$	......

To the Li-triad, $x = 16.3$ In this connection he ascertained that
 „   „   Ca-triad, $x = 24.2$ x of the Li-triad $= \frac{2}{3}$ times the Ca-
 „   „   Cl-triad, $x = 45.8$ triad's x, and x of the Ca-triad $= x$
 „   „   S-triad, $x = 24.1$ of the S-triad.

According to Gladstone, all elements can be assigned to one of these groups, except As and Sb which, although resembling P, do not form a triad with this element.

P	As	Sb
31	75	129

He pointed out lastly that Sb has an atomic weight about twice that of Te (64.2).

Although this scientist did not speculate on the deeper significance of all the demonstrated relationships, in his opinion "we can scarcely imagine that the intimate constitution of these related elementary bodies will long remain an unfruitful field of investigation".

4.9. *Cooke; 1854–1855*

Cooke's publications [25, 26] show that scientists in the United States were already working in the field of atomic weights.

Josiah Parsons Cooke (1827–1894), appointed professor of chemistry and mineralogy at Harvard University in 1850, considered Dumas, whose lectures in Paris he had attended in 1848, to be the leader in this field and thus evidently did not know of Pettenkofer's work. For didactic reasons Cooke arranged the elements in six groups comprising six series of homologous substances comparable to those found in organic chemistry. He too represented the atomic weights of the elements of these six groups (Fig. 6) by arithmetical progressions.

	O	Fl	CN	Cl	Br	J
determined at. w.	8	18.8		35.5	80	126.9
theoretical at. w.	8	17	26	35	80	125
n =	0	1	2	3	8	13

atomic weight $= 8 + n \times 9$

	O	S	Se	Mo	Te	V	W	Ta
determined at. w.	8	16	39.6	46	64.1	68.5	92	184
theoretical at. w.	8	16	40	44	64	68	92	188
n =	0	1	4	5	7	8	11	23

„ $= 8 + n \times 8$ or $4 + n \times 8$

	O	N	P	As	Sb	Bi
determined at. w.	8	14	31	75	129	208
theoretical at. w.	8	14	32	74	128	206
n =	0	1	4	11	20	33

„ $= 8 + n \times 6$

	C	B	Si
determined at. w.	6	10.9	21.3
theoretical at. w.	6	11	21
n =	0	1	3

„ $= 6 + n \times 5$

atomic weight $= 4 + n \times 4$ or $2 + n \times 4$

	Gl	Mg	Al	Ca	Ti	Cr	Mn	Fe	Co	Ni	Cu	Zn	Zr	Sr	La	Ce
determined at. w.	4.7	12	13.7	20	25.2	26.7	27.6	28	29.5	29.5	31.6	32.5	33.6	43.8	47	47.3
theoretical at. w.	4	12	14	20	26	26	28	28	30	30	32	32	34	44	48	48
n =	0	2	3	4	6	6	6	6	7	7	7.	7	8	10	11	11

	R	Ru	Pd	Cd	Sn	Th	Ur	Ba	Ir	Os	Pt	Hg	Pb	Ag	Au
determined at. w.	52.1	52.1	53.3	56	58	59.6	60	68.5	98.5	99.4	98.5	100	103.7	108.1	197
theoretical at. w.	52	52	54	56	58	60	60	68	98	98	98	100	104	108	198
n =	12	12	13	13	14	14	14	16	24	24	24	24	25	26	49

Fig. 6   System of Cooke (1854).

His results showed a close resemblance to Pettenkofer's work. The latter pointed out only the differences in atomic weight between elements of a series of homologues, whereas Cooke provided equations for other series. He himself compared his work to that of Gladstone and considered it as an extension of the latter's work. Gladstone had already represented the triads by arithmetical progressions, particularly on the basis of isomorphism, of the components of the elements .Cooke extended these triads to longer series, as a result of which the coefficient of the arithmetical progressions must become smaller. In our opinion, however, a factor of three for a basic number of one does not make much sense. (So the atomic weight $= 1 + 3n$ in the case of hydrogen and the alkali metal group and the magnesium and cadmium groups.) With this kind of progression, almost any atomic weight can be obtained and many elements included, even elements whose properties show that they do not belong to the series. For that matter, the other series also contain strange combinations. Although Cooke tried, he failed to convince anyone of the analogy between vanadium and sulphur, for example. Though he was aware that vanadium and sulphur do not form isomorphous compounds and knew that vanadium and arsenic possess more properties in common, he placed these elements in the same group because they form analogous acids and because the oxides of vanadium are coloured, like the analogous oxides of molybdenum. To him the decisive factor in the arrangement of elements was whether or not they fell into the atomic weight series 8 (or 4) + $n \times$ 8 (see Fig. 6).

The last series shows that Cooke attached great importance to properties of particular significance in analytical chemistry; Al, Cr, Mn, etc. are placed together, without much regard to valence or other factors.

Like Pettenkofer and Gladstone, Cooke regarded the elements carbon, boron and silicon as homologues. Besides fitting into the series 6 + $n \times$ 5, these elements also possess analogous properties. Later, it was to be found that boron did not belong to this series and that carbon and silicon should be compared with titanium, tin and lead. When the correct atomic weights are used (B = 11, C = 12, and Si = 28), the series into which carbon, boron and silicon fitted so well, disappears. Of the three atomic weights, only that of boron was correct. The atomic weight of carbon required doubling and that of silicon to be changed from 21 to 28. The source of these errors lay in the fact that the three above-mentioned investigators[14] used the atomic weights assumed by Liebig and Kopp. A glance at the table of atomic weights (p. 46 ff.) suffices to show that the values given by Gerhardt in 1843 correspond better to those accepted today. Note that in the first series oxygen is consistently placed as the initial element. Apart from similarities to sulphur, oxygen does not belong to this series.

Cooke, whose publications show that he did much work on atomic weights, considered the series he determined to be the expression of divine laws, rather than accidental relationships, and used them to predict the properties of unknown elements. He believed that an analogy with the laws of astronomy would be found. This idea was developed by Hinrichs in 1864 (see *5.6*). If all atomic weights could be described by a single formula, their properties could be predicted theoretically and the alchemist's dream would be realized, if not practically, then at least in theory.

4.10. Lenssen; 1857

1857 was a very important year in the period preceding the discovery of the periodic system. Not only did a considerably larger number of publications on the subject appear in that year than at any previous time, but the size and importance of some of these publications were also greater (see also *4.11* and *4.12*). The twenty-year-old Lenssen [28] (b. 1837), who was assistant to Carl Remigius Fresenius, professor at the Agricultural Institute at Wiesbaden in 1857, believed that all elements except niobium could be assembled in triads (Fig. 7). To do this he had to halve the equivalent weights of N, P, As, Sb, Bi and Au. A striking fact is that Lenssen used half the actual values of almost all atomic weights, which was certainly unusual in those years.

Although Lenssen's statement that he could classify all elements suggests considerable progress, on closer examination his results are somewhat disappointing. It is difficult to accept that many of these triads—e.g., O, N, C, and Si, B, F—justify his

[14] Cooke reported having used the atomic weights given in a list in the 1852 *Jahresbericht* [27] by Liebig and Kopp. The copy available to us contained no table of atomic weights, however.

#	calculated atomic weight		determined atomic weight		
1	$\frac{K + Li}{2} = Na =$	23,03	39,11	23,00	6,95
2	$\frac{Ba + Ca}{2} = Sr =$	44,29	68,59	43,67	20
3	$\frac{Mg + Cd}{2} = Zn =$	33,8	12	32,5	55,7
4	$\frac{Mn + Co}{2} = Fe =$	28,5	27,5	28	29,5
5	$\frac{La + Di}{2} = Ce =$	48,3	47,3	47	49,6
6	Yt Er Tb		32,2	?	?
7	Th Norium Al		59,5	?	13,7
8	$\frac{Be + Ur}{2} = Zr =$	33,5	7	33,6	60
9	$\frac{Cr + Cu}{2} = Ni =$	29,3	26,8	29,6	31,7
10	$\frac{Ag + Hg}{2} = Pb =$	104	108	103,6	100
11	$\frac{O + C}{2} = N =$	7	8	7	6
12	$\frac{Si + Fl}{2} = Bo =$	12,2	15	11	9,5 [*]
13	$\frac{Cl + J}{2} = Br =$	40,6	17,7	40	63,5
14	$\frac{S + Te}{2} = Se =$	40,1	16	39,7	64,2
15	$\frac{P + Sb}{2} = As =$	38	16	37,5	60
16	$\frac{Ta + Ti}{2} = Sn =$	58,7	92,3	59	25
17	$\frac{W + Mo}{2} = V =$	69	92	68,5	46
18	$\frac{Pa + Rh}{2} = Ru =$	52,2	53,2	52,1	51,2
19	$\frac{Os + Ir}{2} = Pt =$	98,9	99,4	99	98,5
20	$\frac{Bi + Au}{2} = Hg =$	101,2	104	100 [**]	98,4 [**]

[*] with the double atomic weight of fluor the triad is :
$$\frac{Bo + Fl}{2} = Si = \frac{11 + 19}{2} = 15$$
[**] in the original table these two atomic weights have been interchanged

Fig. 7   Calculated and determined atomic weights of the elements of the triads, according to Lenssen (1857).

claims. A triad should show not only relationships between atomic weights but also analogies in properties. Lenssen regarded the element mercury, which occurs in both the 10th and 20th triads, as a connecting element (Fig. 8). It is not quite clear what he meant by this, unless he attached particular value to the numbers 10, 20 and 100 (= atomic weight of mercury). The order of the triads is not taken arbitrarily, as shown by the fact that he could construct so-called *enneads*, i.e. groups of three triads showing the same numerical relations mutually as the elements of a triad (Fig. 9). The actual result is the formation of new triads of elements, but these lack any connection whatsoever. If we consider ennead *a*, we find the triad:

$$Zn = \frac{Na + Sr}{2} = \frac{23 + 44}{2} = 33.5,$$

whose elements show little interrelationship. According to him, the enneads themselves must also obey specific laws, but he was unable to demonstrate this convincingly (Fig. 10).

Lenssen not only established triads and enneads, but also believed so strongly in their existence that he took them as the basis for the determination of the atomic weight of elements which, although known, had not yet been demonstrated con-

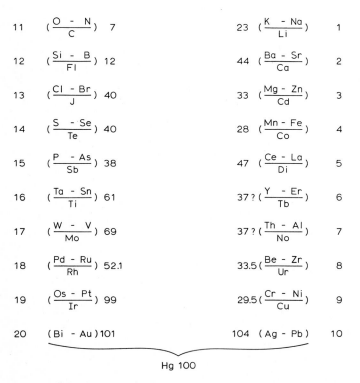

Fig. 8   Triads of Lenssen (1857).

$$a \begin{cases} \text{1e triad} & 23 \\ \text{2e triad} & 44 \\ \text{3e triad} & 33 \end{cases} \quad \frac{23 + 44}{2} = 33.5 \quad \text{i.o.} \quad 33$$

$$b \begin{cases} 7 & 37 \\ 8 & 33.5 \\ 9 & 29.5 \end{cases} \quad \frac{37 + 29.5}{2} = 33.3 \quad \text{i.o.} \quad 33.5$$

$$c \begin{cases} 10 & 104 \\ 20 & 101 \\ 19 & 99 \end{cases} \quad \frac{104 + 99}{2} = 101.5 \quad \text{i.o.} \quad 101$$

$$d \begin{cases} 18 & 52.1 \\ 17 & 69 \\ 16 & 61 \end{cases} \quad \frac{52.1 + 69}{2} = 60.6 \quad \text{i.o.} \quad 61$$

$$e \begin{cases} 15 & 38 \\ 14 & 40 \\ 13 & 40 \end{cases} \quad \frac{38 + 40}{2} = 39 \quad \text{i.o.} \quad 40$$

$$f \begin{cases} 12 & 12 \\ 11 & 7 \\ H & 1 \end{cases} \quad \frac{12 + 1}{2} = 6.5 \quad \text{i.o.} \quad 7$$

Fig. 9   Enneades according to Lenssen (1857).

clusively. The atomic weight of norium, not yet demonstrated as an element, results from triad 7:

$$No = \frac{Th + Al}{2} = \frac{60 + 14}{2} = 37.$$

In triad 6, the atomic weights of Er and Tb are unknown. Lenssen calculated these atomic weights by means of the laws of enneads and triads; the first ennead is indicated as:

$$\frac{\text{triad } 4 + \text{triad } 5}{2} = \text{triad } 6,$$

which corresponds to the triad $\dfrac{Fe + Ce}{2} = Er.$

a	b	c	d	e	f	g
33	34	37	61	40	7	101

d + e = g ; 61 + 40 = 101

a + f = e ; 33 + 7 = 40

$\dfrac{b + e}{2} = c$; $\dfrac{34 + 40}{2} = 37$

Fig. 10   Relations between enneades according to Lenssen

References p. 96

From this equation the known atomic weights give the atomic weight of erbium:

$$\frac{28 + 47}{2} = 37.5.$$

According to Mosander, the atomic weight of Y is 32.2. Therefore, by means of the sixth triad, the atomic weight of Tb can be calculated:

$$Er = \frac{Y + Tb}{2} = \frac{32.2 + 42}{2} = 37.$$

Although consideration of both the triads and the enneads suggests striking relationships, there is no reason to attach great value to these triads and enneads. Many of the relationships have no bearing on homologous elements, and incorrect atomic weights were used. At the end of his first publication Lenssen remarked that in every triad two elements are more closely related to each other than to the third element. He therefore took the third element to be a transitional element to the next triad, which he expressed as follows:

(1)	K	Na	
			Li
(2)	Ba	Sr	
			Ca
(3)	Mg	Zn	
			Cd
(4)	Mn	Fe	
			Co
(5)	Ce	La	
etc.			

In his second treatise [29] this author noted that the spectral colours of the elements of most triads are complementary. This also holds for the colours of the flame when the metals are burnt (in oxygen); e.g.,

K (reddish-blue) + Na (yellow) + Li (bluish-red) = white
Ba (green) + Sr (bluish-red) + Ca (yellowish-red) = white.

Lenssen commented on the fact that the element of the triad with the highest atomic weight usually has a mixed colour, whereas the middle element has the simple colour. An alloy of K, Na and Li would burn with a colourless flame. He believed this conclusion to be justified, because Liebig had demonstrated that a mixture of red manganese of the lowest valence and bluish-yellow ferrous salts become colourless when dissolved.

Lenssen made many factual observations, but did not propose a theoretical basis for them.

4.11. Dumas; 1857

Dumas [42], who was aware of Cooke's work in 1857, continued his comparative studies of the series of elements and the homologous series in organic chemistry (e.g. methyl, ethyl, etc.). He now also regarded the relationships in the light of Prout's hypothesis, although he did not consider it to hold for all the elements, unless the basic unit of matter was taken to be a half or a quarter of an atom of hydrogen instead of one hydrogen atom. He observed that Cl, Br and I do not exactly form a triad according to the series a, $a + d$, $a + 2d$. All four halogens can be placed in the following series, which he, of course, called a *natural family*:

a	$a + d$	$a + 2d + d'$	$a + 2d + 2d' + d''$
F	Cl	Br	I
	35.5	80	127
19	$19 + 16.5$	$19 + 33 + 28$	$19 + 33 + 56 + 19.$

As additional examples, he gave:

a	$a + d$	$a + d + d'$	$a + d + 2d'$	$a + d + 4d'$
N	P	As	Sb	Bi
14	31	75	119	207
	$14 + 17$	$14 + 17 + 44$	$14 + 17 + 88$	$14 + 17 + 176$

C	B	Si	Zr
a	$a + d$	$a + 3d$	$3a + 3d$
6	11	21	33
	$6 + 5$	$6 + 15$	$18 + 15$

O	S	Se	Te
a	$2a$	$5a$	$8a$

or

a	$a + d$	$a + 4d$	$a + 7d$
8	16	40	64
	$8 + 8$	$8 + 32$	$8 + 56.$

Dumas preferred the second of these last two series, because its structure is analogous to the others. Here d is 8. Another example of a case in which $d = 8$, is:

Mg	Ca	Sr	Ba	Pb
a	$a + d$	$a + 4d$	$a + 7d$	$2a + 10d$
12	20	44	68	104
	$12 + 8$	$12 + 32$	$12 + 56$	$24 + 80.$

With $d = 2 \times 8 = 16$:

Li	Na	K
a	$a + d$	$a + 2d$
7	23	39
	$7 + 16$	$7 + 32$.

And furthermore:

Ti	Sn	Ta		Cr	Mo	V	W
a	$a + d$	$a + 2d$		a	$a + d$	$a + 2d$	$a + 3d$
25	59	93		26	48	70	92
	$25 + 34$	$25 + 68$			$26 + 22$	$26 + 44$	$26 + 66$.

Dumas attached great value to these relationships as well as to the fact that the equivalent weights of some elements are either just the same as those of other elements (Mn = Cr = 26) or differ only by a factor of two (S : O = 16 : 8). Berzelius believed these numerical relationships to be merely fortuitous. Dumas nonetheless recognized Berzelius as an expert in the field of atomic weight determinations, and rarely changed an atomic weight on his own authority. Dumas remained what he always considered himself to be, namely an experimentalist, who would not dispute facts on theoretical grounds.

Josiah Parsons Cooke

Jean Baptiste André Dumas

Looking more closely into the matter, we find that this scientist did not really bring to light many new facts or come to any fresh conclusions in 1857. He confined himself to constructing series, but these were already known. Moreover, his arithmetical progressions are sometimes rather arbitrary. The nitrogen series, for instance, contains three unknown factors a, d, and d'. In fact, there is some justification for Pettenkofer's comment [29] upon the publication of Dumas' paper (see 16.2) to the effect that his own work of 1850 had anticipated it.

4.12. Odling; 1857

William Odling [30, 31] (1829–1921), at that time teaching chemistry at Guy's Hospital in London, the English advocate of the new theories of organic chemistry, re-examined the body of facts critically and thus obtained a better idea of the properties of elements. Only when elements had many properties in common did he consider

William Odling

them to be related to each other. To arrive at a *natural classification* in this way, he studied the various established groups to determine analogous properties of the members, also with respect to their compounds.

Like some of his predecessors, Odling was convinced that it was not the simple substance but the element in a compound that carries the properties whose analogies must be studied (see also *3.7*). For this study he used every means available to him, such as atomic theory, the phenomenon of isomorphism, the law of atomic heats, and the regularities of the atomic volumes of elements (also called by him *elementary bodies*).

A striking example is furnished by the three series of acids, derived from chlorine, HCl, HClO to $HClO_4$, the dibasic, H_2S to H_2SO_4, and the tribasic, H_3P to H_3PO_4, each series consisting of 5 compounds mutually representing analogous combinations of chlorine, sulphur or phosphorus.

The construction of the very first of the thirteen groups (Fig. 11) he collected, shows clearly how cautiously Odling proceeded. As he could not find many relations between fluorine and the other halogens, he did not immediately place F, Cl, Br and I in the same group. For the second group he followed the same procedure. He incorporated fluorine in group 1 only with the words "F belongs to Cl, Br and I".

Group 1   F belongs to Cl, Br, and J

Group 2   O belongs to S, Se, and Te

Group 3   N, P, As, Sb, Bi

Group 4   B, Si, Ti, Sn

Group 5   Li, Na, K

Group 6   Ca, Sr, Ba

Group 7   Mg, Zn, Cd

Group 8   Gl, Y, Th

Group 9   Al, Zr, Ce, U

Group 10   Cr, Mn, Co, Fe, Ni, Cu

Group 11   Mo, V, W, Ta

Group 12   Hg, Pb, Ag

Group 13   Pd, Pt, Au

Fig. 11   Groups arranged by Odling (1857). Only the elements without their atomic weights and their relative numerical relationships are shown.

When constructing group 4, Odling encountered difficulties that other investigators usually neglected. Although boron and silicon resemble each other closely, they nevertheless have some properties that disturb the analogy between these two elements. The formulae of their chlorides are BCl_3 and $SiCl_4$, respectively. It is likewise impossible to base a triad on their atomic weights: B = 14.5 (even though incorrect), Si = 28.5, and Ti = 48.4. According to him, this could nevertheless be done by using Berzelius' new value of 29.6 for silicon. Odling noticed another relationship:

$$Sn = 117.6, \quad B = \frac{117.6}{8} = 14.7, \quad \text{and} \quad Si = \frac{117.6}{4} = 29.4.$$

Carbon would belong, because of its resemblance, to the fourth group too.

Odling's great contribution was that he was the first to see a relationship between four groups. He noticed that the first four elements of group 1 through 4, i.e. C, N, O and F, show a remarkable analogy:

C	=	12	forms	H_4C
N	=	14	,,	H_3N
O	=	16	,,	H_2O
F	=	18?	,,	HF.

Actually this analogy could not have been made much earlier, because Kekulé [32, 33] had only stated the quadrivalence of carbon some months before. Odling had already noticed the wide separation between elements with analogous properties with respect to their atomic weights as well as the close proximity of elements with hardly any properties in common. Although he drew no conclusions from this observation, we must assume that he had more insight into its implications than any other investigator up to this time. This opinion gains support from the fact that a few years later, as early as 1864, Odling was able to construct a system of elements independently (see 5.5). He was indeed the only one of the precursors who was able to set up a periodic system on the basis of his own studies.

If we examine the arrangement of the groups (Fig. 11) we see that this scientist did not include magnesium in group 6. He considered this element to be a homologue of zinc and cadmium, as did Cooke and Lenssen (see also 12.4).

Odling was struck by many kinds of relationships between atomic weights, for instance between the elements of the groups 5, 6 and 7.

	Li	=	6.5		Mg =	12.16		Ca =	20
I	Na	=	23.0	II	Zn =	32.50	III	Sr =	43.8
	K	=	39.0		Cd =	55.7		Ba =	68.5
Total			68.5	Total		100.36	Total		132.3

Average value $\dfrac{68.5}{3} = 22.8$ $\qquad \dfrac{100.36}{3} = 33.45$ $\qquad \dfrac{132.3}{3} = 44.1$

Average difference   16   20   24

The three totals, the three means and the three differences form a series, but Odling made no further comment on this point. The averages of the totals, the average values and the average differences are 100.38, 33.45 and 20, respectively. These values are equal to the total, the average and the difference of II, respectively. The three averages, however, are the same as the members of Lenssen's first ennead (Fig. 9, p. 83).

Lenssen pointed out:

atomic weight	Ca	= atomic weight	1st member of III
,,	,, Sr	= ,, ,,	1st + 2nd members of II
,,	,, Ba	= ,, ,,	1st + 2nd + 3rd members of I,

to which the following may be added:

$$Li = \tfrac{1}{2} Mg = \tfrac{1}{3} Ca \text{ and } Na + Zn = Cd \text{ and } Ca + Mg = Zn.$$

The average difference in group 8 is 28; in the next group it is just half this value but, remarkably enough, equal to the atomic weight of Al ($= 13.7$), for which Odling formulated the following additional relations:

$$Al = \tfrac{1}{2} Fe (= 28) = \tfrac{1}{4} U (= 60).$$

We do not believe that these groups were composed very critically. Indeed, even in the opinion of Odling himself, the composition of group II was not based on sufficient analogies. He believed that these elements had not been adequately investigated. The elements of some of the groups[15] composed by him are only analogous in an analytical chemical sense, as in the case of group 12, for instance.

Odling did not attempt to make all elements part of triads, however. He noticed that it is frequently possible to add to a triad an element having an atomic weight equal to half of that of the first member, as well as an element having an atomic weight equal to twice that of the last member.

Gladstone was the only one of his predecessors whom Odling mentioned by name.

4.13. Dumas; 1858

Odling's observation of the connection between the first members of some series and Dumas' discovery [34] in 1858 of the numerical relationship between the members of the fluorine and nitrogen groups and those of the magnesium and oxygen groups, were very important. The difference in atomic weight between the respective members of the fluorine and nitrogen groups is 5 units, and this difference is 4 for the oxygen and magnesium groups. Here Dumas again noticed a correspondence to the radical series in organic chemistry, as shown by the following examples:

$$\begin{array}{llll} F \ = 19 & Cl = 35.5 & Br = 80 & I \ \ = 127 \\ N \ = 14 & P \ \ = 31 & As = 75 & Sb = 122 \end{array} \quad \text{difference: 5}$$

$$\begin{array}{lllll} Mg = 12.25 & Ca = 20 & Sr \ = 43.75 & Ba = \ \ 68.5 & Pb = 103.5 \\ O \ \ = 8 & S \ \ = 16 & Se = 39.75 & Te = \ \ 64.5 & Os = \ \ 99.5 \end{array} \quad \text{difference: 4}$$

$$\begin{array}{llll} NH_4 = 18 & NC_2H_6 = 32 & NC_4H_8 = 46 & NC_6H_{10} = 60 \\ \text{ammonium} & \text{methylammonium} & \text{ethylammonium} & \text{propylammonium} \\ C_2H_3 = 15 & C_4H_5 = 29 & C_6H_7 = 43 & C_8H_9 = 57 \\ \text{methyl} & \text{ethyl} & \text{propyl} & \text{butyl} \end{array} \quad \text{difference: 3}$$

---

[15] Fig. 11 shows the groups given by Odling, but without their atomic weights and interrelationships.

This scientist, who regarded previous classifications of metals as being artificial, arrived at an important discovery, which led to the recognition of a relationship between *all* the elements, after many years of work on the problems in this field. When comparing them with the series of organic radicals, Dumas inclined to the belief that the elements are divisible. In this respect his idea did not differ from Pettenkofer's conclusions. In Dumas' opinion, however, chemists would find great difficulty in furnishing proof of this divisibility. He deduced the possibility of division from Lavoisier's concept of elements. Lavoisier's 32 elements were the final products of analysis. The only truly indivisible elements would be heat, light, oxygen, nitrogen and hydrogen, because each of these belonged to the three kingdoms of nature, the vegetable, the animal and the mineral.

Dumas *[35]*, who had been occupied for the past ten years with his ministerial duties (since 1851 as Minister of Education), in 1859 summarized his share in the discovery of the periodic system in a well-documented 80-page review of his work in this field, to which he added some supplementary material.

After a revision of the atomic weights, the nitrogen and fluorine groups could be described by a regular progression: $a, a + d, a + 2d + d', \ldots\ldots$ This in fact followed from the results given in his previous publication, with the sole addition of 0.5 to the atomic weight of phosphorus. According to Dumas, the atomic weight of phosphorus might well later prove to be 30.5 rather than 31 (Fig. 12).

a	$a+d$	$a+2d+d'$	$a+2d+2d'+d''$	
N	P	As	Sb	Bi
14	31	75	122	210
	$14+16.5+0.5$	$14+33+28$	$14+33+56+19$	$14+33+56+19+88$

F	Cl	Br	J
19	35.5	80	127
	$19+16.5$	$19+33+28$	$19+33+56+19$

Fig. 12 New progressions formulated by Dumas.

The atomic weight of antimony must be taken 7 units smaller than the value determined by Berzelius, which, however, was then exact[16]. If the atomic weight of bismuth is taken as 211 instead of 210, the corresponding term of the progression would become $a + 4d + 4d' + d''$ instead of $a + 2d + 2d' + d'' + d'''$. The correct atomic weight is 209, however.

[16] Schneider *[36]* (1825–1900), who became professor of chemistry at the University of Berlin in 1860, objected to Dumas' modified atomic weights and doubted the validity of the parallel series put forward by Dumas. Schneider himself, be it said, also accepted certain relationships, viz.:

$$Ti = 25$$
$$Nb = 25 + 24 = 49$$
$$Sn = 25 + 24 + 10 = 59$$
$$Ta = 25 + 24 + 2 \times 10 = 69,$$

in which he, too, had changed an atomic weight. Berzelius erroneously determined the atomic weight of tantalum as 92, which differed from the value used by Dumas.

References p. 96

4.14. Mercer; 1858

John Mercer *[37]* (1791–1866) was an English chemist whose work lay in the field of the calico-printing industry, where he made an important contribution by his study of catalysts. A year before his retirement, he made another contribution, this time to the systematization of elements.

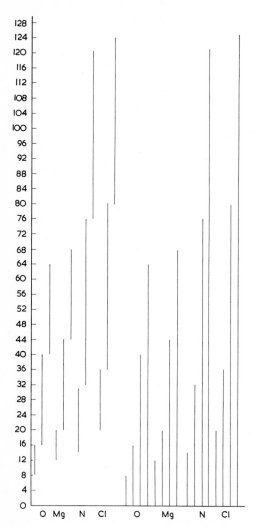

Fig. 13   Parallelism of groups of elements (Mercer; 1858).

Mercer gave a schematic review of the parallels to be drawn between the fluorine, oxygen, nitrogen and magnesium groups (Fig. 13), as already pointed out by Dumas *[28]*. He also found a large number of relationships between atomic weights, almost none of them new, and representing nothing more than the results of additions and subtractions of atomic weights. Mercer also pointed out the similarity to the organic radicals, which he described for the alkali metals as follows:

$$Li = Li = 7 \text{ like } H$$
$$Na = b_2Li_3 = 23 \text{ ,, } C_2H_3$$
$$K = b_4Li_5 = 39 \text{ ,, } C_4H_5, \text{ in which } b = 1.$$

Mercer gave no further explanation and did not say what he meant by b.

4.15. Carey Lea; 1860

A quite new point of view was put foward by Matthew Carey Lea [38] (1823–1897), an American who lived in Philadelphia, and did research in his private laboratory. Like some phlogistians, he considered the possibility of negative weights. He seems to have had a special preference for the number 44 or 45, to which he tried to reduce all kinds of differences and sums of atomic weights.

45	120	Sb =	120.3[*]
44	75	As =	75
45	31	P =	31
45	- 14	N =	14
45	- 59	Sn =	59
45	-104	Pb =	103.5
45	-149	2 As =	150
45	-194	-	-
45	-239	2 Sb =	240.6
45	-284	-	-
44	-328	- =	2 x 164
44	-372	-	-
44	-416	2 Bi =	416

[*] All atomic weights are derived from the Jahresbericht 1857

Fig. 14 Atomic weight relations according to Carey Lea.

Carey Lea also sought some new relationships, because many of the old ones no longer held after new atomic weights were published in 1857. For the nitrogen group the difference between the atomic weights was 44 or 45, with the exception of that between N and P. But if an atomic weight of -14 is used for nitrogen, the difference becomes 45. If 45 or 44 is subtracted several times more from this figure of -14, the results are frequently atomic weights or multiples of atomic weights of elements belonging to the nitrogen group (Fig. 14). This group, however, also includes elements such as Sn and Pb, which bear little resemblance to the elements of the nitrogen group.

Carey Lea assigned the atomic weight 164 to an element yet to be discovered, as well as the atomic weight 152.2, occurring in one of the following examples. He thought the existence of the former element becomes all the more probable because the addition of 44 to the atomic weight 120 of antimony again gives the atomic weight 164; 44 added to 164 makes 208, the atomic weight of bismuth.

Some other examples are:

$$44 \begin{cases} 100 \\ 56 \end{cases} \quad \begin{array}{l} Hg = 100 \\ Cd = 56 \end{array}$$

$$44 \begin{cases} 56 \\ 12 \end{cases}$$

$$44 \begin{cases} 12 \\ -32 \end{cases} \quad \begin{array}{l} Mg = 12 \\ Zn = 32.6 \end{array}$$

$$44.5 \begin{cases} 197 \\ 152.5 \end{cases} \quad \begin{array}{l} Au = 197 \\ -\!-\!= \!-\!- \end{array}$$

$$44.5 \begin{cases} 152.5 \\ 108 \end{cases}$$

$$44.5 \begin{cases} 108 \\ 63.5 \end{cases} \quad \begin{array}{l} Ag = 108 \\ Cu = 63.4. \end{array}$$

Carey Lea continued to elaborate this difference of approximately 45, which sometimes led to complicated relationships for other series of elements; e.g.,

$$\frac{14 + 3 \times 45}{8} = 18.62 \qquad F = 19$$

$$\frac{14 + 6 \times 45}{8} = 35.5 \qquad Cl = 35.5$$

$$Br - Cl = 45$$

$$\frac{14 + 14 \times 45}{8} = 80.5 \qquad Br = 80$$

$$I - Br \sim 45$$

$$\frac{2 \times 14 + 22 \times 45}{8} = 127.25 \qquad I = 127.$$

Other examples, however, yield relationships of no importance, because the elements of the relevant series are not homologous. Some months later, he noticed [39] that many atomic weights show the same ratio as oxygen to nitrogen (4 : 7), or carbon to nitrogen (3 : 7). He was unable to explain these relationships, but he wondered whether there might not be a reason for them. He evidently did not see that, with such a large number of atomic weights involved, it was not surprising that the same proportions occurred more than once. In fact, Carey Lea was not justified in using these relationships for the purpose of correcting atomic weights.

His most important contribution was the suggested inclusion of vanadium as a member of the nitrogen group. In this, Carey Lea agreed with Šafařik [40], professor of chemistry at the Commercial Academy of Vienna.

Carey Lea's work illustrates the extremes to which a scientist might go—even as far as believing in negative atomic weights—to add a few relationships between atomic weights to the existing ones. The publication by this author is all the more remarkable in that not long after its appearance De Chancourtois was to demonstrate conclusively the universal relationship between all the elements.

It is not difficult to understand why Carey Lea's work was so widely criticized. He was justifiably attacked for introducing negative atomic weights without further explanation. He replied *[41]* that he had only pointed out relationships which become apparent when atomic weights with a minus sign are used, and referred to the disputes about negative numbers between Leibnitz and Bernouilli as well as between Euler and d'Alembert, and cited Carnot in vindication of his concepts. In agreement with Carnot, he stated that negative numbers are not smaller than zero and therefore have a real meaning. Carey Lea forgot, however, that their application to atomic weights was faulty. Atomic weights having a minus sign do not express a real situation. Nevertheless, even in 1886 Carey Lea was still working with negative atomic weights (*6.23*).

4.16. Conclusion

In 1817 the concept of elements was generally accepted, but the elements that Döbereiner employed to compose his first triad had been separated in elementary form less than ten years before and the atomic weights of these and many other elements had not yet been correctly determined. His doubt of the elementary character of calcium, strontium and barium can hardly be held against him, because Dalton's atomic theory was too recent to have been conclusively demonstrated, and Prout had put forward his protyle hypothesis two years before.

After Döbereiner had laid the basis in 1817 and 1829 for the determination of relationships between the atomic weights of elements, taking analogous properties into account, Gmelin returned to this subject in 1843 and Pettenkofer in 1850, when they extended Döbereiner's formations of triads. Döbereiner was far ahead of his time, as shown by the fact that he already considered larger groups than those with only three analogous elements.

Pettenkofer wrote the first of the many series of papers on this subject published between 1850 and 1862. This work consolidated the basis for the construction of the periodic system.

Gladstone, Cooke and Lenssen developed this work further. Odling and Dumas saw mutual relationships between the groups of analogous elements; Odling noticed a connection between the four groups of his system, and Dumas established this between the elements of the fluorine and nitrogen groups and those of the oxygen and magnesium groups. Although these relationships were limited, their observation must be considered as a turning-point in the study of connections between elements. The complexity of the elementary substances assumed by some investigators did not actually contribute to the development of the periodic system.

These last views and the ultimately correct interpretations of the concept of atomic weight, which was furthered by the results of the Congress held at Karlsruhe in 1860, were the main pillars on which De Chancourtois was able to construct his system.

Not all the investigators in this field used the atomic weight as the only numerical quantity. Investigations into relationships between atomic weights and other physical properties were not usually conducive to what we see as their purpose, namely the classification of all the chemical elements.

References

1 A. M. AMPÈRE, *Ann. Chim. Phys.* [2], 1 (1816) 295, 373; [2], 2(1816) 5, 105.
2 J. W. DÖBEREINER, *Ann. Physik (Gilbert)*, 56 (1817) 331; see also *Die Anfänge des natürlichen Systems der chemischen Elemente, Ostwalds Klassiker No. 66*, Leipzig 1895.
3 J. W. VAN SPRONSEN, *Chem. Weekblad*, 59 (1963) 352.
4 J. W. DÖBEREINER, *Ann. Physik (Gilbert)*, 57 (1817) 435.
5 J. W. DÖBEREINER, *Lehrbuch der allgemeinen Chemie*, 3 vols., Jena 1811–1812.
6 J. W. DÖBEREINER, *Darstellung der Verhältnisszahlen der irdischen Elemente zu chemischen Verbindungen*, Jena 1816.
7 J. W. DÖBEREINER, *Ann. Physik (Pogg.)*, 15 (1829) 301.
8 J. W. DÖBEREINER, *Grundriss der allgemeinen Chemie*, 3rd ed., Jena 1826.
9 J. W. DÖBEREINER, *Anfangsgründe der Chemie und Stöchiometrie*, Jena 1826.
10 L. GMELIN, *Handbuch der theoretischen Chemie und Stöchiometrie*, 3rd ed., Frankfurt am Main 1827, Vol. I, p. 34 ff.
11 L. GMELIN, *Handbuch der Chemie*, 4th ed., Heidelberg 1843, Vol. I, pp. 52, 456.
12 L. GMELIN, *Handbuch der anorganischen Chemie*, 5th ed., Heidelberg 1852, Vol. I, pp. 48, 463.
13 L. GMELIN, *Handbook of Chemistry*, transl. by H. Watts, London 1849, Vol. II, p. 1.
14 J. J. BERZELIUS, *Lehrbuch der Chemie*, Dresden–Leipzig 1845, Vol. II, p. 1178.
15 H. KOPP, *Ann.*, 32 (1839) 207.
16 H. KOPP, *Ann.*, 36 (1840) 1.
17 See P. EINBRODT, *Ann.*, 58 (1846) 1.
18 M. PETTENKOFER, *Gelehrte Anzeigen der Akademie der Wissenschaften zu München*, 30 (1850) 261, 165; see also *Ostwald's Klassiker No.66*, op. cit. (ref. 2), also published in *Ann.*, 105 (1858) 187.
19 J. DUMAS, *Am. J. Sci.* [2], 12 (1851) 275.
20 J. DUMAS, *Atheneum, Journ. Lit., Sci. and fine Arts*, (1851) 750; *L'Institut, Journal universel des sciences et des sociétés savantes en France et à l'étranger*, Section 1, 19 (1851) 303.
21 J. DUMAS, *Traité de Chimie appliqué aux Arts*, Paris 1828, Vol. I, p. LXXIV ff.
22 P. KREMERS, *Ann. Physik (Pogg.)*, 85 (1852) 37, 246.
23 P. KREMERS, *Ann. Physik (Pogg.)*, 99 (1856) 58.
24 J. H. GLADSTONE, *Phil. Mag.* [4], 5 (1853) 313.
25 J. P. COOKE, *Am. J. Sci.* [2], 17 (1854) 387.
26 J. P. COOKE, *Mem. Am. Acad. Arts Sci.*, New Series, 5 (1855) 235.
27 J. LIEBIG and H. KOPP, *Jahresber. Fortschr. Chem.* (1852).
28 E. LENSSEN, *Ann.*, 103 (1857) 121.
29 E. LENSSEN, *Ann.*, 104 (1857) 177.
30 W. ODLING, *Phil. Mag.* [4], 13 (1857) 423, 480.
31 J. W. VAN SPRONSEN, *Chem. Weekblad*, 60 (1964) 683.
32 A. KEKULÉ, *Ann.*, 101 (1857) 200.
33 J. W. VAN SPRONSEN, *Chem. Weekblad*, 61 (1965) 35.
34 J. DUMAS, *Compt. Rend.*, 47 (1858) 1026; transl. in *Ann.*, 109 (1859) 376.
35 J. DUMAS, *Ann. Chim. Phys.* [3], 55 (1859) 129.
36 R. SCHNEIDER, *Ann. Physik (Pogg.)*, 107 (1859) 619.
37 J. MERCER, *Report Brit. Ass. Adv. Sci. (Transactions)*, (1858) 57, 59.
38 M. CAREY LEA, *Am. J. Sci.* [2], 29 (1860) 98.
39 M. CAREY LEA, *Am. J. Sci.* [2], 29 (1860) 349.
40 A. ŠAFAŘIK, *Sitz.ber. Akad. Wissenschaften (Vienna) Math. Nat. Classe*, 33 (1858) 3.
41 M. CAREY LEA, *Am. J. Sci.* [2], 34 (1862) 387.
42 J. DUMAS, *Compt. Rend.*, 45 (1857) 709; transl. in *Ann.*, 105 (1858) 74.

Chapter 5

Discoverers of the periodic system (1862–1871)

5.1. Introduction

After the foundations of the periodic system had been laid and consolidated in the years up to 1862, a system of elements could be constructed on the basis of the atomic weights. As we have shown in the conclusion of the preceding chapter (*4.16*; see also Chapter 3), in 1862 the time was ripe for a first attempt to classify all the elements. 1862 may rightly be called the year of the birth of the periodic system.

5.2. The *"Vis tellurique"* of Béguyer de Chancourtois [1–3]

It comes as a surprise to learn that the first scientist to compose a periodic system of elements was not a chemist but the mineralogist and geologist Alexandre Emile Béguyer de Chancourtois (1820–1886), since 1848 professor of subterranean typography and since 1856 professor of geology at the École des Mines in Paris. De Chancourtois appears to have been a systematist by nature. He adopted as his basic principle the statement, *Les propriétés des corps sont les propriétés des nombres.*

De Chancourtois applied classification and systematization to mineralogy (e.g. in 1848 he listed a collection of French minerals together with the geologist Le Play), geology (classification of formations, system for description of stones), geography (application of a decimal system), and even to the field of philology (universal alphabet).

He began work on a system of chemical elements after studying a book on volcanic rocks and metal-bearing stones written by his colleague Elie de Beaumont [4], who discussed the geographical distribution of the elements and classified the 59 known elements according to electropositivity. De Chancourtois needed a system of simple substances for his lithological studies. He made use of the mineral system of the American mineralogist James D. Dana [5], which was constructed on the principles then in use—class, order, etc.— according to Rammelsberg, and included the metals in the 5th class.

In 1862 De Chancourtois submitted a system of his own to the Académie des Sciences of Paris, which serialized it in the *Comptes Rendus* [6]. To elucidate the periodicity of the properties of the elements as a function of atomic weights, De Chancourtois proposed a three-dimensional representation (Figs. 15 and 16). Next to a

1862

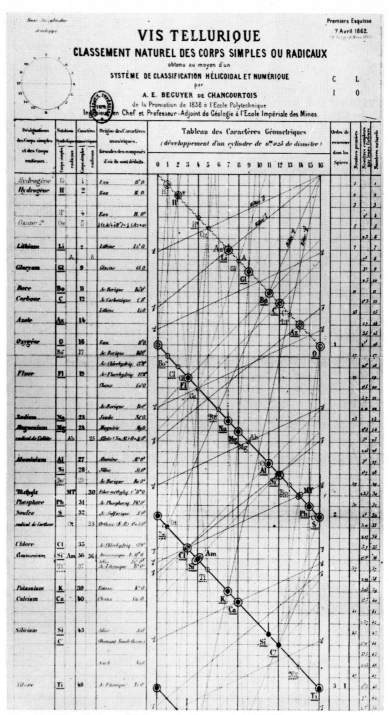

Fig. 15   *Vis tellurique* of Béguyer de Chancourtois (1862)
(upper part; original)

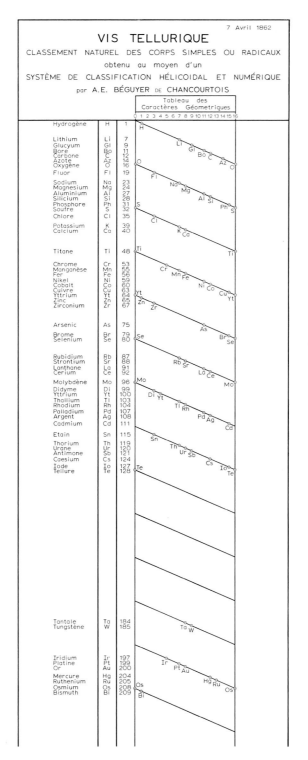

Fig.16 *Vis tellurique* of Béguyer de Chancourtois (upper part; simplified by the present author)

generator of a cylinder, the names of the elements bearing the so-called characteristic numbers (*nombres caractéristiques* or *caractères numériques*) are indicated vertically. For most of these characteristic numbers he used the currently accepted atomic weights, but for a few elements these atomic weights were divided or multiplied by two. De Chancourtois only used whole numbers, knowing that this procedure was supported by both Prout—who [7, 8] conceived protyle[1] or archisom[2] to be a theoretical basic substance with an atomic weight of 1—and Dumas[3]. Most of the atomic weights used by De Chancourtois correspond to those used at present (Table 1). In this system, a spiral or helix encompasses the cylinder, forming a 45° angle with the base and connecting the so-called characteristic points of the elements. These characteristic points are derived by shifting the characteristic number parallel to the base until it reaches a point whose numerical value is the same as that of the characteristic number of the element or that number minus 16 or a multiple thereof, and in such a way as to give a difference equal to or smaller than 16. The base of the cylinder carries the numbers 1 to 16. The length of the base is therefore equal to the atomic weight of oxygen when H = 1 is taken as the standard. De Chancourtois called the resulting helical graphical system *Vis tellurique* after tellurium, which is located at its centre, and because "l'épithète tellurique...... rappelle très heureusement l'origine géognostique, puisque tellus signifie terre dans le sens le plus positif, le plus familier, dans le sens de terre nourricière".

In the *Vis tellurique* the elements resembling each other occur in vertical rows, e.g. O, S, Se and Te. The counterparts of these elements, Mg, Ca, Sr, U and Ba[4], occupy a column located antipodally with respect to the first-mentioned column. In this first publication De Chancourtois limited himself to the description of his *Vis tellurique*. The editors of the *Comptes Rendus* apparently omitted the accompanying illustration [8], for which reason De Chancourtois decided to publish his complete paper independently; this version [9] appeared in 1863. On account of the difficulty involved in representing a cylinder, this was shown in longitudinal cross-section. De Chancourtois used three different representations[5] of the *Vis tellurique* to clarify his concept. These three versions are difficult to understand at first sight, and may cause more confusion rather than lessen it. This, added to the fact that two of the systems show only minor differences and their presentation in print involved great technical difficulties, explains their omission from the *Comptes Rendus*. The lay-out of the graphical reproduction, however, was due to De Chancourtois' oversimplification of the numerical relation-

---

[1] Gr. πρῶτος = first; ὕλη = substance, matter.

[2] Gr. ἀρχή = beginning, origin; σῶμα = body.

[3] Dumas assigned an atomic weight of 35.5 to chlorine, however. Dumas read one paper by De Chancourtois to the *Académie*.

[4] In the text Bi has been added to the former group, but as this element does not occur below Te in the system, its inclusion must have been a printer's error.

[5] The original reproductions were done in colour to emphasize the three alkyl groups: gasolytes, leucolytes and chroïcolytes [9].

ships between elements, as would soon become evident. Further study shows that De Chancourtois [7, 9] incorporated not only all the elements but also such radicals as the ammonium, the alkyl and the alkylammonium, and such compounds[6] as cyanogen, oxides[7], and acids and alloys[8].

This treatment weakened the originality of De Chancourtois' initial idea. At a time when elements, in the sense of simple substances, were clearly distinguished from compounds, De Chancourtois used both in his system. The *Vis tellurique*[9] does show, however, that its creator was the first to realize that the properties of the elements are a function of their atomic weight. This function shows a certain periodicity, which makes it possible to arrange the elements in a graphic representation such that elements with similar properties are located one below the other. However, in his system these elements are not arranged in vertical columns, owing to the choice of the number 16, which determines the periodicity. Consequently, De Chancourtois obtained only a rough approximation by this representation.

The elements of one group, such as the alkali metals, the alkaline earths and the halogens, are approximately in vertical series, however, so that the relationships between the atomic weights of the elements are clearly expressed in the *Vis tellurique*[10]. This description might be taken to imply that De Chancourtois believed that components of nature could be represented by single mathematical equations. Actually, he used mathematics only to describe nature. Later, the relationships appear to be far less simple. Many observations were required to determine their true nature.

De Chancourtois based his system on concrete data, and even avoided using the word atom. He was convinced, however, that the numerical relationships between elements were sufficiently evident to form a basis for classification. He called his *Vis tellurique* a *natural classification*, and used this latter term also as the title of his

---

[6] A curious inclusion is that of silicates, i.e., albite (sodium aluminium silicate) and orthose (the similar potassium compound), with the symbols Ab and Ot in the places of the mean atomic weights of Na and Al, and K and Al, respectively, i.e., 25 and 33. Did the mineralogist De Chancourtois wish to stress these combinations of elements or had he really new elements in mind? He referred to a hypothetical alloy and also stated expressly that his system did not depend on any atomic theory. He used the law of Dulong and Petit, but by advancing the formulas H_2O, HO and HO_2 for water (at any rate in two of his three systems) he showed that he had not yet accepted the consequences of Avogadro's law, at least as far as this case was concerned. Actually, De Chancourtois [7] thought that HO was the most obvious formula. The atomic weight of hydrogen therefore became 2. As he did not include atoms in his theory, we may certainly not call him a Daltonian.

[7] Because the first helix crosses the point with the atomic weight 5, De Chancourtois thought a substance should be placed there. He chose ozone (incorrectly called oxone in the system) and considered this substance to be a kind of oxide of nitrogen; $\frac{1}{6}(N + O)$ has indeed an equivalent weight of 5. What De Chancourtois meant by the symbol A with the atomic weight of 8 is not clear.

[8] e.g. emetique $= \frac{1}{4}KSb_2$, with the equivalent weight 86.

[9] The many compounds adjoining the main spiral impede a comprehensive view of the system. Because of the printing difficulties presented by the *Vis tellurique*, including its great length and the small and complicated print it requires, we have represented it here in a somewhat simplified form, which, however, takes the original plan into account (Fig. 16). A fragment of one of the original systems is reproduced in Fig. 15.

[10] For the omission of a few elements, see 2.5.

References p. 144

chronological classification of the geological formations *[10]* he published in 1874. As a geologist, he was primarily interested in minerals, to which his own words testify: "mais de l'imitation matérielle jusqu'à la formule mathématique, tout n'est qu'approximation, approximation grossière dans ses plus exquises finesses, mesquine dans ses plus vastes portées". These are the words he used in the dedication—almost the creed of a scientist—he inscribed in the copy of the *Vis tellurique* presented to Prince Napoleon.

We also refer here to another *classification tellurique* by De Chancourtois, which is a complete classification, in great detail, of the movements involved in such phenomena as earthquakes, ground waves and slow oscillations.

1863   In 1863 De Chancourtois *[7]* reported on several applications of his *Vis tellurique* to the theory of steel, which represent some ordinary numerical relationships. He found that the elements responsible for the hardness of steel belong to either group *a* or *b*, as follows:

a) elements having an atomic weight equal to a multiple of 11, for instance B (11), Mn (55), Zn (66), As (77), Sb (121), W (187), Ir (198); or

b) elements with an atomic weight equal to a multiple of 7, for instance Si (28), Ti (49), Fe (56), As (77), I (126), V (137 to 140), W (189).

Hardly any attention was paid to De Chancourtois' work, although it was quoted by Kopp and Will in their *Jahresbericht [11]* and referred to by Sainte-Claire Deville[11], who mentioned the *Vis tellurique [12]* in connection with a paper on a system of elements of his own based on mineralogy, published *[13]* as early as 1855. The chemists, however, did not pay attention to De Chancourtois' work until Hartog in England and Lecoq de Boisbaudran and De Lapparent in France claimed priority for De Chancourtois (see *16.7*). The reason for this long neglect is to be sought, on the one hand, in the omission of the illustrations from the *Comptes Rendus*, which made the paper difficult to understand, and, on the other hand, in the fact that De Chancourtois, being primarily a geologist in spite of his broad knowledge of chemistry—which enabled him to continue Dumas' work—never again published any reference to his *Vis tellurique*.

5.3. Newlands' law of octaves *[14]*

John Alexander Reina Newlands (1837–1898), who at the outset of his career left England to fight under Garibaldi (Newlands' mother was Italian), had not attended the Congress of Karlsruhe. He was unaware of the fundamental modifications adopted there in respect of the atomic weights, after being expounded so convincingly by the Italian scientist Cannizzaro. It is hardly surprising that Newlands, who was back in

---

[11] Charles Joseph Sainte-Claire Deville (1814–1876), a mineralogist and geologist who was appointed professor at the Collège de France in 1852, and was the brother of the chemist Henri Etienne Sainte-Claire Deville.

John Alexander Reina Newlands   *Alexandre Emile Béguyer de Chancourtois*

London in 1863 working in Way's laboratory, could not publish a periodic system of elements in his first paper *[15]*. This first paper on the relationships between the chemical elements appeared a year after his description *[16]* of a table giving the composition of and the relationships between organic substances, which shows that Newlands had begun to work on the problem of classification immediately after his return from Italy. He had constructed such a system a few years before, but only published it much later because nothing else on the subject had appeared in the literature. The same classification was also favoured by Foster[12] and Schiel. The long interval of preliminary work Newlands apparently required is easily explained. Being unacquainted with De Chancourtois' recently published articles on his *Vis tellurique* and unaware of the correct atomic weights, Newlands *[15]* could not do much more in 1863 than form some groups of elements with analogous properties and observe that many pairs of elements with analogous properties differed in atomic weight by 8 units or roughly by a multiple thereof. As his precursor, Newlands *[15, 17]* mentioned only Dumas[13], though he was aware that many chemists had pointed out relationships between the equivalent weights of elements belonging to the same natural family. Newlands' collection of eleven groups of elements (Fig. 17) bears a close resemblance

---

[12] See also p. 111.
[13] In 1865 Newlands concluded that all the relationships set up by Dumas fitted into his system.

to Odling's work (Fig. 11, p. 88), although with differences due partly to greater insight and partly to the discovery of several more elements or better determinations of atomic weights. The atomic weights used by Newlands are the equivalent weights accepted by Gerhardt (*3.2, 3.5*).

Newlands' collection of elements is not ideal, however. Some elements are placed in groups into which they do not fit at all, because of their properties; e.g., osmium in the nitrogen group (Group VII). Group IV, whose structure is rather heterogeneous, is headed by magnesium, while this element is also placed, correctly, in the group of the alkaline earth metals. His classification was accompanied by some relationships between the atomic weights.

In the relationships:

$$1 \text{ Li} + 1 \text{ K} = 2 \text{ Na}$$
$$1 \text{ Li} + 2 \text{ K} = 1 \text{ Rb}$$
$$1 \text{ Li} + 3 \text{ K} = 1 \text{ Cs}$$
$$1 \text{ Li} + 4 \text{ K} = 163$$
$$1 \text{ Li} + 5 \text{ K} = 1 \text{ Tl,}$$

the atomic weight of a still-undiscovered metal would be 163. The fourth relationship

Group I   : Li 7, Na 23, K 39, Rb 85, Cs 123, Tl 204

Group II : Mg 12, Ca 20, Sr 43.8, Ba 68.5, $Sr = \dfrac{Ca + Ba}{2}$

Group III : Be 6.9, Al 13.7, Zr 33.6, Ce 47, La 47, Didydium 48, Th 59.6 in which
$Al = 2 Be = \dfrac{Be + Zr}{3}$ (also $Al = \frac{1}{2} Mn$) $Zr + Al = Ce$; $Zr + 2Al = Th$; La en Di $\sim$ Ce

Group IV : Mg 12, Cr 26.7, Mn 27.6, Fe 28, Co 29.5, Ni 29.5, Cu 31.7, Zn 32.6, Cd 56 in which
$Zn = \dfrac{Mg + Cd}{2}$; Co $\sim$ Ni ; $Fe = \frac{1}{2} Cd$; $Mn = \dfrac{Fe + Cr}{2}$

Group V   : F 19, Cl 35.5, Br 80, J 127 in which $Br = \dfrac{Cl + J}{2}$

Group VI : O 8, S 16, Se 39.5, Te 64.2 in which $Se \sim \dfrac{S + Te}{2}$

Group VII : N 14, P 31, As 75, Os 99.6, Sb 120.3, Bi 121.3 in which
$As = \dfrac{P + Sb}{2}$; $Os = \dfrac{As + Sb}{2}$; $Os = \dfrac{Bi - N}{2}$; $Bi = 1Sb + 3P = 120.3 + 93 = 213.3$

Group VIII : C 6, Si 14.2, Ti 25, Sn 58 in which $Sn - Ti = 3(Ti - Si)$

Group IX : Mo 46, V 68.6, W 92, Ta 184 in which $V = \dfrac{Mo + W}{2}$; $W = 2 Mo$ en $Ta = 4 Mo$

Group X   : Rh 52.5, Ru 52.2, Pd 53.3, Pt 98.7, Ir 99 in which
$Rh = Ru \sim Pd$; $Pt = Ir = 2Rh$; $Pt (98.7) = \frac{1}{2} Au (197)$

Group XI : Hg 100, Pb 103.7, Ag 108 in which $Pb = \dfrac{Hg + Ag}{2}$

Fig. 17   Element groups of Newlands (1863).

is valid, thanks to an incorrect determination of the atomic weight of caesium, but one which Newlands should not have overrated.

Another prediction concerns an element to be placed between iridium and rhodium because of the difference in atomic weight between these two elements, which is 46.8. Newlands attached much importance to the difference of about 48 because it occurs so frequently between the extreme elements of a triad, for example between S and Te and between Ca and Ba (Fig. 18).

				difference					difference
Li	7	K	39	32	Mg	12	Ca	20	8
Mg	12	Cd	56	44	O	8	S	16	8
Mo	46	W	92	46	C	6	Si	14.2	8.2
S	16	Te	64.2	48.2	Li	7	Na	23	16
Ca	20	Ba	68.5	48.5	F	19	Cl	35.5	16.5
P	31	Sb	120.3	89.3	N	14	P	31	17
Cl	35.5	J	127	91.5					

Fig. 18 Differences between the extreme elements of Newlands' triads (1863).

Fig. 19 Differences in atomic weights according to Newlands (1863).

The predicted element has never been discovered, because iridium and rhodium are not extreme elements of a triad at all, but, as appeared later, belong next to each other in the series of the homologues of cobalt. It is strange that Newlands should not have seen that this prediction was not soundly based and was therefore premature. He knew that there is also a difference in atomic weight of about 48 between antimony and arsenic, and between bromine and iodine. Surely he did not believe in the existence of yet other elements between these pairs of elements.

The fact that the first two elements of certain groups differ in atomic weight by 8 or 16 (Fig. 19) must have given Newlands food for thought in 1863. With the incorrect atomic weight (half the actual value) of the elements of the carbon, oxygen and magnesium groups, this difference is not always 16, but half the value does occur. As soon as he used the correct values for the atomic weights a year later [18], he saw a more convincing relationship between the atomic weights. Although Newlands had by then informed himself about the research of his predecessors, his contribution in 1863 had not yet reached the level of those of Dumas and Odling, who noticed the relationships between the different groups.

Newlands did not succeed in further systematization until a year later [18], when he used the new atomic weights given by authors like Williamson [19]. With these values the difference in atomic weight between the first two elements of each group becomes 16 (Fig. 20), a number equal to the equivalent weight of oxygen and comparable as a unit to the equivalent weight of hydrogen = 1 used by Prout[14] in his hypothe-

1864

[14] Newlands did not accept Prout's hypothesis. See also his booklet *On the Discovery of the Periodic Law* [20].

sis. The difference in atomic weight between many of the outer elements of the triads is a multiple of 16 (Fig. 21). Newlands could then place the elements in a more comprehensive form. This collection of elements is not yet a periodic system: not all the elements were arranged according to increasing atomic weight, the exceptions being Mo, V, W, and a few others. It is a system (Fig. 22) based on triads, many of which have been extended to an arithmetical progression of 4 or more elements.

			difference	
			H = 1	O = 1
Mg	24	Ca	40	16	1
O	16	S	32	16	1
Li	7	Na	23	16	1
C	12	Si	28	16	1
F	19	Cl	35.5	16.5	1.031
N	14	P	31	17	1.062

Fig.20 Differences in atomic weights according to Newlands (1864).

				difference	
				H = 1	O = 1
Li	7	K	39	32	2
Mg	24	Cd	112	88	5.5
Mo	96	W	184	88	5.5
P	31	Sb	122	91	5.687
Cl	35.5	J	127	91.5	5.718
K	39	Cs	133	94	5.875
S	32	Te	129	97	6.062
Ca	40	Ba	137	97	6.062

Fig.21 Differences in atomic weights according to Newlands (1864).

Newlands made some predictions on the basis of this system (see *7.3*). Because he considered lithium to be a transition element, he included it in both the alkali and the alkaline earths groups. Although beryllium had been discovered as early as 1828 and its properties had long been known, he did not see that this element, rather than lithium, belonged at the head of the group of the alkaline earths.

Newlands made effective use of Quin's discovery *[21]* of a second triad in the alkali group. The equivalent weight of rubidium is almost equal to the average of those of caesium and potassium:

$$83.85 = \frac{128.4 + 39.3}{2}.$$

In the final series we find elements differing by about 70 in atomic weight from the last terms of the triads and sharing properties, in Newlands' opinion, with these elements. In our view, there are few analogies between osmium and tellurium and between barium and lead; they are more easily demonstrated between caesium and thallium. In the series of halogens the final term is lacking. Magnesium is a homologue of calcium as well as of zinc (see also *12.4*).

So far, Newlands had not made more progress than the investigators of the 1850s. However, less than a month after this paper was presented on August 20th, 1864, he surpassed them with a greatly improved system *[22]* (Fig. 23). A cursory glance might raise doubts about its value, as it included about a third of the elements then known and made no mention of atomic weights or differences in atomic weight. For all that, Newlands' system in this form had very great merit. He had abandoned precisely what the chemists before him considered to be so important, viz., the arithmetical progressions of the atomic weights. The discovery of a general system was retarded by the attempts to find series, partly because not all the correct atomic weights were known and partly because a simple numerical relationship between the atomic weights does not exist. The analogous properties of elements were often relegated to a secondary place, although they are primary in classification.

			triad			
			lowest term	mean	highest	
I		Li 7	+17 = Mg 24	Zn 65	Cd 112	
II		B 11				Au 196
III		C 12	+16 = Si 28		Sn 118	
IV		N 14	+17 = P 31	As 75	Sb 122	+88 = Bi 210
V		O 16	+16 = S 32	Se 79.5	Te 129	+70 = Os 199
VI		F 19	+16.5 = Cl 35.5	Br 80	J 127	
VII	Li 7	+16 = Na 23	+16 = K 39	Rb 85	Cs 133	+70 = Tl 203
VIII	Li 7	+17 = Mg 24	+16 = Ca 40	Sr 87.5	Ba 137	+70 = Pb 207
IX				Mo 96	V 137	W 184
X				Pd 106.5		Pt 197

Fig.22   Groups of elements according to Newlands (1864).

		no.		no.		no.		no.		no.
Group	a	N 6	P	13	As	26	Sb	40	Bi	54
	b	O 7	S	14	Se	27	Te	42	Os	50
	c	Fl 8	Cl	15	Br	28	J	41	-	-
	d	Na 9	K	16	Rb	29	Cs	43	Tl	52
	e	Mg 10	Ca	17	Sr	30	Ba	44	Pb	53

Fig.23   First system of Newlands (1864).

Further progress was then made by Newlands, who in 1864 introduced an order number (named by him ordinal number); as a result his articles appeared in rapid succession. Although it was not until 1865 that he propounded the *law of octaves [23]* (Fig. 24) according to which the properties were repeated after each series of seven elements, this law had already been inherent in his preceding system (Fig. 23). Not all the elements were included, places being left open for a great number of the omitted elements: H (1) through C (5), 18 through 26, and so on.

It is nevertheless strange that Newlands should not have immediately classified the candidate elements for these gaps; he knew them and where they should be placed. He remarked that several of the omitted elements, e.g. Mn, Fe, Co, Ni and Cu were the middle terms of triads whose extreme terms were still missing. In the Pt-group, however, it is the middle term that is missing. He thought that this might also be so for Ag and Au. Numbers 11 and 12 were not assigned a place in the system, but the element phosphorus (No. 13) was located exactly one octave further than nitrogen (No. 6), etc.

	No.		No.		No.		No.		No.		No.		No.		No.
H	1	F	8	Cl	15	Co & Ni	22	Br	29	Pd	36	I	42	Pt & Ir	50
Li	2	Na	9	K	16	Cu	23	Rb	30	Ag	37	Cs	44	Tl	53
G	3	Mg	10	Ca	17	Zn	25	Sr	31	Cd	38	Ba & V	45	Pb	54
Bo	4	Al	11	Cr	19	Y	24	Ce&La	33	U	40	Ta	46	Th	56
C	5	Si	12	Ti	18	In	26	Zr	32	Sn	39	W	47	Hg	52
N	6	P	13	Mn	20	As	27	Di &Mo	34	Sb	41	Nb	48	Bi	55
O	7	S	14	Fe	21	Se	28	Ro&Ru	35	Te	43	Au	49	Os	51

Fig.24   Law of octaves of Newlands (1865).

The distance between the arsenic period and the antimony period is two octaves, the same distance as that between the latter and the bismuth period. It is a noteworthy fact that the distance between the phosphorus and arsenic periods is only 6 elements. In 1865 this difference was also adjusted to the law of octaves. He related this law to the laws of musical scales by which a tone recurs an octave higher after every seven tones.

Strangely enough, Newlands was not yet ready in 1864 to assume analogies between, for instance, beryllium and magnesium, boron and aluminium, and lithium and sodium, although they would have provided an excellent extension of the law of octaves. Yet he did immediately solve the problem of the Te–I inversion (see 8.2).

Closer examination of Newlands' system shows that not all the elements were correctly placed with respect to their analogous properties. This was of course impossible, because strictly speaking the law of octaves only holds for the first two periods. Nevertheless, Newlands should not have classified the elements lead, thallium and osmium in the very same way as he had in his earlier grouping (Fig. 22), especially as this resulted in their incorrect incorporation in regard not only to their properties but also to their ordinal numbers.

A classification into main groups, sub-groups and transition elements, which represents a step further towards a perfect periodic system, was adopted in the same year by Odling [24]. Even much later, Newlands was still unwilling to accept a system in which the elements were classified according to main groups and sub-groups. In 1878 he extended most of the periods to include 10 elements [25], but this created a large number of free intervals and even a whole free period, bringing the total number of elements to 105 (Fig. 25) (see also 16.2).

1865   A critical examination of Newlands' law of octaves [23] (Fig. 24) appeared in 1865, a year after the publication of Odling's first periodic system and Meyer's first attempts at systematization. It makes the point that, although the final period had been enlarged to include, not 5, but 7 places—including 8 elements, of which Pt and Ir share one place—, there was no resulting increase as regards analogies in the properties of the elements. The placings of thorium and mercury had now increased the number of incorrect arrangements already mentioned. A more critical evaluation of some analogies would have provided Newlands with a better classification. Neither vanadium nor lead fits into the third group. If he had only maintained the correct sequence of atomic weights for the elements of the final series, some of the elements would have been cor-

No.	No.	No.	No.	No.	No.	No.	No.	No.	No.	No.	No.
	2 Li 2.95	9 Na 9.70	16 K 16.50	26 Cu 26.75	36 Rb 36.03	46 Ag 45.57	56 Cs 56.12	66 –	76 –	86 Au 83.12	96 –
	3 Be 3.97	10 Mg 10.13	17 Ca 16.88	27 Zn 27.51	37 Sr 36.96	47 Cd 47.26	57 Ba 57.81	67 –	77 –	87 Hg 84.39	97 –
	4 B 4.64	11 Al 11.56	18 –	28 Ga 29.49	38 Y 37.13	48 In 47.85	58 Di 58.23	68 –	78 Er 75.11	88 Tl 85.91	98 –
	5 C 5.06	12 Si 11.81	19 Ti 21.09	29 –	39 Zr 37.81	49 Sn 49.79	59 Ce 59.07	69 –	79 La 75.95	89 Pb 87.34	99 Th 99.16
	6 N 5.91	13 P 13.08	20 V 21.60	30 As 31.65	40 Nb 39.66	50 Sb 51.48	60 –	70 –	80 Ta 76.79	90 Bi 88.61	100 –
	7 O 6.75	14 S 13.50	21 Cr 22.03	31 Se 33.50	41 Mo 40.51	51 Te 54.01	61 –	71 –	81 W 77.64	91 –	101 U 101.27
	8 F 8.02	15 Cl 14.98	22 Mn 23.21	32 Br 33.76	42 –	52 I 53.59	62 –	72 –	82 –	92 –	102 –
1 H 0.422			23 Fe 23.63	33 –	43 Rh 44.05	53 –	63 –	73 –	83 Pt 83.29	93 –	103 –
			24 Ni 24.81	34 –	44 Ru 44.05	54 –	64 –	74 –	84 Ir 83.54	94 –	104 –
			25 Co 24.81	35 –	45 Pd 44.98	55 –	65 –	75 –	85 Os 84.05	95 –	105 –

Fig. 25   System of Newlands in which the atomic weights are divided by 2.37 to make them equal to the ordinal number of the elements (1865).

rectly placed immediately, e.g. mercury in the series of the bivalent elements, thallium in the aluminium group and lead in the series of the quadrivalent elements. It was to take another year before Newlands [26] realized this.

Other incorrectly arranged elements were Cr, Fe, In, Ro, Ru, Mn, Au, Di, Mo, W, U and Ta, Mn even being placed as a homologue of P, and Fe as a homologue of S, etc. Platinum and iridium are acceptable as homologues of palladium, cobalt and nickel, but to consider them as resembling the halogens is rather far-fetched. Here, as well as in other places where two elements occur in the same space—Ba and V, Ce and La, etc.—the consequences of the law of octaves are felt. The use of periods of 8 elements would have avoided the forced character of the arrangement. But Newlands' creation of more space in 1878 led to an excessive number of free intervals (Fig. 25).

Newlands' ordinal or atomic number could be found, in his opinion, by dividing the atomic weight by a constant number; this constant varies slightly for the various groups:

atomic number	constant
4–17	2.5
18–34	2.75
35–46	3
47–56	4

It follows that, when three atomic numbers form a triad, the atomic weights do more or less the same thing, even though the elements may not resemble each other, for instance:

$$\text{atomic number} \qquad \text{atomic weight}$$

$$\text{Ti} = \frac{\text{F} + \text{Se}}{2} = 18 = \frac{8 + 28}{2}; \qquad 50 = \frac{19 + 79.5}{2},$$

a conclusion which seems rather pointless.

1866   In his next publication, which reported an improved system and was in fact a version of a paper read in 1865 before the Chemical Society, Newlands [26] claimed precedence for his discovery of the law of octaves (see 16.3). At this meeting, Gladstone, who himself had a share in the development of the classification of elements (4.8), raised the objection that Newlands' system left no free intervals for any undiscovered elements. He argued that four elements had been discovered during the past few years: thallium, indium, caesium and rubidium. Newlands' reply was beside the point; he stated that the metals in the final vertical column resembled both each other and the corresponding elements in the horizontal series.

G. C. Foster[15] asked humorously whether Newlands could not have arranged the elements in alphabetical order[16]. Foster believed that a system like Newlands' could result from a coincidence, as demonstrated by the fact that in the system Fe is placed too far from Co and Ni, and Mn too far from Cr. According to Foster, such a separation of elements was unsatisfactory. Newlands took this attack seriously and replied, that he had also attempted other arrangements, for instance by specific weights, but the classifications based on Cannizzaro's atomic weights had given him the best results.

After this meeting, Newlands answered his critics further by a publication [29] in which he referred again to the outstanding points of his arguments that no system could reproduce the analogies between Cr, Mn, etc. better, and that the distance of Mn from Cr, and Fe from Co and Ni, is exceptionally large. He answered Gladstone more directly by showing that a period need not necessarily consist of 7 elements, but might also include 8 or 9. The resulting free intervals would serve for any elements discovered later. He took as an example the changes resulting from the development of his first (Fig. 23) into his second system (Fig. 24): P and As were separated initially by 13 numbers, later by 14. This was the inevitable consequence of the discovery and classification of indium [30, 31].

This reasoning, however, implied that all newly-discovered elements would lead to one or more new series. This seemed very unlikely, but Newlands' ideas were actually substantiated by the discovery of the noble gases (see Chapter 9), which required the addition of a new column to the periodic system.

5.4. Discussions with Studiosus

In 1864 a discussion started between Newlands and a man who preferred to conceal his identity under the pseudonym Studiosus. The latter [32] asserted that the atomic weights of all elements were a multiple of 8 or roughly 8 (Fig. 26) (see also 7.4). Others joined in as this discussion went on. Newlands [18] remarked, quite correctly, that the value of *about 8* was taken rather loosely, and furthermore that Studiosus was also unable to incorporate all the elements into his scheme. Newlands stated that if the atomic weights were 8 or a multiple thereof, the differences should also be 8 or a multiple thereof. Studiosus [33] objected that this need not hold for deviations from the value of 8. In this case, indeed, the differences in atomic weight may deviate further from 8 than the atomic weights themselves. Studiosus gave as an example I (127) and Te (129), each of which differs by one unit from a multiple of 8 (128) and between which the difference is two units.

This so-called *law of Studiosus* naturally attracted criticism from many sides. Noble [34] criticized its limited validity. However, he also considered that Newlands had no right to comment on Studiosus'

---

[15] George Carey Foster (1835–1919) had just been appointed professor of physics at University College, London. The nomenclature of chemistry was his main interest. Starting in 1857, he studied that of organic chemistry, and in 1865—then reader in natural philosophy at Anderson University, Glasgow—he extended this study to the field of inorganic chemistry [27], in which his opinions were opposed to those of his former teacher, Williamson [28] (see also p. 105).

[16] Bunsen was another scientist who showed reluctance to commit himself where the periodic system was concerned by remarking that one could just as well classify the stock exchange reports in a system.

work because Newlands himself accepted excessive deviations; e.g.,

$$Zn\,(65) = \frac{Mg\,(24) + Cd\,(112)}{2} \quad \text{and} \quad V\,(137) = \frac{Mo\,(96) + W\,(184)}{2} \quad \text{(see Fig. 22, p. 107).}$$

An author calling himself Inquirer [35], who attempted to mediate in the argument, reproved Noble for attacking the other two in that manner for the way they rounded off atomic weights. Dumas, Mercer and others agreed, arguing that this procedure was acceptable because the atomic weights were not always known with equal accuracy.

Li	8						8	
O	16	; N	14				16	= 8 x 2
Mg	24	; Na	23				24	= 8 x 3
S	32	; P	31				32	etc.
Ca	40	; K	39				40	
Ti	50						48	
Ni	58.5	; Co	58.5	; Mn	55	; Fe 56	56	
Cu	63.5	; Zn	65	; Y	64		64	
Se	79.5	; Br	80				80	
Zr	89.5	; Sr	87.5				88	
Mo	96	; Di	96				96	
Ru	104	; Ro	104				104	
Cd	112						112	
Sn	118	; U	120	; Sb	122		120	
Te	129	; J	127				128	
Ta	138	; V	137	; Ba	137		136	
W	184						184	
Os	199	; Hg	200				200	
Pb	207	; Bi	210				208	
Th	238						240	

Fig.26   Atomic weights as multiples of 8 (according to Studiosus).

5.5. William Odling, pioneer and discoverer [36]

William Odling was the only one of the discoverers of the periodic system who had played a part in the preparatory period. In 1857 Odling, then demonstrator of applied chemistry at Guy's Hospital in London, was not far from the discovery of a universal periodicity in both the chemical properties of the elements and their atomic weights (see 4.12).

Dumas and Odling were the scientists who might have been expected to discover the periodic system of the elements. That the discovery does not bear Dumas' name is only partly accounted for by the fact that, during the Second Empire (1852–1870), he served as a senator, as President of the *Conseil Supérieur d'Instruction Publique* and as President of the Municipal Council of Paris *[37]*. He even became Director of the Mint in 1867. Nevertheless, his many publications during these years show that he remained a keen chemist, albeit one whose interests necessarily lay in the field of applied chemistry. In any event, he found time to publish such books as the collected works of Lavoisier. Not until France became a republic in 1870 did Dumas have more time for purely scientific work, but by then the period leading up to the discovery of the periodic system was over.

Odling was among those who read papers at the Karlsruhe Congress. In his interesting talk he emphasized the fact that an element cannot have more than one atomic weight. Like Kekulé and Wurtz, he argued for a unitary system and for uniformity in the scale of atomic weights. Odling had long supported the unitary theory of Gerhardt and Laurent; in 1855 he had translated Laurent's *Méthode chimique [38]*, and had performed scientific investigations with Gerhardt in Paris.

Partly owing to his preparatory work and his intense interest in chemical systems **1864** and nomenclature, particularly in the inorganic field *[39]*, Odling was able, as early as 1864, to draw up an extensive periodic classification of elements, independent of the work done by Newlands[17]. At that time Odling was reader in chemistry at St. Bartholomew's Hospital in London, as successor to Frankland, who shared with him the study of valence. Curiously, Williamson *[40]*, who was also instrumental in deciding Odling to adopt the new molecular theory in spite of the latter's originally sceptical attitude *[41, 42]*, could at that time only classify the elements according to valence. Odling's system *[24]* (Fig. 27) is also more universal than that of Newlands, who, by groping his way from series of 5 elements over a number of octaves, arrived at a system in which not all the elements were correctly placed. Odling made a clear distinction between the main group and sub-group elements and was thus the first to see that some series of analogous elements had to be split in two to obtain a pure periodicity of their properties.

Although Odling did not give a theoretical explanation of the relationships he found between the elements, he believed in a general law.

Analysis of Odling's system shows that it included 57 of the 60 elements known in 1864. Newlands' first *periodic* system had provided for only 24. Odling's 57 elements were, in principle, arranged in the order of increasing atomic weight, but where required by the principle of analogy, Odling made exceptions. As a result, this investigator was the first to introduce the inversion of the elements tellurium (129) and iodine (127). He noted only one exception to the order of increasing atomic weight. Cobalt and nickel, the first pair, differ only slightly in atomic weight, and the second elements of each of the two other pairs, potassium–argon and thorium–protactinium, had not yet been discovered. Argon was not discovered until 1894, protactinium not

---

[17] Newlands, who also arrived at a periodic system in the same year, was born in the same quarter of London, Southwark, as Odling.

until 1917. Gold was assigned its correct place below silver, although this did not follow from the atomic weight then in use. It is to be regretted that the elements Fe through Cu were not assigned their respective places next to Ro (rhodium) through Ag, which now seems obvious. Thallium, which had been discovered by Crookes in 1861 and had already been classified by De Chancourtois, was placed correctly by Odling in 1864 between Hg and Pb but not as a homologue of aluminium, because he apparently did not know what to do with the copper and zinc groups and did not solve that problem until 1868 (Fig. 28). This correct arrangement is the more striking because of the close resemblance between thallium and the alkali metals as far as their properties are concerned. Even Meyer (Fig. 37, p. 128) and Mendeleev (Fig. 44, p. 135), each in his first publication, placed it below caesium. At this time its homologue gallium had still to be discovered and the newly-discovered indium (1863) had not been sufficiently analyzed to permit classification.

					Ro	104	Pt	197
					Ru	104	Ir	197
					Pd	106,5	Os	199
H	1	"		"		Ag	108	Au	196,5
"		"		Zn	65	Cd	112	Hg	200
L	7	"		"		"		Tl	203
G	9	"		"		"		Pb	207
B	11	Al	27,5	"		U	120	"	
C	12	Si	28	"		Sn	118		
N	14	P	31	As	75	Sb	122	Bi	210
O	16	S	32	Se	79,5	Te	129	"	
F	19	Cl	35,5	Br	80	I	127	"	
Na	23	K	39	Rb	85	Cs	133		
Mg	24	Ca	40	Sr	87,5	Ba	137		
		Ti	50	Zr	89,5	Ta	138	Th	231,5
		"		Ce	92	"			
		Cr	52,5	Mo	96	V	137		
		Mn	55			W	184		
		Fe	56						
		Co	59						
		Ni	59						
		Cu	63,5						

Fig.27   System of Odling (1864).

Odling opened the report on his discovery by stating that there were a number of lacunae in the regular progression of the atomic weights, e.g. between 40 and 50 and between 65 and 75. If, in his system, the symbol   ,,   indicates an element still to be discovered, then Odling predicted Ga and Ge, among other elements. He did not leave a free interval for Sc, which must have a place between Ca and Ti. That Odling did not put V in the free interval between Ti and Cr was due to his use of the old value for its atomic weight. Newlands, too, used the value 137, rather than the correct value 51.

Odling derived his system from the observed phenomena of chemical analogies and then deduced from it other analogies not yet observed. Thus, he examined the identical series of hydrogen compounds $HCl \rightarrow HClO_4$, $H_2S \rightarrow H_2SO_4$, $H_3P \rightarrow H_3PO_4$ [43]. From among these, he decided to accept the series $H_4Si \rightarrow H_4SiO_4$.

Odling also provided a basis for a system of silicates [44, 45] on the analogy of the ortho-, meta- and pyro-phosphates. In this connection it is interesting to repeat the story of an examination held by him, during which he asked a student what he meant by a normal acid and received the reply, "Ortho-nitric acid H_3NO_4, is a normal acid". After a brief silence, Odling remarked, "I cannot contradict you" [45]. In all his scientific work he aimed at a comprehensive approach and always tried to systematize wherever possible. He also worked on classification and nomenclature in organic chemistry [46], one of his publications dating from 1862.

1868

It is nevertheless strange that Odling[18], who at that time knew of Dumas' preparatory work, should publish, four years later (1868), without any explanation, a system

		Triplet groups							
H	1					Mo	96	W	184
						-		Au	196,5
						Pd	106,5	Pt	197
L	7	Na	23	-		Ag	108	-	
E	9	Mg	24	Zn	65	Cd	112	Hg	200
B	11	Al	27,5	-		-		Tl	203
C	12	Si	28	-		Sn	118	Pb	207
N	14	P	31	As	75	Sb	122	Bi	210
O	16	S	32	Se	79,5	Te	129	-	
F	19	Cl	35,5	Br	80	J	127	-	
		K	39	Rb	85	Cs	133	-	
		Ca	40	Sr	87,5	Ba	137	-	
		Ti	48	Zr	89,5	-		Th	231
		Er	52,5	-		V	138	-	
		Mn	55						

Fig.28   System of Odling (1868).

with incomplete classification of chromium, the transition metals and gold (Fig. 28), both in Watts' *Dictionary of Chemistry* [47] and in a book of his own [48]. Several other elements, such as lithium, thallium, lead and tungsten, were given a better place in this new system. Odling reduced the total number of elements to 46. The elements Fe, Co, Ni, Ro (= Rh), Ru, Ir, Os, Cu, Ce, U and Ta were omitted, evidently becaues he did not know what to do with them, which is surprising. Cu could have been placed next to Ag and Au, which in his second system occupied an isolated position but had been classified correctly by him in 1864. Another retrogressive step was the inclusion of Mo, now as a homologue of W but no longer as that of Cr. On the whole,

[18] Odling had just been appointed professor of chemistry at the Royal Institution of Great Britain, as Faraday's successor.

Odling was less successful in the composition of this system than in that of 1864, although the over-all picture is somewhat clearer and some of the above-mentioned elements had received better placing.

5.6. Hinrichs' natural system

Scientists in the United States were also seeking numerical relationships between elements at this time. Cooke and Carey Lea had been active in the 1850s. Hinrichs knew of their work when he composed *[49]* his series of elements in 1866, a year before he succeeded in forming a periodic classification *[50]*, although he mentioned only Carey Lea by name. He was also acquainted with the results of Dumas who, as we have already noted, published many papers in American periodicals.

Gustavus Detlef Hinrichs (1836–1923) was born in Lunden (Holstein), then part of Denmark, and studied at the University of Copenhagen until 1861, when he left his native country to go to the United States of America *[51]*. There, only a year later, he became professor of natural philosophy, chemistry and modern languages at the University of Iowa at Iowa City. In 1871 he exchanged this professorship for one in the physical sciences. Hinrichs' first professorship shows the breadth of his interests and it is therefore not surprising to learn that he also studied and worked in astronomy. Regularities he noticed in the distances between the planets led him to make comparisons between them and values distances of the atomic weights of the elements.

1864   In 1864, when he attempted to estimate the relative age of the planets *[52]* he made use of the following series of distances between the planets and the sun as well as the differences in those distances:

distance to the sun		difference
Mercury	60	
Venus	80	$20 = 1 \times 20$
Earth	120	$40 = 2 \times 20$
Mars	200	$80 = 4 \times 20$
Asteroid	360	$160 = 8 \times 20$
Jupiter	680	$320 = 16 \times 20$
Saturn	1320	$640 = 32 \times 20$
Uranus	2600	$1280 = 64 \times 20$
Neptune	5160	$2560 = 128 \times 20$

As the formula for these distances, Hinrichs gave the following series: m, $m + n$, $m + 2n$, $m + 4n$, $m + 8n$, etc. It must be borne in mind here that m is different for each body. The differences in distance can be described by the formula $2^x \times n$, in which n is the difference between the distances of Venus and Mercury from the sun, and $x = 0, 1, 2$, etc.

Plato[19] [53] himself concluded that a simple relationship was to be found in the distances of the planets, the moon and the sun, to the earth. In the numerical series of Moon 1, Sun 2, Venus 3, Mercury 4, Mars 8, Jupiter 9, Saturn 27 he distinguished two series: 1, 2, 4, 8, with the general formula 2^x, and 1, 3, 9, 27, with 3^x.

In 1766, Titius [56], professor of physics at the University of Wittenberg, found the series 0, 3, 6, 12, 24, 48, etc., each term of which plus 4 represented the distances between the planets and the sun. The results corresponded fairly well to the real value, the only missing planet being for the fifth term, i.e. between Mars and Jupiter. Ceres, one of the many planetoids, was discovered in 1801 and proved to fit in this place. For the last term $(4 + 64 \times 3 = 196)$, Uranus was discovered in 1781. It is interesting to note that Hinrichs visualized a system of elements in the form of a spiral, because Plato [57] also saw the celestial bodies as describing spiral orbits.

Hinrichs' interest in astronomy never flagged, as is shown, for instance, by his detailed study of the meteorite that fell in Esterville in 1879. His study of the line spectra of the elements in 1864 must be considered as a second preliminary study of the relationships between elements. Immediately after the discovery of spectral analysis by Kirchhoff and Bunsen in 1859, Hinrichs thought that the dark spectral lines of the elements would be found to conform to simple laws, from which information about the relative dimensions of the atoms could be derived. Bunsen himself never accepted any relationships between the spectral lines and the chemical properties of atoms. Even in his last lectures [58] he made no mention of either the hypothesis of Avogadro or the periodic system discovered by two of his own students, Meyer and Mendeleev.

Hinrichs [59] found the differences between the wavelengths of the spectral lines of an element to be whole multiples of the smallest difference. Thus, for instance, the ratios 1 : 2 : 4; 3 : 6 : 1, etc. occur for calcium. He came to the conclusion that the lines must be determined by the size of the atomic particles. Their optics would therefore show the form and size of their atoms; their chemistry, the weight. The mean difference in wavelength between the spectral lines of Ca (4.8 units) and Ba (4.4 units) led Hinrichs to the daring assumption that one dimension of the atoms of these two elements must be the same or almost the same in both.

Two years later Hinrichs [49] elaborated his theory, again showing himself to be a **1866** Platonist[20]. His hypothesis stated that there must be a correlation between the di-

---

[19] Before him the Pythagoreans, among others, and after him Plato 's pupil Aristotle, saw a relationship between the harmonics of musical intervals and that of the distances between planets [54].

Kepler [55] elaborated on this concept in his *Harmonice Mundi*. He related the ratios of the planetary velocities to the intervals of the musical scale as follows:

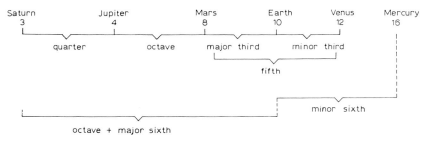

[20] Plato had also represented the elements by geometrical figures: the earth as a cube, water as an icosahedron, air as an octahedron, and fire as a tetrahedron.

mensions of the atoms and these intervals, as the greater the intervals of the dark spectral lines are, the smaller are the atomic weights of the elements of a group. The atoms would have the shape of a rectangular prism whose volume could be written as $A = a \cdot b \cdot c$, in which A is the atomic weight and a, b and c represent the lengths of the sides, the base being the same for atoms of the same group. A prism with a square base,

oxygen group; quadratic formula $A = n \cdot 4^2$

	n	A	calc.	det.	difference
oxygen	1	$1 \cdot 4^2$	= 16	16	0.0
sulfur	2	$2 \cdot 4^2$	= 32	32	0.0
selenium	5	$5 \cdot 4^2$	= 80	80	0.0
tellurium	8	$8 \cdot 4^2$	= 128	128	0.0

alkali metals group; quadratic with pyramid $A = 7 + n \cdot 4^2$

				difference
lithium	0	7	7	0.0
sodium	1	$7 + 1 \cdot 4^2 = 23$	23	0.0
potassium	2	$7 + 2 \cdot 4^2 = 39$	39	0.0
rubidium	5	$7 + 5 \cdot 4^2 = 87$	85.4	- 1.6
cesium	8	$7 + 8 \cdot 4^2 = 135$	133	- 2.0

chlorine group; quadratic formula $A = n \cdot 3^2 \pm 1$

				difference
fluorine	2	$2 \cdot 3^2 + 1 = 19$	19	0.0
chlorine	4	$4 \cdot 3^2 - 1 = 35$	35.5	+0.5
bromine	9	$9 \cdot 3^2 - 1 = 80$	80	0.0
iodine	14	$14 \cdot 3^2 + 1 = 127$	127	0.0

alkaline earth group; quadratic formula $A = n \cdot 2^2$

				difference
magnesium	3	$3 \cdot 2^2$	= 12	12	0.0
calcium	5	$5 \cdot 2^2$	= 20	20	0.0
strontium	11	$11 \cdot 2^2$	= 44	43.8	- 0.2
barium	17	$17 \cdot 2^2$	= 68	68.5	+ 0.5

Fig.29   Atomic weight relations according to Hinrichs (1866).

has $A = a^2 \cdot c$. Some prisms have a pyramidal appendix, in which case the formula for the atomic weight becomes $A = a \cdot b \cdot c + k$ (k being a constant). When $c = n$ (n being a whole number), the atomic weights become:

$$A = n \times a \cdot b \quad \text{or} \quad A = n \times a^2$$

In this way Hinrichs was able to find correlations between the elements of a group of homologues as well as between the groups (Fig. 29). He also showed that the product $n \times d$ (in which d is the interval of the dark spectral lines) is almost constant per group.

He based his considerations on the existence of a protosubstance, the atoms of which would be four times lighter than hydrogen. Hinrichs gave no explanation of this assumption, but he was well aware that it was in contradiction to the elementary nature of the chemical elements. He hoped to be able to prove that, just as there is a unit of force, there is a unit of substance, together forming an all-pervading being. This concept placed Hinrichs in the realm of philosophy, whereas it was his intention to preserve the physical nature of his hypothesis on the primary substance so that it could provide a basis for theoretical mechanical deductions about the properties of the elements. Although Hinrichs' hypothesis concerning the shape of the atoms was based on very shaky grounds, his considerations may be regarded as a preliminary study of the periodic relationships of the elements.

Only a year later *[50]* he published a lithographic reproduction of a handwritten **1867** book entitled *Programm der Atomechanik oder die Chemie eine Mechanik der Pan-atome*[21,22].

That Hinrichs had not given enough thought to his hypothesis on the proto-substance is shown by the change in this paper of the atomic weight of the proto-atom from $\frac{1}{4}$ to $\frac{1}{2}$, with H = 1 as the standard. Hinrichs was convinced that the Creator had only made elements with whole atomic weights, and therefore rejected any experimental methods leading to results conflicting with this persuasion. As chlorine incontestably had the atomic weight 35.5, he took H = $\frac{1}{2}$ as the unit. The protosubstance, here called pantogen, was so important to Hinrichs that he made it the basis for the atomic weights of all the elements, which are consequently about twice the old values. He suggested the names *atograms* or *Hinrichs' atomic numbers*, indicated by *g*. In determining these atomic weights, Hinrichs assumed the relationships found by him to be correct, which eliminated any difference between the real value of the atomic weights (= numbers) and the theoretical values deduced from the formula.

We may take as one instance Genus II, Chloroid, for which species the relationship $g = 1 + m(p)$ holds, in which g is Hinrichs' atomic weight of an element of the group, m the length and (p) the surface of the atomic prism:

---

[21] Hinrichs continued to entertain a special preference for Europe *[51]*, which explains the fact that he published mainly in French periodicals and, for instance, wrote this work in German, adding only an abstract in French. It is therefore hardly surprising that Hinrichs was better known in Europe than in the country he lived in. The fact that Hinrichs himself brought out this volume as well as other books and pamphlets shows that he wanted to have complete freedom to publish, which more or less forced him into the conflict on precedence with Dana, chief editor of the *American Journal of Mining*.

[22] A few months later he presented to the scientific world a French abstract in which he announced that an English abstract would appear in the *American Journal of Mining*. This abstract appeared as a pamphlet in July, 1868, but was certainly not a straight translation of the French text. The copy of the *Atomechanik* in the library of the University of Utrecht is bound together with some pamphlets whose contents chiefly concern a fierce controversy with Dana, in which William Crookes and Hugo Fleck supported Hinrichs.

	A	α	g	m	p	s	S
Fl	19	−1.0	36	5	⑦	—	—
Cl	35.5	0.0	71	10	⑦	1.33	53.4
Br	80	−1.5	157	12	⑬	2.97	53.8
Io	127	0.0	254	11	㉓	4.95	57.2

s = specific weight, S = atostere (atomic volume) and α = deviation from the relationship $A = \frac{1}{2} g$.

Strangely enough, Hinrichs had abandoned the relationship established for the group of halogens $A = n \cdot 3^2 \pm 1$, which was simpler (Fig. 29). It is evident from the foregoing that Hinrichs composed his system of elements, which he called a natural classification, in the classic manner. He distinguished two orders: I, *Trigonoid*, which

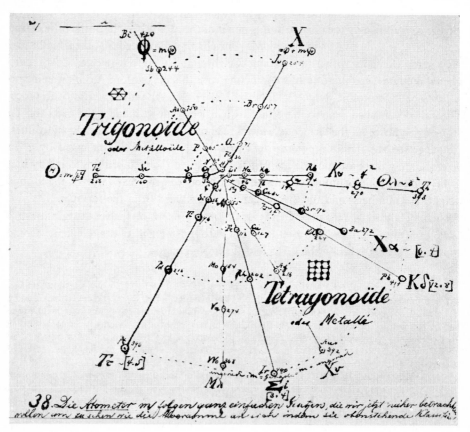

Fig.30   System of Hinrichs (1867).

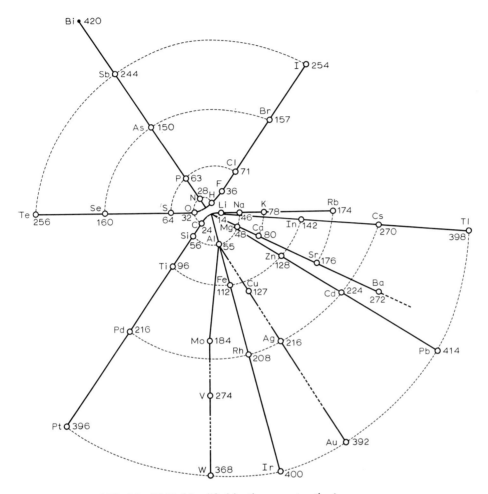

Fig.30a System of Hinrichs (1867) (simplified by the present author).

comprised the metalloids, and II, *Tetragonoid*, which comprised the metals. These
orders are subdivided into 4 and 7 genera, respectively, each of the groups consisting
of species, i.e. the elements (Fig. 30; see also Chapter 2). The term *trigonoid* refers to
the atomic structure formed by panatoms in a triangular arrangement; the panatoms
of the *tetragonoid* class are arranged in squares.

Hinrichs did not mention whether he chose the shape of the system into which he
wanted to incorporate all the elements because of its resemblance to a natural form—
for instance the shell of *Solarium perspectivum* Linné—, or for its similarity to the
solar system[23], but it is a fact that the spiral or star shape gives an excellent representation of the analogies between elements (Figs. 30 and 30a). The arrangement is not

[23] Which Plato visualized as having the shape of a screw.

entirely according to increasing atomic weight; the Φ (nitrogen group) and Θ (oxygen group), for instance, should have exchanged places; but Hinrichs' system may justifiably be called periodic: all the elements occur in groups (genera) and a mutual relationship is established among almost all the elements.

Genera			Species		
x =	1	2	3	4	5
Υ	H				
Kα	Li	Na	Kα	Rb	
Xα	-	-	Ca	Sr	Ba
Kδ	-	Mg		Zn	Cd	Pb	**)
·Υγ	-	-.	-	-	Hg
Kν	-	-	Cu	Ag	Au
			Co		
			Ni		
Σι		Al	.Fe		
			Mn	Rh	Ir
			Cr		
Tτ	C	Si	Ti	Pd	Pt
				Sn	
φ	N	P*)	As	Sb	Bi
Θ	O	S	Se	Te	-
X	Fl	Cl	Br	Io	-
Υ	H				

*) In original one reads T

**) Outline by the present author

Fig.31   Classification of elements (Hinrichs; 1869).

Two years later *[60]*, he improved the form of his classification (Fig. 31). It is to be regretted that Hinrichs did not know what to do with Be and B. He was unable to assign them to any of his groups, which may explain their absence from a system in which there was actually a good place available for them.

Hinrichs' pamphlet contains other material as well, some of it very interesting, including a discussion—beyond the scope of this study—of the smallest particles, which reveals his adherence to the Neo-Democritean school in his attempt to explain many physical properties of the elements and their interrelationship[24].

1869   In 1869 Hinrichs published another paper on his system, in which he stated clearly

[24] Hinrichs *[60–63]* also gave several discussions of his so-called *atomechanical* explanation in 1869.

that in his natural classification[25] the elements are arranged according to increasing atomic weight, but he did not mention their numerical values (Fig. 31). An analysis of the system shows that many elements (Fig. 32) are arranged according to decreasing rather than increasing atomic weight. The sequence of the elements that follow Ti,

								x	calc.	det.	diff.
	Zn		Cd		Pd			. . . .	1	12	-	-
	-		-		Hg			magnesium	2	24	24	0
	Cu		Ag		Au			-				
	Co											
	Ni							zinc	3	64	65.2	+1.2
	Fe							cadmium	4	112	112	0
	Mn		Rh		Ir			lead	5	208	207	-1
	Cr											

Fig.32   A part of the classification of elements (Hinrichs; 1869).

Fig.33   A series of atomic weights calculated by Hinrichs.

viz. Cr, Mn, etc. Zn, should have been reversed to provide them with a place between Ti and As; after the latter, the progression proceeds normally. The same holds for the elements Pd through Cd, and Ir through Pb. The places of Pd and Rh are reversed with respect to their atomic weights, however, perhaps because Hinrichs classed Rh with the Al-group and Pd with the C-group on account of the relationships he indicated (here differing numerically from those in his book) between the atomic weights of the elements of the individual groups. Hinrichs gave a far from simple formula for each group, including those not composed entirely of homologues. Free intervals for elements still to be discovered were left open. Only one of these "predicted" atomic weights was confirmed later, i.e. the element after iodine (see 7.5).

The argument shows that Hinrichs remained a follower of Prout; there must be a common divisor for the atomic weights (perhaps $\frac{1}{2}$). The exceptional position occupied by the "parallelogram" Cr–Zn–Pd–Ir (Fig. 32) might be explained by the fact that Hinrichs wanted in any case to see the elements Zn, Cd and Pb classified as a group next to Mg, and therefore these elements are represented in this group by the formula: $6 \times 2^x + 16$ (omit the constant for $x = 2$) in which $x = 2, 3, 4$ and 5 (Fig. 33). To continue the analogy, Hinrichs could also have classified most of the other elements of this "parallelogram" between the alkali metals and the alkaline earth metals.

The element beryllium was not placed because of the difficulty of deciding whether to consider it as the first member of the iron or of the cadmium group, thus placing it as a homologue of aluminium or magnesium, respectively.

Hinrichs considered the relationships he found to be an extension of those of Dumas. He stated [65] that his natural classification in no way resembled those of

[25] Hinrichs included this system with scarcely any modifications in his chemistry textbook [64], which appeared in 1871. The "chromium-parallelogram" is only partially filled in. In this textbook the division into genus and species is again applied.

NAME	GROUP	SYMBOL	OF	THE	ELEMENTS			
Pantoïds	Υ	H	—	—	—	—		
Kaloïds	Kζ	Li	Na	Ka	Rb	—		
Chalcoïds	Xζ	—	—	Ca	Sr	Ba		
Cadmoïds	Kδ	(Be?)	Mg	Zn	Cd	Pb		
Hydrargoïds...	Υγ	—	—	—	—	Hg		
Cuproïds	Kν	—	—	Cu	Ag	Au		
Ferroïds	Σν	(Be?)	Al	Co-Ur Ni Fe	Rh	Ir		
Molybdoïds....	Mλ	Bo	—	Mn Cr	Mo	Wo		
Titanoïds	Tτ	C	Si	Ti	Pd	Pt		
?					Sn			
Nioboïds	Nβ	—	—	Va	Nb	Ta		
Phosphoïds....	Φ̇	N	P	P	As	As	Sb	Bi
Sulphoïds	⊕	O		S		Se	Te	—
Chloroïds......	Χ	Fl		Cl		Br	Io	—
Pantoïds	Υ	H	—	—	—	—		

Fig. 34   System of Hinrichs (1869).

Thenard and Graham [66]. Graham had based his classification on isomorphism. Another table of Hinrichs (Fig. 34) is again a periodic classification. It should be mentioned that Hinrichs omitted his spiral system, in order to facilitate publication; but he certainly attached much value to the latter system, which he frequently used in his classes. Although the elements beryllium and boron were included, Hinrichs had evidently not yet made his final choice between placing beryllium at the head of the magnesium, or at that of the aluminium group. The elements Cr, Mn, Fe, Ni, Co and Ur he again considered to be varieties of a subgroup. The other elements of the "parallelogram" (Figs. 31 and 32) have received normal places here. Mo and Wo were also classified. It is strange, however, to find that P occurs twice (once as the last term of the sodium period and once as the first term of the period consisting only of the elements P, S and Cl), which results in the unnecessary division of the period Na → Cl into two parts. This was Hinrichs' last publication on his system. In 1893 he reported a system of atomic weights [67], but only to illustrate the advantages of C = 12 as the standard.

5.7. Lothar Meyer's contributions

Most of the scientists who participated in the discovery of the periodic system regarded Dumas and, in some cases, Pettenkofer as their precursors. Julius Lothar Meyer (1830–1895), professor of chemistry at the Institute of Physiology of the University of Breslau (Wroclaw) since 1859, was one of them. When writing *Die modernen Theorien der Chemie* in 1862, the work of the precursors as well as his attendance at the Karlsruhe Congress, made it possible for him to review the numerical relationships between elements, which he also compared with the radicals of a series of organic homologues. This work [68, 69], which did not appear until 1864 and was based on both Dalton's atomic theory and Avogadro's hypothesis, bears the stamp of Meyer's solid theoretical insight and his systematic mind. Meyer, like Mendeleev, thought it necessary that his textbook should be based on a classification of elements, although he did not succeed in incorporating all the elements into a single system. In addition to the system (Fig. 35) based on the valence[26] of the elements (which did not include all the elements), Meyer described seven groups (Fig. 36) based predominantly on a

1864

Julius Lothar Meyer   *Gustavus Detlef Hinrichs*

valence of 4. The last group comprises elements with different valences[27]. Valence was not the only factor to determine the arrangement. Meyer observed that the elements in the group of analogous elements, whether included in the system or not, differ in atomic weight by 16 or a multiple thereof; the values *c.* 46 and 87–90 play the most important role.

Both regularities had already been demonstrated in the groups of analogous elements, as we have already seen, and Meyer knew of them. At first sight, Meyer's classification resembles a periodic system, but this is misleading. Firstly, not all the elements are included; secondly, the B–Al group, which is required for even a short

[26]   This term was first used by Meyer.
[27]   For the classification of some of these elements, see Chapter 12.

	4-werthig	3-werthig	2-werthig	1-werthig	1-werthig	2-werthig
	-	-	-	-	Li = 7.03	(Be = 9.3?)
Differenz =	-	-	-	-	16.02	(14.7)
	C = 12.0	N = 14.04	O = 16.00	Fl = 19.00	Na = 23.05	Mg = 24.0
Differenz =	16.5	16.96	16.07	16.46	16.08	16.0
	Si = 28.5	P = 31.0	S = 32.07	Cl = 35.46	K = 39.13	Ca = 40.0
Differenz =	$\frac{89.1}{2}$ = 44.55	44.0	46.7	44.51	46.3	47.6
	-	As = 75.0	Se = 78.8	Br = 79.97	Rb = 85.4	Sr = 87.6
Differenz =	$\frac{89.1}{2}$ = 44.55	45.6	49.5	46.8	47.6	49.5
	Sn = 117.6	Sb = 120.6	Te = 128.3	J = 126.8	Cs = 133.0	Ba = 137.1
Differenz =	89.4 = 2 x 44.7	87.4 = 2 x 43.7	-	-	(71 = 2 x 35.5)	-
	Pb = 207.0	Bi = 208.0	-	-	(Tl = 204?)	-

Fig.35   System of Meyer (1864).

	4-werthig	6-werthig	4-werthig	4-werthig	4-werthig	2-werthig	
	Ti = 48	Mo = 92	Mn = 55.1 Fe = 56.0	Ni = 58.7	Co = 58.7	Zn = 65	Cu = 63.5
Differenz	42	45	{ 49.2 48.3	45.6	47.3	46.9	44.4
	Zr = 90	Vd = 137	Ru = 104.3	Rh = 104.3	Pd = 106.0	Cd = 111.9	Ag = 107.94
Differenz	47.6	47	92.8 = 2 x 46.4	92.8 = 2 x 46.4	93.0 = 2 x 46.5	88.3 = 2 x 44.2	88.8 = 2 x 44.4
	Ta = 137.6	W = 184	Pt = 197.1	Ir = 197.1	Os = 199.0	Hg = 200.2	Au = 196.7

Fig.36   Elements next to Meyer's 1864 system.

periodic system, is lacking; and lastly, Meyer did not refer to the point of increasing atomic weight. He remarked only "Nachstehende Tabelle giebt solche Relationen für sechs zusammengehörig wohl charakterisirte Gruppen von Elementen".

Two aspects of Meyer's work should be noted here. Firstly, Meyer, as an atomist, spoke of "sog. Atome", "so-called atoms", in comparing the series of elements with the homologous series of organic radicals and molecules, and in fact preferred to consider the elements as compounds of atoms of higher order, just like composite radicals. Secondly, in his system we find a free interval between silicon and tin. Although Meyer did not refer to this, it can be deduced from his calculation of the differences in atomic weights that he took into account an element belonging in this space. This is confirmed by the fact that he mentioned the possibility that, according to the laws already found, the atomic weights would show better agreement in their mutual differences after their weights had been determined more accurately. Therefore, Meyer certainly expected a law to be found, although he advised great caution in arbitrarily altering the existing atomic weights.

Arija's contention [70] that Meyer based his system on purely empirical grounds and did not have sufficient courage to make predictions is unjustified, as Meyer's own words show: "Je mehr die systematische Ordnung der Chemie sich befestigt, desto mehr wird es erlaubt sein, die Speculationen dem Empirismus gleichberechtigt an die Seite zu stellen'' *[71]*. In 1870 his opinion on this point had not changed.

Meyer's second system was ready in 1868, when he prepared the second edition of his textbook *[72]*, which he based on the atomic theory and systematization of elements. This volume was not published until 1872. The delay in publication may have been caused by his move from Eberswalde, where he had been professor at the Academy of Forestry, to Karlsruhe, where he was appointed professor at the Polytechnic Institute in 1868 to succeed Weltzien, the organizer of the 1860 Congress. A sketch of the relevant system, intended to serve as section 91 of his textbook (the same numbering as in the 1864 edition), was reproduced in *Ostwald's Klassiker* in 1895—the year in which Meyer[28] died—by Karl Seubert (1851–1921), Meyer's colleague[29] at the University of Tübingen since 1885. In July, 1868, Meyer gave this manuscript to Adolf Remelé, his successor as professor of chemistry at Eberswalde, who in turn passed it on to Seubert.

1868

A study of the system (Fig. 37) shows that Meyer succeeded in including all the elements in main and sub-groups and thus in forming a periodic system. The only exception to the classification according to increasing atomic weight, is the placing of aluminium as well as the elements Mo, Vd and W in column 15. Regarding column 15, Meyer may have been confused by the incorrect determination of the atomic weight of vanadium, which was given the same value as tantalum and barium; Berzelius had in 1835 given it the value of 137, but this was changed to 69 by Cannizzaro (correct value 51). Like most of his predecessors, Meyer evidently did not know what to do with aluminium, which even appeared twice. The difficulty was due to the fact that scandium and gallium, in the third group, had not yet been discovered, and thallium was often considered to be a homologue of the alkali metals. Only Odling (Fig. 28, p. 115) classified B, Al and Tl in the same column. Meyer could have placed aluminium between columns 7 and 8, next to Si, or between 13 and 14, like the other elements of the third group. In his first article *[74]* on the periodic system of elements, published in a periodical in 1870, these elements were treated in this way (Fig. 38) (see also Chapters 11 and 12).

When Meyer wrote this separate article on the periodic system of elements he had read the German abstract, printed in the *Zeitschrift für Chemie [75]*, of Mendeleev's first publication *[76]*, which was in Russian. As Meyer arrived at his discovery of the periodic system of elements independently and he and Mendeleev did not influence each other, it is interesting to compare the systems of these two scientists. The great resemblance between them is evident (Figs. 38 and 39). Meyer did remark that ''Die nachstehende Tabelle ist im Wesentlichen identisch mit der von Mendelejeff gegebenen'', for which he paid dearly in the battle for priority (see *16.4*). That Meyer worked independently and did not simply copy Mendeleev's system, is shown by several superior solutions in Meyer's system. In contrast to Mendeleev's first system,

1870

---

[28] From 1876 until his death, Meyer was professor at the University at Tübingen, as successor to Fittig.

[29] In 1883 Meyer and Seubert published a textbook on the determination of atomic weights *[73]*.

1	2	3	4	5	6	7	8	9	10	11	12	13	14	15	16
		Al - 27.3	Al = 27.3*)												
		28.7/2 = 14.3													
							C = 12.00	N = 14.04	O = 16.00	Fl = 19.0	Li = 7.03	Be = 9.3			
							16.5	16.96	16.07	16.46	16.02	14.7			
							Si = 28.5	P = 31.0	S = 32.07	Cl = 35.46	Na = 23.05	Mg = 24.0			
							89.1/2 = 44.55	44.0	46.7	44.51	16.08	16.0			
Cr = 52.6	Mn = 55.1	Fe = 56.0	Co = 58.7	Ni = 58.7	Cu = 63.5	Zn = 65.0	89.1/2	As = 75.0	Se = 78.8	Br = 79.97	K = 39.13	Ca = 40.0	Ti = 48	Mo = 92	
	49.2	48.3	47.3		44.4	46.9		45.6	49.5	46.8	46.3	47.6	42	45	
	Ru = 104.3	Rh = 104.3	Pd = 106.0		Ag = 107.94	Cd = 111.9	Sn = 117.6	Sb = 120.6	Te = 128.3	J = 126.8	Rb = 85.4	Sr = 87.6	Zr = 90	Vd = 137	
	92.8 = 2×46.4	92.8 = 2×46.4	93 = 2×46.5		88.8 = 2×44.4	88.3 = 2×44.15	89.4 = 2×44.7	87.4 = 2×43.7			47.6	49.5	47.6	47	
	Pt = 197.1	Ir = 197.1	Os = 199.0		Au = 196.7	Hg = 200.2	Pb = 207.0	Bi = 208.0			Cs = 133.0	Ba = 137.1	Ta = 137.6	W = 184	
											71 = 2×35.5				
											?Tl = 204?				

*) Crossed in its original and made valid again by placing small points under them

Fig.37 An unpublished system of Meyer (1868).

I	II	III	IV	V	VI	VII	VIII	IX
	B = 11.0	Al = 27.3	–	–	–	?In = 113.4	–	Tl = 202.7
	C = 11.97	Si = 28				Sn = 117.8	–	Pb = 206.4
			Ti = 48	–	Zr = 89.7			
	N = 14.01	P = 30.9		As = 74.9		Sb = 112.1		Bi = 207.5
			V = 51.2		Nb = 93.7		Ta = 182.2	
	O = 15.96	S = 31.98		Se = 78		Te = 128?	W = 183.5	–
			Cr = 52.4		Mo = 95.6			
	F = 19.1	Cl = 35.38		Br = 79.75		J = 126.5		–
			Mn = 54.8		Ru = 103.5		Os = 198.6?	
			Fe = 55.9		Rh = 104.1		Ir = 196.7	
			Co=Ni = 58.6		Pd = 106.2		Pt = 196.7	
Li = 7.01	Na = 22.99	K = 39.04		Rb = 85.2		Cs = 132.7	Au = 196.2	–
			Cu = 63.3		Ag = 107.66			
?Be = 9.3	Mg = 23.9	Ca = 39.9		Sr = 87.0		Ba = 136.8	Hg = 199.8	–
			Zn = 64.9		Cd = 111.6			

Differenz von I zu II und von II zu III ungefähr = 16.
Differenz III zu V, IV zu VI, V zu VII schwankend um 46.
Differenz VI zu VIII, von VII zu IX =88 bis 92.

Fig.38   Periodic system of Meyer (1870).

Meyer did not separate the elements of the main and sub-groups but he did include the so-called transition metals Fe through Ni, etc. He thus formed a more distinct transition group in his system although, like Mendeleev, he included Mn among them, and Ni and Co had to occupy the same place (see also *11.3*). Meyer classified Hg as a homologue of Cd, Mendeleev classified it as a homologue of Ag; Pb is given as a homologue of Sn rather than of Ba; Tl is included in the boron group, and not in the group of the alkali metals. Meyer classified In with the correct atomic weight in this same third group; Mendeleev, however, was misled by the incorrect weight (75.6 instead of 113.4), having assumed that this element was bivalent instead of tervalent (see p. 218).

			Ti = 50	Zr = 90	? = 180
			V = 51	Nb = 94	Ta = 182
			Cr = 52	Mo = 96	W = 186
			Mn = 55	Rh = 104.4	Pt = 197.4
			Fe = 56	Ru = 104.4	Ir = 198
			Ni =Co = 59	Pd = 106.6	Os = 199
			Cu = 63.4	Ag = 108	Hg = 200
H = 1					
	Be = 9.4	Mg = 24	Zn = 65.2	Cd = 112	Au = 197?
	B = 11	Al = 27.4	? = 68	Ur = 116	
	C = 12	Si = 28	? = 70	Sn = 118	Bi = 210
	N = 14	P = 31	As = 75	Sb = 122	
	O = 16	S = 32	Se = 79.4	Te = 128?	
	F = 19	Cl = 35.5	Br = 80	J = 127	
Li = 7	Na = 23	K = 39	Rb = 85.4	Cs = 133	Tl = 204
		Ca = 40	Sr = 87.6	Ba = 137	Pb = 207
		? = 45	Ce = 92		
		?Er = 56	La = 94		
		?Yt = 60	Di = 95		
		?In = 75.6	Th = 118?		

Fig.39   Periodic system of Mendeleev (1869).

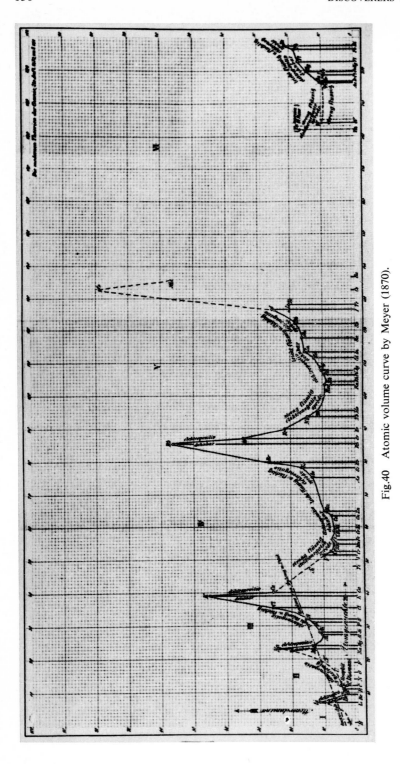

Fig.40   Atomic volume curve by Meyer (1870).

In both systems the elements V, Mo and W are arranged correctly. In Meyer's system we do not find uranium and thorium, whose atomic weights were so uncertain that they could not be used as a basis for any classification. Mendeleev placed these elements incorrectly. Inconsistently, Meyer did not classify osmium, tellurium and gold according to increasing atomic weight. His arrangement of these elements was intended to place them next to their homologues, which was also suggested by the regularity of the curve of the atomic volume (Fig. 40). Later, only the reversal of the sequence of tellurium and iodine appeared to be well-founded (see 8.2). The atomic weights of osmium, platinum and gold were determined incorrectly (see also 11.3). Meyer could not understand why the elements of the columns showed such wide mutual differences when in principle their properties recur after 16, 46, or 88 to 92 units of atomic weight, and are thus a periodic function of the atomic weights. Only the valence shows regularity:

$$1 \longrightarrow 4 \longrightarrow 1$$

The nature of the elements as a function of their atomic weight must be discovered by studying the changes in their properties as functions of the atomic weight. As a case in point, Meyer took the atomic volumes (i.e. atomic weight/density, using hydrogen as the unit of atomic weight and water as the unit of density) of solid elements (only chlorine being a liquid here)[30]. The periodic changes in the atomic volume as a function of the increasing atomic weight are expressed in a graph. For the application of this graph to the prediction of elements, see 7.7.

Meyer continued to doubt the indivisibility of the atoms, as he had in 1864. The regularity of the relationships found between elements led him to assume mass particles of a higher (third) order forming aggregates as atoms. He remained cautious about changing atomic weights on the basis of regularities in their relation to each other, but considered that deviations should be taken as an incentive to make new determinations. Meyer, unlike Mendeleev, always kept in mind the possible existence of a primary substance underlying the composition of elements. The primary matter would account for the similarity between the properties of many elements *[77]*. We may consider this publication as closing the period of discovery. The second edition of Meyer's textbook *[72]* appeared in 1872. In this volume his system (Fig. 41) is shown as a spiral; the Be-group must be attached to the B-group to form a cylinder. Meyer took this system as the basis for a comparative examination of various properties of the elements, including the specific weight, the melting point, and the coefficient of expansion. This textbook was therefore the prototype of a long series of books in which the periodic system of elements was chosen as the basis for the study of inorganic chemistry.

1872

---

[30] Unknown volumes are indicated by small dots.

References p. 144

	1	2	3	4	5	6	7	8	9	10
I									H I ; Li 7,01	Be 9,3
II	B 11,0	C 11,97	N 14,01	O 15,96	F 19,1					
III	Al 27,3	Si 28	P 30,96	S 31,98	Cl 35,37				Na 22,99	Mg 23,94
IV	? 47?	Ti 48	V 51,2	Cr 52,4	Mn 54,8	Fe 55,9	Co 58,6	Ni 58,6	K 39,04	Ca 39,90
V	? 70?	? 72?	As 74,9	Se 78	Br 79,75				Cu 63,3	Zn 64,9
VI	? 88?	Zr 90	Nb 94	Mo 95,6	? 98?	Ru 103,5	Rh 104,1	Pd 106,2	Rb 85,2	Sr 87,2
VII	In 113,4	Sn 117,8	Sb 122	Te 128	J 126,53				Ag 107,66	Cd 111,6
VIII	? ·173?	? 178?	Ta 182	W 184,0	? 186?	Os 198,6	Ir 196,7	Pt 196,7	Cs 132,7	Ba 136,8
IX	Tl 202,7	Pb 206,4	Bi 207,5						Au 196,2	Hg 199,8

Fig.41   Spiral system of Meyer (1872).

5.8. Dmitri Ivanovitch Mendeleev

1869   Dmitri Ivanovitch Mendeleev (1834–1907), professor at the University of Leningrad since 1865 of chemistry, and since 1867 only of inorganic chemistry, was the last, but the most consistent, of the group of discoverers. The records [90] show that Mendeleev discovered the periodic system on February 17th, 1869. Because he was ill, it was Menschutkin who reported this discovery on March 6th at a meeting of the Russian Chemical Society. On March 1st, however, Mendeleev had already sent a pamphlet on the subject to the printer, copies of which he intended to send to scientists of his acquaintance in Russia and other countries. His first publication appeared in 1869 in Russian [69, 76], but an abstract in German [75] was published shortly afterwards. Mendeleev, like Meyer, prepared an elementary textbook. He had already completed a textbook on organic chemistry in 1861, not long after attending the Karlsruhe Congress, and thought that the basic principles of inorganic chemistry also required systematic treatment. In the final chapter on the relationships between elements he wanted to do more than cite the numerical relationships given by Kremers, Cooke, Pettenkofer, Odling (1857), Dumas and Lenssen, though he certainly appreciated the significance of their work. One of Mendeleev's objections was that only Lenssen had included all the known elements in a single system. As we have seen (*4.10*), the list Lenssen published in 1857 is far from being a periodic system. In the first place, the triads required modification. The esteem in which Mendeleev held Lenssen's work is consistent with the innate courtesy he often showed, and which stood him in good stead during the later conflicts about precedence, although it did not prevent him from claiming credit for his share in resolute terms (see, e.g., *16.4*).

Dmitri Ivanovitch Mendeleev

According to his own statement *[78]*, Mendeleev had not seen the much more important publications of De Chancourtois, Newlands, Odling (1864 and 1868), Meyer and Hinrichs. It was not until the meeting of the Russian Chemical Society in March, 1869, when Mendeleev presented his results, that Savshchenkov told him of his translation of Odling's *A Course of Practical Chemistry [79]*. Mendeleev *[78]*, who later studied this work, concluded that because Odling had not added an explanatory note to his table of elements and atomic weights, he had not seen its true implications (see also *16.5*).

Mendeleev needed a textbook to instruct his own students. He himself had received his education at Leningrad, where one of his teachers was Voskressensky, a pupil of Liebig. Voskressensky discussed the problems of inorganic chemistry in great detail, and investigated the elements W, Os, V and Ir. Consequently, Mendeleev was well-prepared when he began his research. During his attempts to find relationships between elements he was struck by the great accuracy with which the atomic weights of most of the elements had been determined. Even so, Mendeleev refused to accept any hypothesis about either the primary substance or the complexity of elements

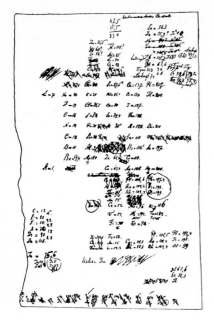

Fig.42   Rough draught of a system of Mendeleev (1869).

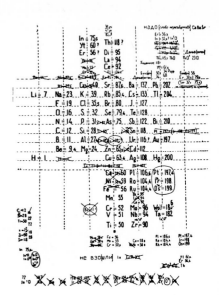

Fig.43   Rough draught of a system of Mendeleev, made more legible (1869).

[80, 81]. He also refused to take the concept of atom as a starting point, because the term atomic weight could then be substituted for elementary weight *[78]*.

What he would accept *[82]* in 1871 was Marignac's suggestion *[83]* that an exception to the law of the conservation of matter entered into the origination of the elements[31] *[84]*.

Mendeleev arrived at the periodic system by puzzling[32] with little cards on which the names of the elements were written *[87, 88]*, a method which Kedrov *[89, 90]* very characteristically called *patience*. But he could not include it in the textbook *[91]* he published in the same year (1869).

A comparison of Meyer's system of 1870 with Mendeleev's first system (Fig. 39, p. 129) immediately shows that the system of the Russian scientist encompasses almost all the facets of a true periodic system of elements and should therefore be seen as the culmination of the period of discovery. Special mention may be made of certain aspects: the division into main and sub-groups, the vacant spaces left for undiscovered elements together with the prediction of some of their properties, i.e., the homologues

---

[31] Marignac thus in a sense anticipated later theories and experiments related to this problem, including Einstein's law *[85]*.

[32] The sketches *[86]* by Mendeleev reproduced in Figs. 42 and 43 show that the construction of the periodic system was indeed something like solving a puzzle.

of aluminium and silicon (see also *7.3*), the classification of the transition metals, and the reversal of tellurium-iodine (see *8.2*).

It should be clearly understood that Mendeleev's work was so brilliant precisely because he knew nothing of the studies of his contemporaries which had been published on the same subject since 1862. If he had known of them, Mendeleev's work could only be considered as a useful summary and continuation of that of his predecessors. It should be noted that, in those days, because of poor communication and publication facilities, the scientific responsibility of an author was not yet taken to include thorough literature searches.

Mendeleev rightly broke with the tradition of those investigators who still classified the elements solely according to a few analogous properties, in particular valence, which they considered the most important property. It is obvious that this criterion could not be maintained, for the simple reason that many elements can have two or more valences. Mendeleev drew on all the available chemical knowledge, the rule of isomorphism, and the law of Dulong and Petit, to give his system as sound a basis as possible and to decide in doubtful cases of classification. He considered the division into main and sub-groups to be of fundamental importance; for instance, both Mg and Be should be placed in the series of Zn and Cd. He also deliberately separated Cu and Ag from the alkali metals (see *12.4*).

Although Mendeleev also considered a system without a division into main and sub-groups, he believed that this could only be done by denying natural analogies. Such a division conflicts, however, with the spiral or, strictly speaking, screw-shaped system (Fig. 44), which admits the incorporation of main and sub-group elements in

Li	Na	K	Cu	Rb	Ag	Cs	-	Tl
7	23	39	63.4	85.4	108	133	-	204
Be	Mg	Ca	Zn	Sr	Cd	Ba	-	Pb
B	Al	-	-	-	Ur	-	-	Bi?
C	Si	Ti	-	Zr	Sn	-	-	-
N	P	V	As	Nb	Sb	-	Ta	-
O	S	-	Se	-	Te	-	W	-
F	Cl	-	Br	-	J	-	-	-
19	35.5	58	80	100	127	160	190	220

Fig.44   Mendeleev's spiral system (1869).

the same groups. The whole development of this system, which may be considered to be a preliminary study, makes an odd impression. The elements Ce, Mn, Fe, Ni and Co (52–59) must form the transition from the bottom of column 3 to the top of column 4, which also applies to the elements Mo, Rh, Ru and Pd of column 5 → 6, and Au, Pt, Os, Ir and Hg of column 8 → 9. It is difficult to see why Cr, Mn, Mo, Au and Hg were not placed within this system (see also *7.6*).

Mendeleev's firm belief in the validity of the periodicity in the system of elements led him to suspect that the atomic weight of tellurium had been incorrectly determined. He thought that a value of 124 to 126 instead of 128 would be reasonable, and that this would also place tellurium and iodine (see *8.2*) in the natural sequence.

Further study of the first system (Fig. 39, p. 129) makes it very clear that Mendeleev could not find a solution to the problem of the rare earths (see *10.3*). Certain elements would seemingly be simple to place—Hg (below Cd), Au (below Ag), Pb (below Si), and perhaps Tl in the series of B—but these elements are located in places other than those which might be expected from their atomic weights.

Mendeleev observed that the atomic weight of most of the elements widely distributed in nature is below 60.

At the end of his first publication the author gave a survey in eight points which, together with the system shown in Fig. 39, forms the contents of the German *[75]* abstract[33]. These eight points include the fact that the analogous elements in the system differ in atomic weight from left to right by a constant value, but that analogous elements in a vertical sequence also occur. The atomic weights of these elements are almost the same (Pt, Ir, Os, etc.). Mention is also made of elements still to be discovered, such as the homologues of Al and Si.

A second publication in Russian, comprising a report of the meeting of Russian scientists held on August 23rd, 1869 in Moscow, which had reached only a small audience partly because the periodical was not widely circulated, presented a system (Fig. 131, p. 287) in which the transition elements are described explicitly *[94, 95]*.

Mendeleev was not satisfied with his systems or proposals for systems. He thought that a cubical system might represent the situation better, but he was unable to construct one himself (*6.17*).

1870 Partly as a result of Meyer's paper *[74]*, Mendeleev *[96]* discussed some details of his system again a year later. He considered the structure of the system to be in principle completed. What remained to be done was to improve the location of a few elements whose classification was not yet satisfactory. This was the main purpose of the paper. In the first place, the inclusion of uranium in the boron group seemed to him incorrect; his original opinion had changed in the same year in which his first publication had appeared *[93, 94]*. He now placed uranium, with an atomic weight of 240 instead of the currently accepted 116, in the chromium group below tungsten, as had already been done in his textbook *[97]*. Like Meyer, Mendeleev also changed the atomic weight of indium from 75 to 113, after which this element fitted into the boron group. Cerium, with a new atomic weight one and a half times greater (138 instead of 92), no longer disturbed the regularity of the system: as a quadrivalent ele-

---

[33] Mendeleev's first paper was summarized and published by the Russian correspondent Von Richter in the second volume of the *Berichte der deutschen chemischen Gesellschaft [92]*, and the *Journal für praktische Chemie [93]* mentioned the system only. The German chemists therefore had sufficient opportunity to learn of Mendeleev's work.

REIHEN	Gruppe I R^2O	Gruppe II RO	Gruppe III R^2O^3	Gruppe IV RH^4 RO^2	Gruppe V RH^3 R^2O^5	Gruppe VI RH^2 RO^3	Gruppe VII RH R^2O^7	Gruppe VIII RO^4
1	H=1							
2	Li=7	Be=9.4	B=11	C=12	N=14	O=16	F=19	
3	Na=23	Mg=24	Al=27.3	Si=28	P=31	S=32	Cl=35.5	
4	K=39	Ca=40	-=44	Ti=48	V=51	Cr=52	Mn=55	Fe=56, Co=59 Ni=59, Cu=63
5	(Cu=63)	Zn=65	-=68	-=72	As=75	Se=78	Br=80	
6	Rb=85	Sr=87	?Yt=88	Zr=90	Nb=94	Mo=96	-=100	Ru=104, Rh=104 Pd=106, Ag=108
7	(Ag=108)	Cd=112	In=113	Sn=118	Sb=122	Te=125	J=127	
8	Cs=133	Ba=137	?Di=138	?Ce=140	-	-	-	- - - -
9	(-)	-	-		-		-	
10	-	-	?Er=178	?La=180	Ta=182	W=184	-	Os=195, Ir=197 Pt=198, Au=199
11	(Au=199)	Hg=200	Ti=204	Pb=207	Bi=208	-	-	
12	-	-	-	Th=231	-	U=240	-	- - - -

Fig.45   System of Mendeleev (1871).

ment it became a homologue of zirconium. Lastly, the place of thorium was also changed: with an almost doubled atomic weight, it now assumed its proper place in the fourth column. With one exception, all these classifications are in agreement with what we accept today. Only uranium has another location at present, which could not have been foreseen in Mendeleev's time.

Mendeleev could say little about the classification of yttrium, erbium, didymium and lanthanum, since their atomic weights were not yet known with certainty. Mendeleev, like Meyer, assumed that the atomic weights of tellurium and osmium were lower than those of iodine and platinum, respectively (see 8.2 and 11.2).

This system, in which in Mendeleev's opinion the analogies between the elements in both the horizontal and the vertical series are clearly apparent, is almost identical to that (Fig. 45) of his next publication [78], in which the main and sub-group elements belong to the same groups.

This paper was 96 pages long and written in German. Mendeleev dealt exhaustively **1871** with the properties of inorganic substances, such as the transition of acid-forming to base-forming elements, to demonstrate the fundamental aspect of the periodic system in inorganic chemistry. He called his system *the periodic law*. Two tables (Figs. 45 and 46) serve to clarify his argument. In one (Fig. 45), in which no main and sub-groups are distinguished, the columns of homologous elements are placed vertically. The copper group occurs in both the first and eighth column. Many elements alternate in the columns; thus, fluorine is above manganese rather than above chlorine. The other system (Fig. 46) consists of short and long periods. Mendeleev called the elements hydrogen through fluorine typical elements, partly because of the small differences between their atomic weights and those of the elements of the following period. This in contrast to

			K =39	Rb = 85	Cs =133	–	–
			Ca=40	Sr = 87	Ba=137	–	–
			–	?Yt = 88?	Di =138 ?	Er = 178?	–
			Ti =48?	Zr = 90	Ce=140?	?La= 180?	Th=231
			V =51	Nb= 94	–	Ta = 182	–
			Cr =52	Mo= 96	–	W = 184	U =240
			Mn=55	–	–	–	–
			Fe =56	Ru =104	–	Os= 195 ?	–
			Co=59	Rh =104	–	Ir = 197	–
Typische Elemente			Ni =59	Pd =106	–	Pt = 198 ?	–
H =1	Li = 7	Na=23	Cu=63	Ag=108	–	Au= 199 ?	–
	Be = 9.4	Mg=24	Zn =65	Cd=112	–	Hg=200	–
	B =11	Al =27.3	–	In =113	–	Tl =204	–
	C =12	Si =28	–	Sn =118	–	Pb=207	–
	N =14	P =31	As =75	Sb =122	–	Bi =208	–
	O =16	S =32	Se =78	Te =125?	–	–	–
	F =19	Cl =35.5	Br =80	J =127	–	–	–

Fig.46   Another system of Mendeleev (1871).

the differences in atomic weights and properties between the last-mentioned elements and the last seven elements of the other periods. He applied the differences again when he made a further distinction according to odd and even elements (Fig. 47).

Mendeleev supported his statement that the usefulness of a system increases with the number of its applications, by recapitulating a few of them. The system can be applied:

1) as a classification of elements;
2) to determine the atomic weights of elements not sufficiently analyzed;
3) to examine properties of unknown compounds;
4) to correct erroneously determined atomic weights; and
5) to collect information about the properties of compounds.

Mendeleev could indeed make all these applications; in this respect his system is superior to the others. Mendeleev stated that he had introduced no hypotheses and did

typical elements

I	II	III	IV	V	VI	VII
H						
Li	Be	B	C	N	O	F
Na						

even elements							VIII			odd elements						
I	II	III	IV	V	VI	VII				I	II	III	IV	V	VI	VII
–	–	–	–	–	–	–				–	–	–	–	–	–	–
							–			–	Mg	Al	Si	P	S	Cl
K	Ca	–	Ti	V	Cr	Mn	Fe	Co	Ni	Cu	Zn	Ga	–	As	Se	Br
Rb	Sr	Yt	Zr	Nb	Mo	–	Ru	Rh	Pd	Ag	Cd	In	Sn	Sb	Te	J
Cs	Ba	La	Ce	–	–	–	–	–	–	Au	Hg	Tl	Pb	Bi	–	–
–	–	Er	Di(?)	Ta	W	–	Os	Ir	Pt	–	–	–	–	–	–	–
–	–	–	Th	–	U	–	–	–	–	–	–	–	–	–	–	–

Fig.47   System of Mendeleev (1879).

Predictions	Determinations
Eka[*] - aluminium	Gallium
	(discovered in 1875 by Lecoq de Boisbaudran)

	Eka[*] - aluminium	Gallium
at. w.	68	69.9
sp. w.	6.0	5.96
at. vol.	11.5	11.7

	Ekaboron	Scandium
		(discovered in 1879 by Nilson)
at. w.	44	43.79
oxide Eb_2O_3 sp. w. 3.5		Sc_2O_3 sp. w. 3.864
sulphate $Eb_2(SO_4)_3$		$Sc_2(SO_4)_3$
bisulphate not isomorphous with alum		small narrow columns

	Ekasilicon	Germanium
		(discovered in 1886 by Winkler)
at. w.	72	72.3
sp. w.	5.5	5.469
at. vol.	13	13.2
oxide	EsO_2	GeO_2
sp. w. oxide	4.7	4.703
chloride	$EsCl_4$	$GeCl_4$
boil. pnt. chloride	$< 100°$	$86°$
density chloride	1.9	1.887
fluoride	EsF_4	$GeF_4 \cdot 3H_2O$
not gaseous		white solid mass
ethyl compound	$EsAe_4$	$Ge(C_2H_5O)_4$
boil. pnt. ethyl compound	$160°$	$160°$
sp. w. ethyl compound	0.96	a little < 1

[*] Eka = Prefix being the Sanskrit numeral one

Fig.48   Prediction of elements by Mendeleev (1871).

not accept Prout's, but this does not mean that he excluded the interpretation of observations or the formulation of laws if their meaning was properly understood. Because
he was consistent, he also refused to accept a primary substance (see also *16.5*). This
explains why Mendeleev's contribution is so remarkably concrete and practical.

For Mendeleev's treatment of the rare earths, his predictions and his approach to the
tellurium–iodine problem, the reader is referred to *10.3, 12.2* and *7.6*, respectively.
Mendeleev not only predicted elements but also many of their properties and the
compounds to be formed with them (see Fig. 48).

The curious fact remains that a paper Mendeleev published in Russian during the
same year *[82]* still contains the earlier system (Fig. 39, p. 129) with some uncertain
atomic weights.

5.9. *Contributions made by other investigators during the period of discovery*

In the last years of this period Kremers *[98]*, who still considered the atomic
weights to be dependent on temperature[34] (*4.7*), published a graphical representation

[34] As late as 1890, Palmer *[99]* still ascribed the defects of the periodic system to comparison of the
elements at, e.g., room temperature. In 1904 Martin *[100]* proposed comparing the elements at other
temperatures, and referred to Gay-Lussac's modification of Boyle's law. His view is also subject to
criticism.

and a diagram of a system (Fig. 49) containing vacant spaces between the beryllium and carbon groups for the transition elements. This arrangement, however, bears little resemblance to Meyer's system of 1868 (Fig. 37, p. 128).

Kremers' system is not clearly presented and also contains numerous incorrect locations of elements, such as Be, C and B, due to incorrect determinations of the atomic weights. As a result, Kremers succeeded only partially in arranging the series of the transition elements and the group Cu, Ag, Au according to increasing atomic weight (*11.3*). As we know, these elements were not classified better by either Meyer or Mendeleev.

Just after Mendeleev's publications, Baumhauer [101] also published a system (Fig. 50). This investigator knew the work of both Mendeleev and Meyer and was the first to publish a system with a spiral representation. Although it contains the homologous elements in groups, many elements could have been classified more correctly, such as the transition metals and most of the elements of the subgroups. Baumhauer was also unable to provide definite places for the elements Y (61.7), Er (112.6), In (37.8), Ru (104.4), Rh (104.4), Di (95), La (92), Ce (92), Th (231.5) and jargonium, partly because of their dubious atomic weights. Baumhauer saw the atomic weights of the elements as functions of those of the other members of the same group. For this reason he considered the elements to be very closely related and refused to rule out the existence of a primary element in the composition of all elements. Baumhauer's contribution did not lie in his arrangement of elements but in the fact that he was the first to completely develop a spiral system.

	I	II	III	IV	V	VI	VII	VIII	IX	X	XI	XII
1)	Li		Be					C	B	N	O	Fl
	7		7					12	14.6	14	16	19
2)	Na	Mg	Al						Si	P	S	Cl
	23	24	27.4						28.4	31	32	35.4
3)	K	Ca							Ti		Cr	
	39	40							50		52.2	
4)		Zn	Y	Cu	Co	Ni	Fe	Mn		As	Se	Br
		65.2	61.7	63.4	60	58	56	55		75.2	79.4	79.8
5)		Cd	E	Ag	Pd	Rh	Ru	U	Sn	Sb	Te	J
		112	112.6	107.7	106.6	104.4	104.4	120	118	122	128	126.5
6)	Rb	Sr							Nb	V	Mo	
	85.4	87.6							94	102.6	96	
7)	Cs	Ba										
	133	137.2										
8)				Au	Pt	Ir	Os			Bi		
				197	197.4	198	199.2			210		
9)									Ta		W	
									182		184.2	
10)	Hg	Tl	Pb						Th			
	200	203.6	206.4						234			

Fig.49   System of Kremers (1869/70).

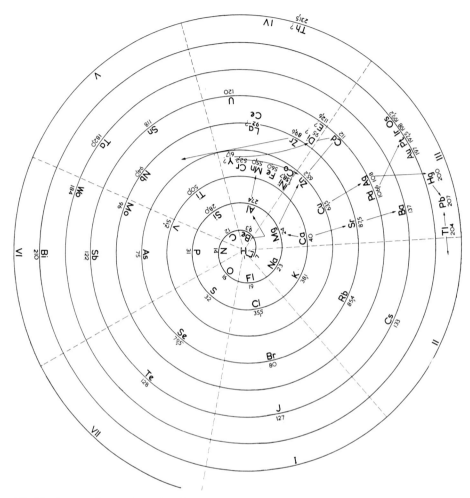

Fig.50   System of Baumhauer (1870).

5.10. Problems of nomenclature[35]

A controversy having arisen later between Rudorf [102] and Schmidt [103] about the historical naming of the classification of elements, we should now examine the nomenclature used by Mendeleev and Meyer.

In his first publication [76], which appeared early in 1869 Mendeleev spoke of a *system of elements*. He also used this designation in the title[36] of the system he sent to

---

[35] See also 2.7.

[36] The full title is *Essai d'un système des éléments d'après leurs poids atomiques et fonctions chimiques*. The system with this title (translated into German) was published in 1869 in the *Journal für praktische Chemie* [93].

some colleagues *[104]*. In his rough notes (Figs. 42 and 43, p. 134) he first called his system *classification of elements and division of elements*. The designation of the system of elements, however, was changed in his 1871 treatise *[105]*: "Ich bezeichne als *periodisches Gesetz*, die weiter zu entwickelnden gegenseitigen Verhältnisse der Eigenschaften der Elemente zu deren Atomgewichten, welche auf alle Elemente anwendbar sind; diese Verhältnisse besitzen die Form einer periodischen Funktion". In a paper *[106]* read before the Russian Chemical Society on December 3rd, 1870, Mendeleev used the term *natural system*, which had appeared on page 163 of the publication mentioned above. During the conflict with Odling over precedence he remarked that the law of periodicity is presented by a *periodic system [107]*.

In 1867 Hinrichs called the system a *natural classification* (Figs. 30 and 30a, pp. 120 and 121).

In December of 1869, Meyer *[74]* was already stating "dasz die Eigenschaften der Elemente grossentheils *periodische* Funktionen der Atomgewichte sind". Note that Meyer did not assign a name to his 1870 system, even in 1872. In his 1880 publication he spoke of *periodische Atomistik*. In 1893, for a lecture on this subject *[108]*, he preferred the name *natural system of elements* to *periodic system*, but in his textbook *[109]*, published in that same year, he favoured the latter designation. The general impression given is that Meyer did not consider a special term necessary. In the 6th edition of his book *Die Atome und ihre Eigenschaften [110]* (1896), he left it at *Systematische Zusammenstellung*.

Many years later, however, Nickel *[111]* objected to the name *periodic system* on the grounds that, mathematically speaking, there are no periods. Although, for example, Li–Na and Na–K differ by 16 in atomic weight, this is not so in the case of K–Rb. Instead of rubidium, manganese should occupy this place, but the latter element does not belong there where analogy is concerned. Nickel therefore preferred the term *natural system*.

Schmidt *[103]* in 1917 believed that, up to then, the law defining periodicity could only be imperfectly represented by *periodic system*. Newlands continued to call his system the *law of octaves* or *periodic law*.

5.11. Conclusion

This chapter has shown how six investigators were able to construct systems, once the structural foundations had been laid. It was in France that the groundwork for a periodic classification of elements was first completed. It is a striking fact that systems were constructed in almost all the principal countries of Europe, in most cases by scientists who were evidently unaware of each other's work. In those same years, the periodicity of the elemental system was also discovered in the United States.

We wish to emphasize here that we do not consider only one or two scientists (Mendeleev, Meyer) to be the discoverers of the periodic system of elements and the other investigators mentioned in this chapter as their predecessors. We recognize six independent discoverers: Béguyer de Chancourtois, Newlands, Odling, Hinrichs, Meyer and Mendeleev, in chronological order. This is not to say, of course, that all these scientists made equal contributions to the discovery and development of the periodic system, but rather that each of them played his own special part. It could be hardly expected that a geologist like De Chancourtois, who arrived at the periodic system several years before the other discoverers, could influence its development as much as the last discoverer, Mendeleev, who, moreover, taught chemistry and published a chemical textbook. This is the main reason why we do not agree with Kedrov [112]—any more than with Leicester [113]—who recognized only Mendeleev as the discoverer of the periodic system and classed the other investigators among the predecessors.

In our opinion, Kedrov, despite his efforts to do so, failed to show that there was an essential difference between Mendeleev's approach and the inductive methods which had dominated work in the natural sciences since the seventeenth century. The other discoverers also used the method of induction in constructing their systems from the available chemical data. The fact that some of these scientists explained or predicted more than others by making deductions from their system is of secondary importance, no matter how highly Mendeleev's predictions of new elements have been esteemed.

In this respect we may compare the discovery of the periodic system with other scientific discoveries, in which the protagonists' claims proved not to be absolutely watertight. Did not Copernicus, the discoverer of the heliocentric universe, think of immobile stars and circular courses of the planets? Did not Lavoisier class heat and electricity among the elements? Did not Dalton, who introduced atomic weights, determine their numerical values rather poorly himself? And did not Chadwick, the discoverer of the neutron, misconceive the true nature of this elementary particle?

Early hypotheses on atomic structure can hardly be considered to have made contributions towards the understanding of relationships between the elements. Indeed, Mendeleev avoided doing so as a matter of principle. Hinrichs, and to a lesser extent Meyer, regarded speculations as indispensable. They were of the opinion that the numerical relationships between the atomic weights of related elements could be reduced to simple relations between the components of the atom.

However, many positive results were obtained even without assuming atoms, as Wilhelm Ostwald demonstrated. Both Mendeleev and De Chancourtois, in fact, believed that they could also dispense with the concept of atom. Dalton's atomic theory—in which the atoms are realities—had hardly any direct influence on the systematization of elements. In this connection it may be noted that, in general, empiricism and positivism had such a profound influence on scientists that few of them hazarded mere speculations about natural phenomena. Remarkably enough, however, Mendeleev became one of them towards the end of his life.

Essentially, the approach to systems remained pragmatic. Meyer and Mendeleev, being teachers, both needed a system of elements for their lectures and textbooks.

Meyer's approach to systematization, published in 1864, and Mendeleev's rough notes found in the archives of Leningrad University, prove this beyond doubt. Kedrov called Mendeleev's system the result of a jig-saw puzzle, solved by *patience*. Odling, too, approached the system of elements from the angle of practical chemistry. The mineralogist De Chancourtois was by nature a systematist, and introduced systematization into many other fields. Hinrichs, as a result of his interest in astronomy sought relationships resembling those among the distances between the planets.

The discoveries resulting in the periodic system proved that the relationships previously found between the elements were correct and therefore also proved the correctness of the assumption that the element is the end product of chemical analysis. The *practical* element could then be identified with the *theoretical* element. The system as it finally took shape in 1871 represented the achievement of a sound classification applicable to chemistry, even though the form had not yet been entirely perfected.

The problems still remaining included the incorporation of the rare earths, the deviations from the principle of arrangement according to increasing atomic weight, and the prediction and addition of several elements. These subjects will be discussed in Part II.

Apart from these problems, scientists were also trying to devise new forms for the periodic system and to establish a numerical relationship between atomic weights, both of which will be discussed in the next chapters.

References

1 J. W. VAN SPRONSEN, *Chem. Weekblad*, 58 (1962) 576.
2 J. W. VAN SPRONSEN, *L'Histoire de la Découverte du Système périodique des Eléments chimiques et l'Apport de Béguyer de Chancourtois*, Paris 1965.
3 E. FUCHS, *Ann. Mines* [8], 9 (1887) 505; *Notice nécrologique sur M.A.E. Béguyer de Chancourtois*, Paris 1887.
4 J. B. A. L. L. ELIE DE BEAUMONT, *Bull. soc. géol. France* [2], 4 (1847) 1249.
5 J. D. DANA, *A System of Mineralogy comprising the most recent Discoveries*, 2nd ed., New York–London 1844.
6 A. E. BÉGUYER DE CHANCOURTOIS, *Compt. Rend.*, 54 (1862) 757, 840, 967; summary in *Bull. soc. géol. France* [2], 20 (1862/63) 647.
7 A. E. BÉGUYER DE CHANCOURTOIS, *Compt. Rend.*, 56 (1863) 253, 479.
8 A. E. BÉGUYER DE CHANCOURTOIS, *Vis Tellurique, Classement naturel des Corps simples ou radicaux, obtenu au moyen d'un Système de Classification hélicoïdal et numérique*, Paris 1863.
9 A. E. BÉGUYER DE CHANCOURTOIS, *Compt. Rend.*, 55 (1862) 600.
10 A. E. BÉGUYER DE CHANCOURTOIS, *Compt. Rend.*, 79 (1874) 89.
11 H. KOPP and H. WILL, *Jahresber. Fortschr. Chem.*, 15 (1862) 6; 16 (1863) 14.
12 C. SAINTE-CLAIRE DEVILLE, *Compt. Rend.*, 54 (1862) 782.
13 C. SAINTE-CLAIRE DEVILLE, *Compt. Rend.*, 40 (1855) 177.
14 J. W. VAN SPRONSEN, *Chymia*, 11 (1966) 125.
15 J. A. R. NEWLANDS, *Chem. News*, 7 (1863) 70.
16 J. A. R. NEWLANDS, *J. Chem. Soc.*, 15 (1862) 36.
17 J. A. R. NEWLANDS, *Chem. News*, 12 (1865) 94.
18 J. A. R. NEWLANDS, *Chem. News*, 10 (1864) 59.
19 A. W. WILLIAMSON, *J. Chem. Soc.*, (1864) 211.
20 J. A. R. NEWLANDS, *On the Discovery of the Periodic Law*, London 1884, p. 32.

21 C. W. QUIN, *Chem. News*, 4 (1861) 253.
22 J. A. R. NEWLANDS, *Chem. News*, 10 (1864) 94.
23 J. A. R. NEWLANDS, *Chem. News*, 12 (1865) 83.
24 W. ODLING, *Quart. J. Sci.*, 1 (1864) 642.
25 J. A. R. NEWLANDS, *Chem. News*, 37 (1878) 255.
26 J. A. R. NEWLANDS, *Chem. News*, 13 (1866) 113.
27 G. C. FOSTER, *Phil. Mag.* [4], 29 (1865) 262; [4], 30 (1865) 57.
28 A. W. WILLIAMSON, *Phil. Mag.* [4], 29 (1865) 464.
29 J. A. R. NEWLANDS, *Chem. News*, 13 (1866) 130.
30 J. A. R. NEWLANDS, *ibid.*, 10 (1864) 95.
31 J. A. R. NEWLANDS, *ibid.*, 10 (1864) 240.
32 STUDIOSUS, *ibid.*, 10 (1864) 11.
33 STUDIOSUS, *ibid.*, 10 (1864) 95.
34 J. NOBLE, *ibid.*, 10 (1864) 120.
35 INQUIRER, *ibid.*, 10 (1864) 156.
36 J. W. VAN SPRONSEN, *Chem. Weekblad*, 60 (1964) 683.
37 J. B. DUMAS, *La Vie de J. B. Dumas, 1800–1884*, publ. in 1924.
38 A. LAURENT, *Chemical Method*, transl. by William Odling, London 1855.
39 W. ODLING, *Phil. Mag.* [4], 16 (1858) 37; 26 (1863) 380; 27 (1864) 119.
40 A. W. WILLIAMSON, *Notices of Proceedings at the Meetings of the Members of the Royal Institute of Great-Britain*, 4 (1864) 247; *Z. Chem. Pharm.*, 7 (1864) 697.
41 W. ODLING, *A Manual of Chemistry*, descriptive and theoretical, London 1861, Vol. I pp. 2, 3.
42 See also H. BROCK and D. M. KNIGHT, *Isis*, 56, No. 183 (1965) 5.
43 W. ODLING, *op. cit.* (ref. *41*), p. 376.
44 W. ODLING, *Chem. News*, 15 (1867) 164, 230.
45 H. B. DIXON, *Proc. Roy. Soc. (London)*, A100 (1922) i.
46 W. ODLING, *Guy's Hospital Reports* [3], 8 (1862) 278.
47 W. ODLING, in H. WATTS, *A Dictionary of Chemistry*, London 1868, Vol. III p. 975.
48 W. ODLING, *A Course of practical Chemistry*, 3rd ed., London 1868, p. 226.
49 G. HINRICHS, *Am. J. Sci. Arts* [2], 42 (1866) 350.
50 G. HINRICHS, *Programm der Atomechanik oder die Chemie eine Mechanik der Panatome*, Iowa City 1867.
51 C. C. WYGLIE, *The Palimpsest*, 11 (1930) 193.
52 G. HINRICHS, *Am. J. Sci. Arts* [2], 37 (1864) 36.
53 PLATO, *Timaios*, see e.g. PLATON, *Sämtliche Werke*, Hamburg 1959, Vol. V § 8, *Die Zusammenfügung der Weltseele*, pp. 158, 159; see also J. L. E. DREYER, *A History of Astronomy from Thales to Kepler*, 2nd ed., New York 1953, pp. 62, 63.
54 T. HEATH, *Aristarchus of Samos, the ancient Copernicus*, Oxford 1913, p. 105 ff.; ARISTOTLE, *De Caelo*, 290 b 12 e.g. in *Aristotle on the Heavens*, The Loeb Classical Library, Gr.-Eng., London–Cambridge 1939, p. 192; ARISTOTLE, *Ueber den Himmel*, Paderborn 1958, p. 87.
55 J. KEPLER, *Harmonice Mundi*, Linz 1619; J. KEPLER, *Weltharmonik*, transl. and introd. by M. CASPER, Munich–Berlin 1939, pp. 23, 294 ff.
56 See e.g. J. C. POGGENDORF, *Geschichte der Physik*, Leipzig 1879, p. 160.
57 PLATO, *op. cit.* (ref. *53*), p. 161; DREYER, *op. cit.* (ref. *53*), p. 69.
58 J. R. PARTINGTON, *A History of Chemistry*, London 1964, Vol. IV p. 283; see also H. DEBUS, *Erinnerungen an Wilhelm Bunsen*, Cassel 1901, p. 144 ff.
59 G. HINRICHS, *Am. J. Sci. Arts* [2], 38 (1864) 31.
60 G. HINRICHS, *Proc. Am. Ass. for Advancement of Science*, 18 (1869) 112.
61 G. HINRICHS, *ibid.*, 17 (1869) 209.
62 G. HINRICHS, *ibid.*, 17 (1869) 223.
63 G. HINRICHS, *ibid.*, 18 (1869) 100.
64 G. HINRICHS, *The Elements of Chemistry and Mineralogy*, Davenport–Leipzig 1871.
65 G. HINRICHS, *The Pharmacist*, 2 (1869) 10. In the copy of *Atomechanik* (University Library, Utrecht) used by us, this article had been bound in.
66 T. GRAHAM, *Elements of Chemistry*, London 1842, p. 142 ff.
67 G. HINRICHS, *Compt. Rend.*, 117 (1893) 663, 1075.

68 L. MEYER, *Die modernen Theorien der Chemie und ihre Bedeutung für die chemische Statistik*, Breslau (Wroclaw) 1864, p. 135.

69 *Das natürliche System der chemischen Elemente, Abhandlungen von Lothar Meyer und D. Mendelejeff*, ed. by K. Seubert, Ostwald's Klassiker No. 68. Leipzig 1895.

70 S. ARIJA, *Erforschung der Atome und Medelejews periodisches System der chemischen Elemente*, Markneukirchen, no date, p. 36.

71 L. MEYER, *op. cit.* (ref. *68*), Breslau (Wroclaw) 1864, p. 144.

72 *Ibid.*, 2nd ed., Breslau (Wroclaw) 1872, p. 294.

73 L. MEYER AND K. SEUBERT, *Die Atomgewichte der Elemente aus den Originalzahlen neuberechnet*, Leipzig 1883.

74 L. MEYER, *Ann.*, Suppl. VII (1870) 354.

75 D. MENDELEJEFF, *Z. Chem.*, 12 (1869) 405.

76 D. MENDELEEV, *Zhur. Russ. Khim. Obshch.*, 1 (1869) 60.

77 L. MEYER, *op. cit.* (ref. *71*), 5th ed., Breslau (Wroclaw) 1884, p. 129, 134.

78 D. MENDELEJEFF, *Ann.*, Suppl. VIII (1871) 133.

79 W. ODLING, *Kurs prakticheskoi Khimii*, St. Petersburg 1867.

80 D. MENDELEJEFF, *Grundlagen der Chemie*, Leipzig 1892, p. 24.

81 See also F. A. PANETH, *Scientia*, 58 (1935) 219, 272.

82 D. MENDELEEV, *Zhur. Russ. Khim. Obshch.*, 3 (1871) 25.

83 C. MARIGNAC, *Arch. sci. phys. nat.*, 9 (1860) 97.

84 D. F. LARDER, *Educ. in Chem.*, 2 (1965) 271.

85 J. H. SCOTT, *J. Chem. Educ.*, 36 (1958) 64.

86 B. M. KEDROV, *Filosofskii Analiz Pervych Troedov D. I. Mendeleeva o Perioditseskom Zakone (1869–1871)*, Moscow 1959, photocopy III.

87 P. WALDEN, in G. BUGGE, *Das Buch der grossen Chemiker*, Weinheim 1961, Vol. II p. 241.

88 B. M. KEDROV, *Cahier d'histoire mondiale*, 6 (1960) 644.

89 B. M. KEDROV, *op. cit.* (ref. 86), e.g. Chap. II.

90 O. N. PISARZHEVSKY, *Dmitri Ivanovich Mendeleev, his Life and Work*, Moscow 1954.

91 D. MENDELEEV, *Osnovy Khimii*, 2 vols., St. Petersburg 1869–1871.

92 D. MENDELEJEFF, *Ber.*, 2 (1869) 553.

93 D. MENDELEJEFF, *J. prakt. Chem.*, 1 (1869) 251.

94 D. MENDELEEV, *Proceedings 2nd meeting of scientists* (Russ.), 23 aug. 1869, p. 62.

95 B. N. MENSCHUTKIN, *Nature*, 133 (1934) 946.

96 D. MENDELEJEW, *Bull. Acad. Imp. Sci. (St. Petersburg)* [3], 16 (1870) 46.

97 D. MENDELEEV, *Osnovy Khimii* 1871, Vol. II p. 382.

98 P. KREMERS, *Physikalisch-chemische Untersuchungen*, Vol. I, *Unzerlegbare Körper und Verbindungen erster Ordnung*, Wiesbaden 1869/70, p. 15 ff.

99 C. K. PALMER, *Proc. Colorado Sci. Soc.*, 3 (1890) 287.

100 G. MARTIN, *Chem. News.* 90 (1904) 175, 189.

101 H. BAUMHAUER, *Die Beziehungen zwischen den Atomgewichte und der Natur der chemischen Elemente*, Brunswick 1870.

102 G. RUDORF, *Das periodische System der Elemente, seine Geschichte und Bedeutung für die chemische Systematik*, Hamburg–Leipzig 1904, p. 74.

103 C. SCHMIDT, *Das periodische System der Elemente*, Leipzig 1917, p. 96.

104 B. M. KEDROV, *op. cit.* (ref. 86), photocopies III and IV.

105 D. MENDELEJEFF, *Ber.*, 3 (1870) 990.

106 D. MENDELEJEFF, *Ann.*, Suppl. VIII (1871) 139.

107 D. MENDELEJEFF, *Ber.*, 4 (1871) 348.

108 L. MEYER, *Ber.*, 26 (1893) 1230.

109 L. MEYER, *Grundzüge der theoretischen Chemie*, 2nd ed., Leipzig 1893, p. 53 ff.

110 L. MEYER, *Die Atome und ihre Eigenschaften*, 6th ed., Breslau (Wroclaw) 1896, p. 127.

111 E. NICKEL, *Naturwiss. Wochenschr.*, 6 (1891) 528.

112 B. M. KEDROV, *op. cit.* (ref. 86), e.g. Chap. IV.

113 H. M. LEICESTER, *Chymia*, 1 (1948) 67.

Chapter 6

Contributions to the development of the periodic system from 1871 to the present[1]

6.1. Introduction

Since its discovery, new forms or types of the periodic system have been found again and again. It has been cast into three-dimensional forms (screw, spiral, cone, sphere, to mention only some) as well as many two-dimensional types. Although none can justly lay claim to being the only correct form, there is one deserving preference, viz., a two-dimensional system in which all the elements have their rightful place by reason of their properties. We shall therefore first sketch the development of the two-dimensional system, to which Bayley, Bassett, Thomsen and Werner were the principal contributors. It took several decades to complete this system, which occupies such an important place in the teaching of chemistry. This relatively long period of development and, in particular, the ultimate recognition of the most favourable form, were not determined solely by the lack of sufficient elements, especially in the very difficult task of classifying the series of the rare earths. It was also the result of the attitude of the chemists, who were sometimes too conservative and sometimes too progressive, being conservative in the matter of the correct arrangement of all the known rare earth metals, and progressive in their attitude to proposed forms of periodic systems which, although ingenious, did not express the periodicity more effectively[2].

The discovery of more rare earth metals and the noble gases, as well as the formulation of a new atomic theory and the quantum theory helped materially towards the completion of a more comprehensive system.

Although the discovery of the radioactive substances threatened to retard both the development of the concept of element and the classification of elements, the two latter ultimately prevailed. Just as the periodic system had evolved by trial and error, so did its further development progress during the next few decades without the support of theory. It is all the more astonishing, therefore, that Werner should have succeeded in finally completing a comprehensive system of elements in 1905, just under ten years before the publication of Bohr's atomic theory.

---

[1] See also Chapter 16 for the contribution of the discoverers.
[2] After this chapter had been drafted, the work of Mazurs *[1]* appeared. This book gives a large number of systems arranged according to form, but does not discuss them in detail.

References p. 209

6.2. Bayley; 1882

The first system to appear in this new stage, a system in which all the elements were considered as equal components of matter, was that of Thomas Bayley *[2]* (b. 1854), demonstrator in chemistry of the Mining School at Bristol. Being an experienced teacher, he was able to create a pre-eminently comprehensive system (Fig. 51), all the while keeping in mind Meyer's graph, in which the atomic volume is set against the atomic weight (Fig. 40, p. 130). Bayley's system made a clear distinction between main groups and sub-groups[3], and left vacant spaces (although not yet enough) for the rare earth metals (see *10.5*). Connecting lines indicate that properties analogous to those of the elements of the first two periods are to be found in the elements of both the main and sub-groups.

Bayley suspected that the elements of the fourth period (not including the transition metals) also have two homologues each, and this caused him to leave some vacant spaces for the rare earth metals. The same suspicion led him to assume the existence of a second series of homologues of the elements Ag through I, which would then belong to the period next to Bi, in which only thorium had been placed. This element, however, was classified arbitrarily, and no place at all was reserved for uranium. Bayley's prediction did not prove to be correct and was in any case based on considerations of symmetry involving faulty analogy. The elements Rb through the element after Mo are not homologues of the elements following didymium.

We might conclude that Bayley designed his system on primarily aesthetic grounds: the governing principles are comprehensibility and a suitable position for each element. The result, however, is a system in which an analogous division has been carried out besides the usual division into elements of the main groups and sub-groups in order to provide a place for the rare earth metals. But this result has a negative side as Bayley considered these elements to be indirect homologues of other elements, e.g., zirconium, niobium, etc. On the other hand, he did treat the transition metals as a separate group.

6.3. Bassett; 1892

Bassett *[3]* realized that, with a system analogous to Mendeleev's, which recognized only a division into main groups and sub-groups, it would be impossible to arrive at the proper sequence of the rare earth metals. Bassett therefore did not treat these as homologues of other elements (Fig. 52), but as independent elements, to which a second group including Th and U could perhaps be added later (see *10.5*). It is regrettable that for some reason he did not completely break with the then current system, but continued to adhere to a relatively closed form. As a result, the alkali metals, for example, had to be distributed over three columns (in this case horizontal-

[3] For the concept of main and secondary groups, the reader is referred to Chapter 12.

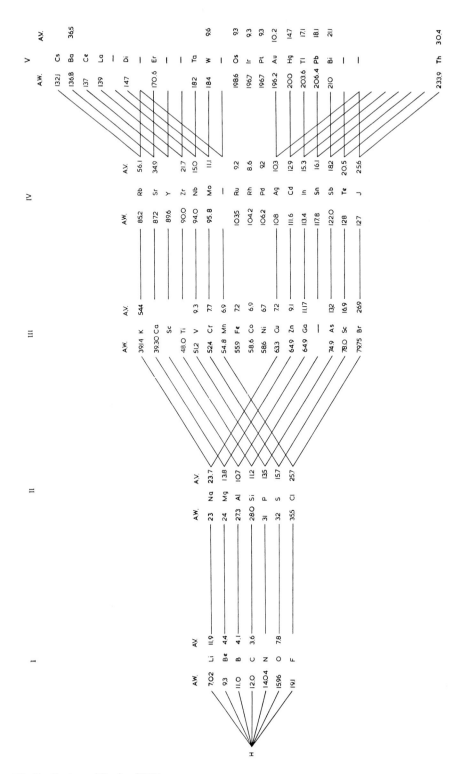

Fig.51   System of Bayley (1882).

ly), and the copper and zinc groups were considered as groups of elements homologous to sodium and magnesium. Bassett wanted to classify the rare earth metals at some terminal point of the periodic system and therefore placed Cs, Ba and La and the respective elements of the next period out of the field of their homologues K, Ca

```
                                        Cs  133          226 ?
                                        Ba  137          ?
                                        La  138.2        ?
                                        Ce  140.2   Th  232.6
                                        Ndy 140.8        ?
                                        Pdy 143.6   U   239.6
                                            148?         241 ?
                                        Sm  150          ?
                                            ?            ?
                                            ?            ?
                                            154          248?
                                            ?            ?
                                        Tb  159.5        ?
                                        Ho  162          ?
                                            ?            ?
                                        Er  166.3        ?
                                            169?         263?
                                        Tm  170.4
                                            ?
                                        Yb  173
                                            174?
                                            ?
                                            ?
                                            ?
                        K   39.1   Rb  .85.5    Ta  182.6
                        Ca  40     Sr  87.6     W   184
                        Sc  44     Y   89.1     189?
                        Ti  48     Zr  90.6
                        V   51.4   Nb  94
                        Cr  52.1   Mo  96
                        Mn  55         100?
                        Fe  56     Ru  101.6    Os  191.7
                        Ni  58.7   Rh  103.5    Ir  193.1
                        Co  59     Pd  106.6    Pt  195
Li  7    Na  23    Cu  63.4   Ag  107.9   Au  197.3
Be  9    Mg  24.3  Zn  65.3   Cd  112     Hg  200
B   11   Al  27    Ga  69     In  113.7   Tl  204.7
C   12   Si  28.4  Ge  72.3   Sn  119     Pb  207
N   14   P   31    As  75     Sb  120     Bi  208.9
O   16   S   32.1  Se  79     Te  125         ?
F   19   Cl  35.5  Br  80     I   126.9       216?
```

Fig.52   System of Bassett (1892).

and Sc, and Rb, Sr and Y. Another result of this re-grouping is that below the three last elements spaces became vacant in which Bassett could only put a question mark. That he attached real significance to these spaces in his system is proved by the numerical value of the atomic weight related to one of these intervals. A few other vacant spaces are also occupied by question marks, with or without an atomic weight (see *7.8*).

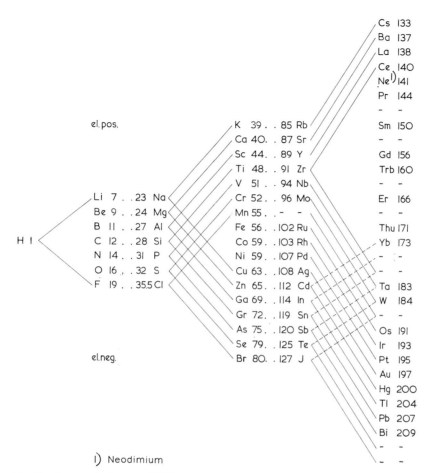

1) Neodimium

Fig. 53   System of Thomsen (1895).

As a result of this mode of classification, the transition metals do not occur as a separate group as in Bayley's system. The merits of Bassett's system, however, should not be underestimated.

6.4. Thomsen; 1895

Three years later, Hans Peter Jörgen Thomsen [4] (1826–1909), professor of chemistry at the University of Copenhagen from 1866 to 1891 and known especially for his thermo-chemical studies, succeeded in forming what for his time was a rather ideal system (Fig. 53). This system is very similar to Bayley's and Bassett's, from which Thomsen eliminated most of the disadvantages or faults mentioned above. The rare earth metals, most of which are not related to other elements, occur as a separate group. Only Ce, Yb and the element following Yb in atomic weight which Thomsen may or may not have considered to be a rare earth metal (this is not clear) are also seen as a homologue of Zr, Cd and In, respectively (*10.5*).

The unknown element adjoining Yb, Ta, W and another unknown element (the future rhenium), are likewise connected with Sn, Sb, Te and I, respectively, by dotted lines indicating analogies in property. But Thomsen appears to have considered the analogy with Zr, Nb, Mo and the later discovered technetium to be more certain, to judge from the unbroken lines connecting these elements. The first-mentioned unknown element, the homologue of zirconium, later proved to be hafnium. It is regrettable that Th and U, so logically classified by Bassett, are absent.

An incidental fact is Thomsen's remark that the elements of the groups can be arranged as follows:

							number	
			1				1	1st group H
		3	1	3			7	2nd group Li → F
	5	3	1	3	5		17	3rd and 4th group K → Br and Rb → I
7	5	3	1	3	5	7	31	5th group Cs → 2 elements after Bi.

In principle, Thomsen succeeded in constructing a comprehensive system that satisfied the demands of analogy. In this respect, it represents a perfection of Bayley's system. But the demands of the principle of analogy mean that, with respect to the arrangement of the elements, no simple system of groups can be set up, because many elements have two homologues. Bassett preferred to assign only one homologue to each element, which obviously could lead to a uniform system. We have already pointed out that this method sometimes gave weaker analogies a preferential position. Both kinds of system found followers. Thomsen built on Bayley's work, whereas Werner and later Romanoff favoured Bassett's principle of classification (see below).

6.5. Werner; 1905

Alfred Werner *[5]* (1866–1919), professor of chemistry at Zurich, who created the inorganic coordination theory and later received the Nobel Prize, gave the system a form which may be called almost ideal (Fig. 54). His special merit is that he applied this system to the whole of inorganic chemistry. There are, of course, some lapses from what we might call a perfect system. Where we now take the electron configuration as a basis, Werner could only use comparisons of the analogies in properties. His intuition certainly stood him in good stead. Only uranium and actinium were incorrectly placed, while the final series includes the addition of radioactive La, Bi and Te isotopes. Just before the year in which Werner's system was published, some investigators thought they had discovered these substances as new elements (see *13.3*).

The incorrect placing of uranium is therefore particularly to be regretted, because

H 1.008																	He 4
Li 7.03	Be 9.1											B 11	C 12	N 14.04	O 16.00	Fl 19	Ne 20
Na 23.05	Mg 24.36											Al 27.1	Si 28.4	P 31.0	S 32.06	Cl 35.45	A 39.9
K 39.15	Ca 40.1	Sc 44.1	Ti 48.1	V 51.2	Cr 52.1	Mn 55.0	Fe 55.9	Co 59.0	Ni 58.7	Cu 63.6	Zn 65.4	Ga 70	Ge 72	As 75.0	Se 79.1	Br 79.96	Kr 81.2
Rb 85.4	Sr 87.6	Y 89.0	Zr 90.7	Nb 94	Mo 96.0		Ru 101.7	Rh 103.0	Pd 106	Ag 107.93	Cd 112.4	Jn 114	Sn 118.5	Sb 120	Te 127.6	J 126.95	X 128
Cs 133	Ba 137.4	La 138		Ta 183	W 184.0		Os 191	Ir 193.0	Pt 194.8	Au 197.2	Hg 200.3	Tl 204.1	Pb 206.9	Bi 208.5	Te α ?		
	Ra 225	La α ?										Pb α ?		Bi α ?			

Lanthanides / actinides:

Ce 140	Pr 140.5	Nd 143.6		Sa 150.3	Eu 151.79	Gd 156	Tb 160	Ho 162	Er 166	Tu 171
Th 232.5				U 239.5						Ac ?

Yb 173	

Fig. 54 System of Werner (1905).

Werner regarded this element as a member of a second group of rare earth metals, just as he did thorium and actinium (see *14.2*). Lanthanum was classed by Werner with the rare earth metals, which made a total of fifteen and led to a vacant space below yttrium. This error is to be ascribed to Werner's attempt to find a formula for the number of elements per period. The formula for the 7th and 8th periods is $3 + 5 + 2 \times 5 + 3 \times 5 = 33$. This indeed leaves 15 elements, later called the lanthanides, as well as 15 elements, later called the actinides.

A second error was the assumption of two periods, each including three elements, of which only hydrogen and helium were known. Concerning the four vacant spaces beside and above helium and above hydrogen, Werner said nothing (compare Rydberg's formula, see p. 232 and *13.5*).

The encircled pairs of elements in his system indicate reversals in the order of the atomic weights. It is a remarkable fact that the pair consisting of neodymium and praseodymium is also found among these pairs of elements (see *8.6*).

Some scientists had doubts about the effectiveness of this system. Augusto Piccini [6] (1854–1905), professor of pharmacy and chemistry at the University of Florence, who had done much research on inorganic chemistry, went so far as to contend that Werner disturbed the regularity of the existing system without substituting a more acceptable one. Piccini objected that the elements of the main and sub-groups do not occur in the same group. Richard Abegg [7] (1869–1910), professor of chemistry at Wroclaw, was also critical. He, too, was opposed to a division of the system into main and sub-groups, which, in his opinion, conflicted with the principle of the periodic system. According to Abegg, this principle meant that all periods must be equal in length. In addition, he preferred to include the noble gases and the transition metals in one group in order to separate the most negative from the most positive elements. Werner [8] tried to refute Abegg's arguments by emphasizing that in his own system each element occupied a specific position, which gave the best demonstration of analogies in properties between the elements. Abegg [9] was not convinced, however, and believed that some compromise system would result from the various suggestions—an expectation which has not been fulfilled.

Georg Rudorf [10] (b. 1881), who determined several atomic weights and wrote one of the few historical studies of the periodic system [11], was not impressed by Werner's system, either. In his opinion, Werner paid attention only to periodicity, so that the system was composed in an artificial way—a remark hardly to be expected from a scientist of his stature.

6.6. Bohr; 1922

Before discussing a few later systems, we must bear in mind that more than half a century has elapsed since the publication of Werner's system; half a century during which an entirely new atomic theory ultimately explained the relationship between elements. Corrections have therefore been made in the valid systems, but it must be

Niels Bohr

Alfred Werner
[Copyright University Library, Zürich]

emphasized that the modern periodic system had in principle been constructed before the development of the modern atomic theory.

The system (Fig. 55) created by Bohr *[12, 13, 14]* in 1922, and characterized by us as the closing stage of the series of systems of Bayley and Thomsen, did therefore not bring any revolutionary changes. It is simply an up-to-date form of Thomsen's system: all the elements have their places and there are vacant spaces only for undiscovered elements. Bohr left fourteen places vacant for the second group of rare earth metals, though not starting with thorium (see *14.4* and *10.8*).

6.7. Romanoff; 1934

All we have to say about Romanoff's system (Fig. 56) is that, at most, it shows divergence in the division into main and sub-group elements. This system was announced in 1934 by Viatschelaw Iljitsch Romanoff *[15]* (b. 1880), professor of physics at the University of Moscow, on the occasion of the centenary celebration of Mendeleev's birth. Comparison with Werner's system shows that, apart from the addition of elements discovered after 1905 and a few minor modifications, the difference consists in the divisions of the periods. Romanoff made a division after the fragments of the second and third groups Be–Mg and Sc–Y, whereas Werner made it after Li–Na

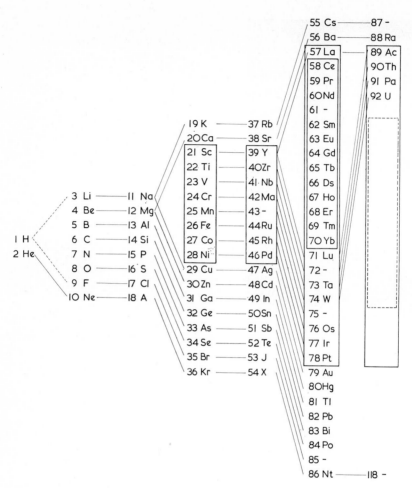

Fig.55   System of Bohr (1922).

and Ca–Sr in order to indicate the elements of the main groups in the 4th and following periods (see also *12.5*).

6.8. Zmaczynski's fan-shaped system; 1937

The system (Fig. 57) constructed by Zmaczynski *[16]*, which is divided into ortho-groups (comprising elements with regularly-filled electron shells) and meta-groups, and had been used since 1930 at the University of Minsk, was called by him a *periodic system of the elements by De Chancourtois–Mendeleev–Werner–Bohr*. This designation shows that Zmaczynski did not underestimate the contribution of De Chancourtois. Instead of Bohr, it would perhaps have been better to mention Thomsen or Bayley,

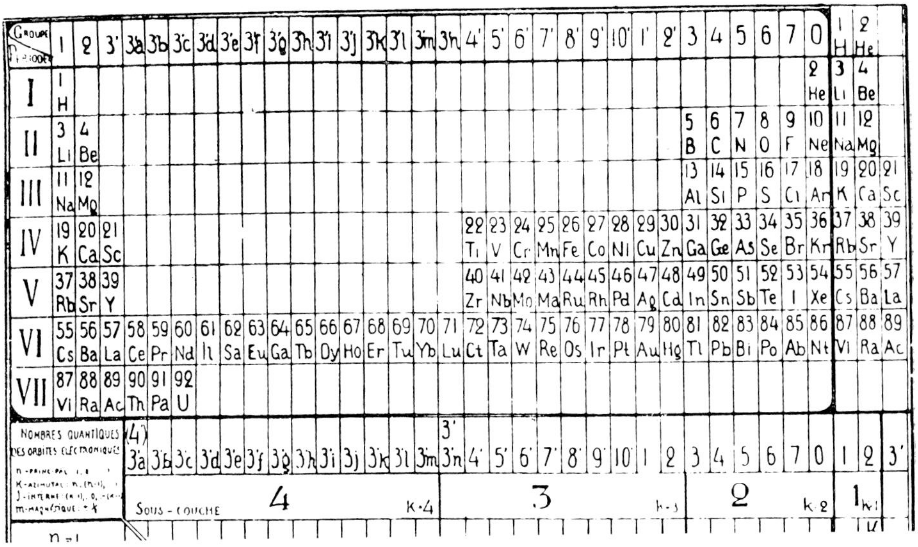

Fig.56   System of Romanoff (1934).

as these investigators had formed systems similar to Bohr's much earlier. The fan-shaped system gave a less obvious division of the first two periods (see *12.5*), and the elements actinium through uranium were not regarded as rare earth metals (see also *10.11*).

Achimov's *[17]* system (Fig. 58) took the form of the cross-section of a pyramid. He, too, based his system on the principle that the lengths of the periods and the analogies in properties between the elements of these periods must be clearly demonstrated.

6.9. *Von Antropoff and Scheele*

A statement by the physical chemist Nernst *[18]*, "Nach meinen Erfahrungen kann die Bedeutung des periodischen Systems nicht hoch genug geschätzt werden", led Andreas von Antropoff *[19, 20]* (b. 1878), professor of physical chemistry at Bonn, who did some of his research on atomic physics, to subject a number of systems to close scrutiny and to set up three systems himself.

These systems should be visualized as being wound around a cylinder and cut through at the group of the noble gases, so that the latter elements occur on both the left and right sides. In these systems hydrogen occupies a middle position at the top. The first system affords the best representation of the situation; it includes the rare earth metals as independent elements. This system is essentially similar to that of Thomsen or Bohr, but Ac, Th, Pa and U are not arranged as homologues of the rare earth metals, as they are in the second system[4] (see also *6.19*).

Scheele *[21]* completed Von Antropoff's system by adding newly discovered elements. His arrangement (Fig. 60) includes four kinds of periods. The first, which

[4] The third system (Fig. 59, see p. 160), which is virtually the same as the second, was brought out commercially, in colour and enlarged, as a wall-chart, and a smaller size was also obtainable. This system was widely used for teaching.

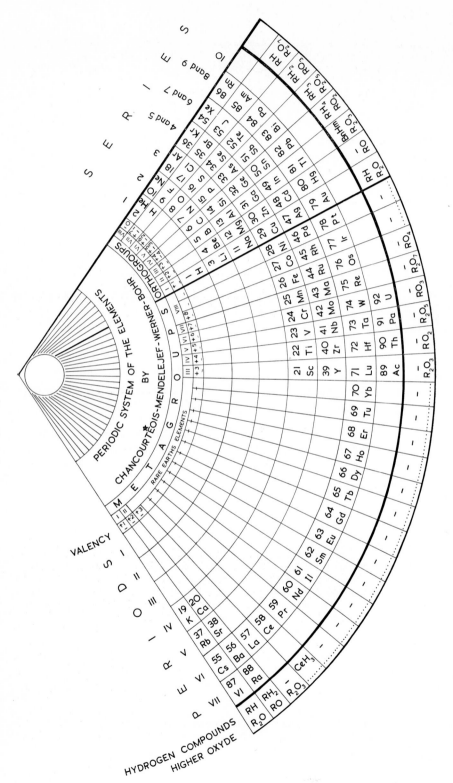

★ Name spelled incorrectly

Fig.57 Fan-shaped system of Zmaczynski (1937).

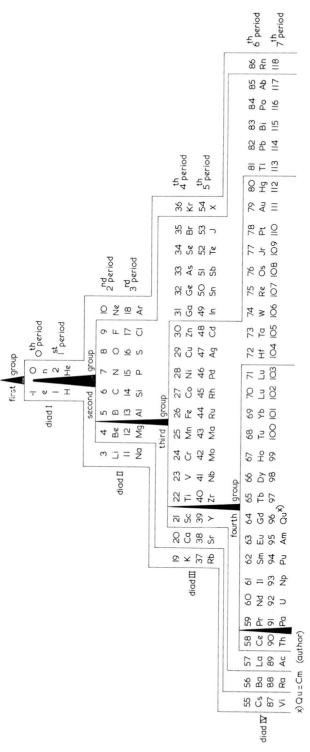

Fig.58 System of Achimof.

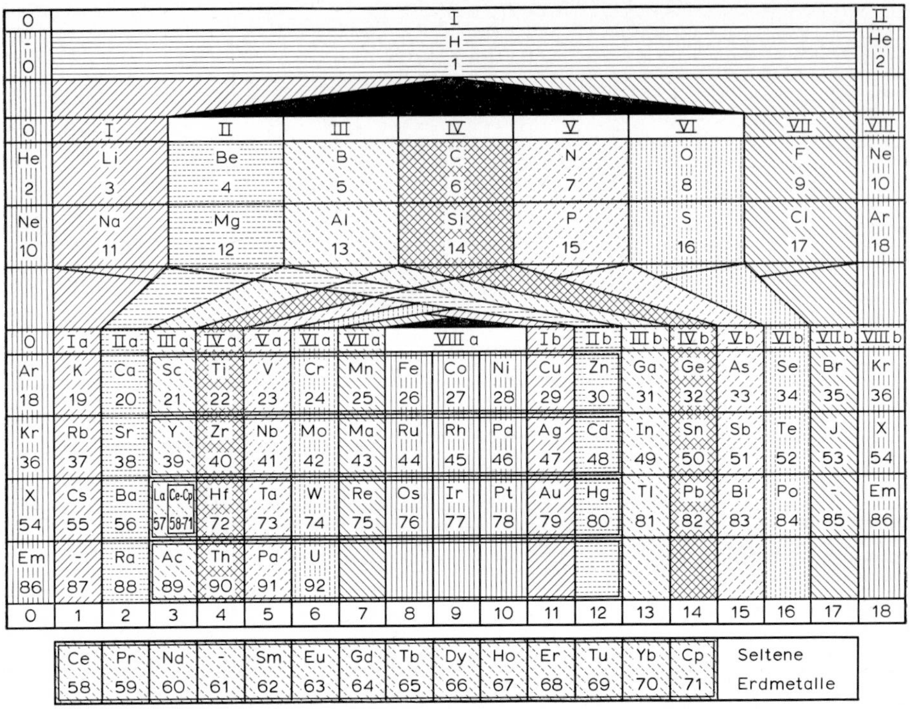

Fig.59   System according to Von Antropoff.

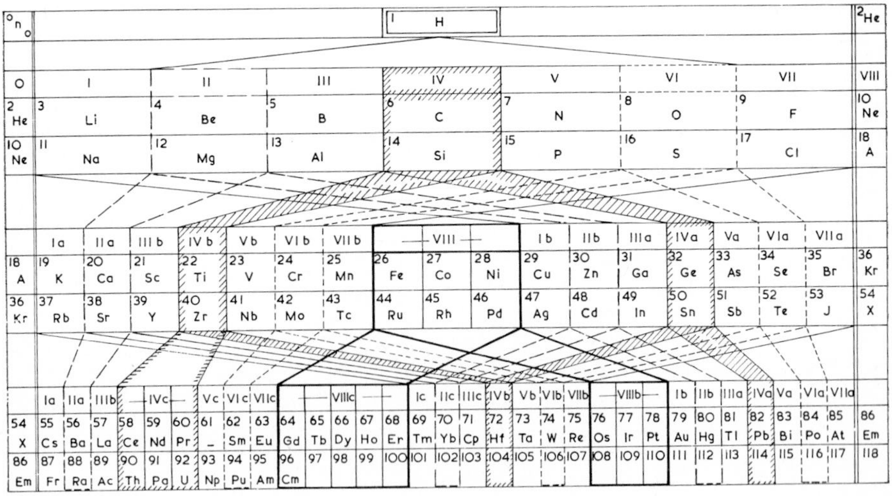

Fig.60   System of Von Antropoff, revised by Scheele (1949).

comprises only one period, is composed of the elements H and He. The next two periods, of the second kind, consist of 8 elements each. Periods 4 and 5 comprise 18 elements, Ar → Br and Kr → I, which are directly related to the elements of periods 2 and 3. The letters *a* and *b* indicate to which element a given element is most closely related. The elements labelled *a* belong to the main group, those with I*b* to VII*b* to the sub-group; group VIII includes the transition metals. Scheele also related the next two periods (6 and 7) to the previous ones; three kinds of elements can be distinguished here, however. Besides the elements of the main and sub-groups, we now see the inclusion of the rare earth metals which are regarded, either separately or in groups, as homologues of the elements of the previous periods. Thus, the element I*c*, i.e., thulium, belongs to rubidium and silver, indicated as I*a* and I*b*. In this sixth period the place IV*c* holds three elements: Ce, Nd and Pr, but the elements 61, samarium and europium each occupy individual places: V*c*, VI*c* and VII*c*. In the 6th period, VIII*c* is occupied by five rare earth metals. However, this extreme division of the system seems exaggerated. The rare earths cannot be classified as homologues of the other elements.

6.10. Spiral and helical systems

These forms were intended to show the periodic system more clearly, but they were seldom successful and if at all, only partly. Some details were occasionally expressed more effectively, but often at the cost of over-all clarity.

In geometry a distinction is made between two figures that are sometimes confused in common usage: the spiral and the screw. A spiral is a two-dimensional figure, whereas a screw has three dimensions. Both have been used to give a clear representation of the periodic system of elements. Both the spiral and the screw-shaped (helical) systems offer the advantage of forming a single unit. There is no gap between the places occupied by the first and last groups of homologous elements, as is the case with a system having a rectangular form. An additional advantage is offered by the fact that the constantly increasing periods of elements are placed more naturally. But as all periods—if we may also assume this for the last two—occur in pairs[5], the spiral representing the system should not be smooth-flowing, i.e. the revolutions should not increase regularly in length. In some spiral systems, however, in order to retain the regularity of the spiral, the periods are not divided into short and long ones. This is illustrated by the systems of Baumhauer, Meyer, Huth and Stoney, for instance.

A screw-shaped system provides an additional means of expressing the analogies in properties and, which is also important, the differences. Because this advantage was initially underestimated, the screw was less frequently used than the spiral. Only after the discovery of the isotopes did some investigators find that the screw shape offered a solution of the problem of including these atomic particles in the periodic system.

[5] The first period (H and He) forms an exception to this *rule*.

The first spiral system dates from the time of the discovery of the periodic system. In 1867 Hinrichs constructed a system (Fig. 30, p. 120) that we have described as star-shaped (5.6).

Without knowing of Hinrichs' work, Mendeleev immediately saw that what he called the spiral representation of the elements (Fig. 44, p. 135) was more satisfactory than a two-dimensional form. We have already seen (5.8) that Mendeleev actually meant a screw-shaped system, which he said he had originated by curving a two-dimensional system over a cylinder and inserting Cr → Co, Mo → Pd, and Au → Hg as transition metals.

In 1870 Baumhauer (5.9) had shown the relationships between the elements in a spiral system (Fig. 50, p. 141), in which he had attempted to place the elements of the main and sub-groups separately.

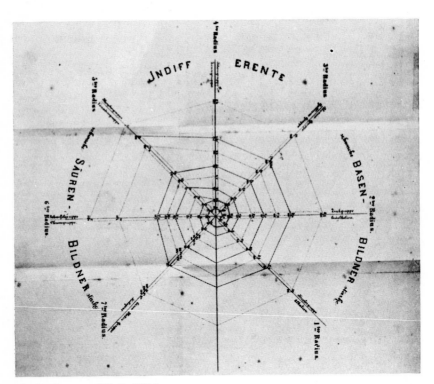

Fig. 61   System of Huth (1884)

The botanist Huth [22] (1845–1897), however, made no distinction in 1884 between the elements of the main and sub-groups (Fig. 61)[6]. He based his system on the work of Baumhauer, who also divided the spiral into only seven sectors, which made it impossible to classify the transition metals separately.

[6] A system proposed by Carnelley [23] in 1886 resembles Huth's.

Meyer considered the possibility of drawing up a system in which the transitions between the various groups of analogous elements would be smoother than in a two-dimensional system. Meyer, like Mendeleev, made no attempt to depict a screw-shaped system. He only indicated how a two-dimensional system could be converted into a screw-shaped one. For this purpose Meyer [24] formed not linear but step-shaped periods, so that the place of the last element of a period was about at the same level as the place of the first element of the next period. If this two-dimensional system is rolled around a cylinder, the result is one uninterrupted series of elements: "Denkt man sich die Tafel auf einen senkrecht stehenden Cylinder so aufgerollt, dass die mit *Bor* und *Aluminium* beginnende Gruppe sich an die der alkalischen Erdmetalle, *Beryl-lium, Magnesium, Calcium,* u.s.w. so anreiht, dass die im Atomgewichte einander folgende Elemente sich unmittelbar an einander schliessen, so erhält man, wie leicht ersichtlich, eine *spiralförmig angeordnete nach der Grösse der Atomgewichte continuir-lich fortlaufende Reihe aller Elemente"*. This system (Fig. 41, p. 132) was published by Meyer in 1872 (see 5.7); he was acquainted with Baumhauer's system but it did not appeal to him, because he thought it was too arbitrary.

The choice between the helix and the spiral is purely a matter of taste. In both forms the elements lie in an unbroken line. A three-dimensional system has the advantage of greater possibilities for representation, but the untrained eye observes analogies more quickly in the two-dimensional version. New spiral and screw-shaped systems continued to appear after the discovery of the periodic system.

A short article by Vincent [25], giving a formula for the relationship between the atomic weight and the order number of the elements 3 up to 60 as $W = 1.21 (n+2)$, led George Johnstone Stoney [26] (1826–1911) to point to the system he had set up in 1888. In a paper read before the Royal Society [27], Stoney, whose work was principally in the field of astrophysics and who introduced the term electron in 1894, had searched for an arithmetical relationship between the atomic weights and the order number of elements. The system he described was never published; nor do reports of his lecture convey Stoney's idea of this relationship. In 1902 Stoney published an improved system (Fig. 62). As far back as 1888 Stoney had left a radius in his spiral vacant, i.e. radius 16, for the noble gases discovered a few years later. The reason for this choice was that the radius carrying the elements of the transition group lay between the radii of the elements of the main and sub-groups. It was therefore necessary[7] to place radius 16 in the production of the former radius, i.e. number 8. The drawback of this system[8] is that the elements of the periods Li through Ne and Na through A are arranged in series. This brings the elements C, N, etc. above Ti, V, etc., while Na, Mg, etc. are placed as homologues of the elements of the sub-group. Neon, for example, would then belong to the iron group, although this was not Sto-

---

[7] The hatched parts in the system have various meanings. The triangular one refers to elements having a large atomic volume in the solid state. The sectors, however, indicate a small atomic volume [28].

[8] Another spiral, for which Stoney did not provide a drawing, illustrates a logarithmic law.

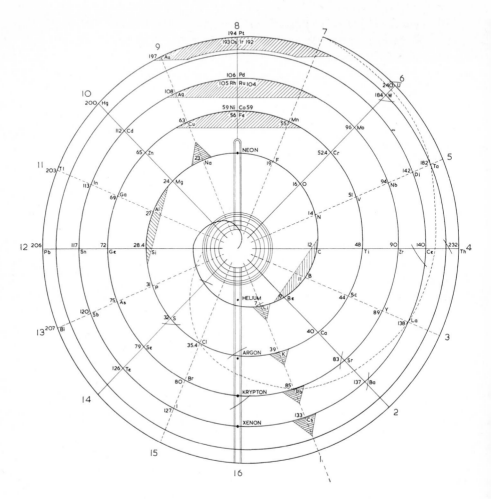

Fig.62   System of Stoney (1902).

ney's intention. These peculiarities are clearly the result of the consistent division of each period into 16 parts[9].

Benjamin Kendall Emerson [30, 31] (1843-1932) found an elegant solution for this defect by placing the small periods in small spirals and the large periods in larger ones. Emerson in fact started from a screw-shaped system (Fig. 63) which he called a helix and which included several hypothetical elements (see 9.5). To clarify this system, in which Emerson expressed the periodicity of the different properties of the elements, he used the projection that assumed the shape of a spiral (Fig. 64). The rare earth metals unfortunately occupy places from which their properties should exclude

[9] Erdmann's system, published two years later [29] and even containing, as homologues of other elements, the rare earth metals mostly omitted by Stoney, suffers from the same difficulty.

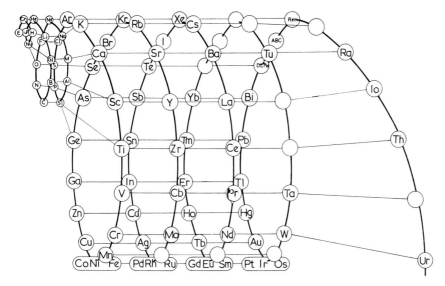

Fig.63   Helical system of B. K. Emerson (1911).

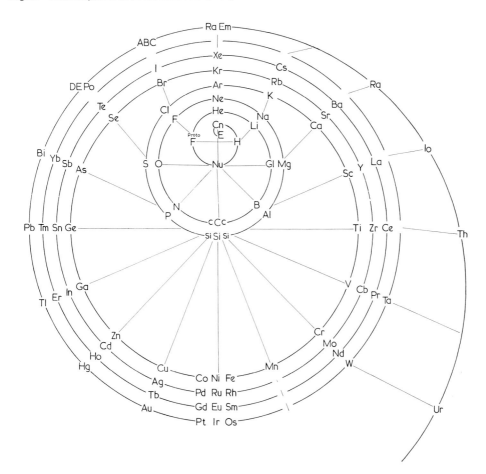

Fig.64   Spiral system of B. K. Emerson (1911).

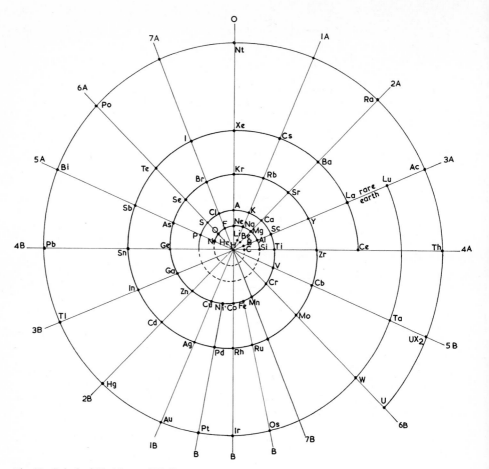

Fig.65   Spiral of Harkins and Hall.

them. It is also surprising to find that the elements carbon and silicon occur three times.

Another solution expressing the difference between small and large periods in a spiral system is found by filling up eight places of the first two periods. One of the first examples of such a system (Fig. 65) was given by R. E. Hall and by William Draper Harkins (1873–1951), professor of chemistry at Chicago, who did much work on molecular and atomic structure in relation to the periodic system *[32]*.

Edgar I. Emerson *[33]* also considered this solution many years later, in 1944. In this system Be and Mg have an exceptional position (Fig. 66)[10]. They are considered as it were, to be elements combining the main and sub-groups (see *12.5*). Hydrogen appears as a homologue of both the alkali metals and the halogens. At the top of the column of the noble gases we find the element 0, the neutron; Emerson believed that

[10] In the same year (1944), Schultze *[34]* constructed an analogous system.

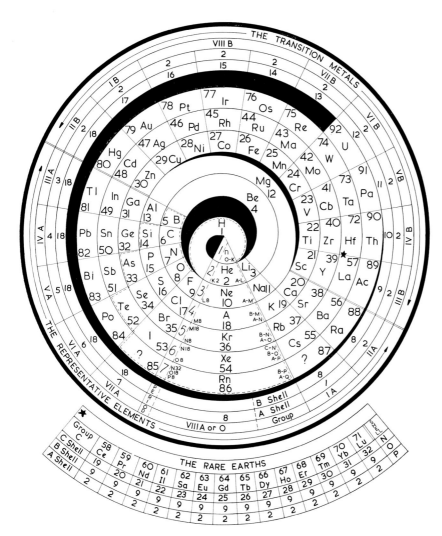

Fig.66 Spiral system of E. I. Emerson.

this nuclear particle had the properties of a noble gas. He made the segment of the circle containing the group of noble gases three times wider than the segment containing the other groups. Emerson did this because he regarded the former group as the prototype of the groups of elements and because its position is diametrically opposite the transition groups (Fe, etc.), which also cover a space of three columns. Emerson assigned a separate place to the rare earth metals.

Another of Harkins' and Hall's systems (Fig. 67) also takes the form of a screw[11]. Starting with the fourth period, i.e. after argon, the revolutions of this screw alter-

[11] Yet another screw-shaped system is that of Monroe and Turner [35].

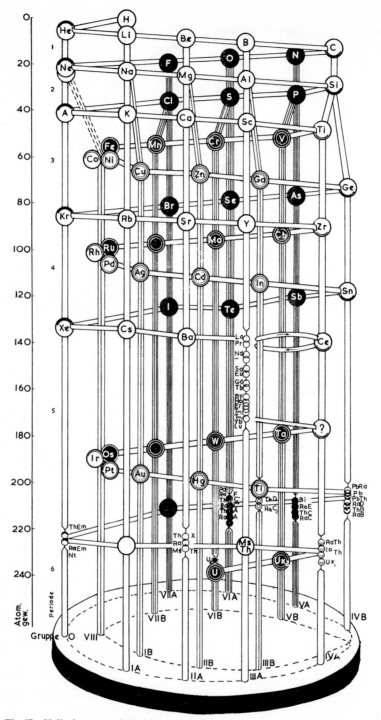

Fig.67   Helical system of Harkins and Hall.

nately contain homologous elements. The authors also used the third dimension to represent the rare earth metals and the newly discovered isotopes. The value of this system was soon realized [36].

The essence of the elements greatly interested Harkins and Wilson [37, 38]. Working together on their hydrogen–helium system, they proceeded from the idea that many elements are built up of hydrogen, possibly in combination with helium, or of helium alone. With this in mind, they tried to classify some elements. The divergence of the atomic weights from whole numbers is caused by the *pack effect* between the hydrogen and helium atoms, which results from the fields of force operating between the atoms[12] [39]. Although neither scientist could provide practical proof of this assumption, they thus anticipated the mass defect that was to be demonstrated later. The extent to which Einstein's theory influenced their thinking is not ascertainable.

The first of these authors, Harkins [41], went more deeply into this theory by stating that all the elements must be considered as intra-atomic—not as chemical—compounds. In his opinion [42], the first phase in the formation of such a complex may be the transformation of hydrogen into helium. Later, this idea was used by Bethe in his explanation of stellar energy and the occurrence of the element helium on the stars. Harkins [43] divided the elements into two series: one series having even atomic weights with n He as the general formula, and one series having odd atomic weights with n He $+$ H or n He $+$ H$_3$ as the formula. Thus, the composition of Li $=$ He $+$ H$_3$; Be $=$ 2He $+$ H; B $=$ 2He $+$ H$_3$; C $=$ 3He, etc. At a time when the atomic number was not yet generally accepted as a principle of classification, Harkins did so. In his metaphysical considerations he also anticipated the actual composition of the atoms in form of what was later known as protons and neutrons. Harkins undeniably had a speculative mind: he arrived at this conclusion about the atom solely on the basis of the practical results of X-ray analyses (diagrams) and radioactivity studies, without any further atomic research. On this basis Harkins and Hall [32] built up their system; they also calculated the number of elements possible in a period, arriving at the formulae 2×2^2, 2×3^2 and 2×4^2 (see *13.5*).

Other ways of representing the elements on a spiral have been suggested[13]. Arnaldo Teofilo Pietro Piutti [45, 46] professor at the University of Naples and investigator of noble gases and radioactivity, divided the spiral into nine segments, thus bringing the elements of main and sub-groups to lie on the same segment (Fig. 68). The rare earth metals suffered from this division; an extremely small space between the third and fourth groups being assigned to these elements. This system also included the

---

[12] Scheringa [40] doubted whether this effect would influence the numbers, as deviations of the atomic weights from whole numbers are not systematic.

[13] To depict a spiral, Caswell [44] made use of polar coordinate paper. The atomic numbers are radial coordinates. Groups of homologous elements form 20° angles with each other. Thus, $\theta = 0°$ applies to the noble gases, $\theta = 20°$ to the alkali metals, $\theta = 340°$ to the halogens. Be and Mg lie at $\theta = 40°$, B and Al at $\theta = 260°$; the rare earth metals lie on a straight line represented by $\theta = 60°$.

CLASSIFICAZIONE PERIODICA degli ELEMENTI

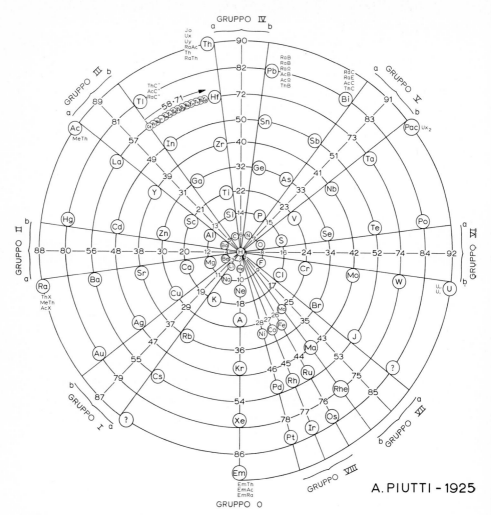

All atomic weights omitted by the present author

Fig.68   System of Piutti.

radioactive disintegration products in the space between the elements having the corresponding atomic numbers.

Nodder's system (Fig. 69) *[47]* shows the rare earth metals to much better advantage, partly because he used a loop[14] with places for the elements of the long periods. Nodder wanted to express the relationship between the elements of the main and sub-groups by showing the groups in parallel. The sub-group starts in this

[14] Monroe and Turner *[35]* used a similar loop in their system, but without enhancing its expressivity.

system only with the vanadium group, which is a marked deviation from the usual division.

6.11. Circular models

Circles have also been used to express the analogies in the properties of the elements. One of the first attempts to draw this kind of system was made in 1875 by Wiik [48] (1839–1909), who was appointed professor of geology and mineralogy at the University of Helsinki in 1877. As the attempt failed to produce a system that in any true sense could be termed periodic, there is no need to discuss it.

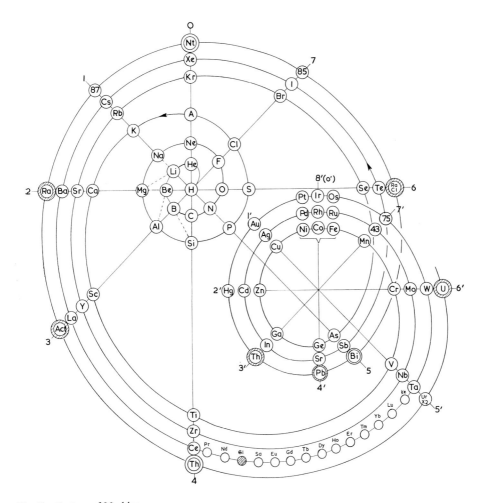

Fig.69   System of Nodder.

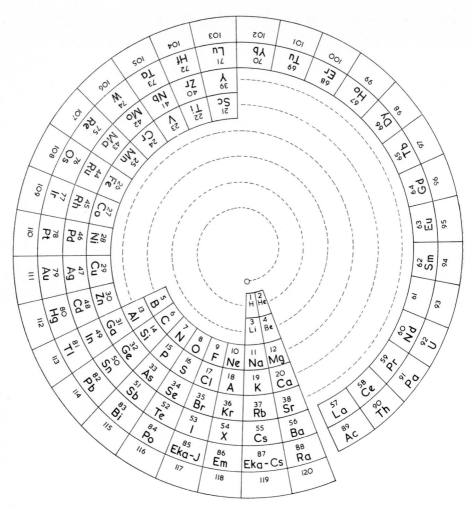

Fig.70   Circular system of Janet.

A system (Fig. 70) far more aptly representing the situation was devised by Janet (1848–1932), an entomologist and geologist [49], though it did not see the light of day until 1929. Janet distinguished four kinds of periods and so assigned places of equal value to all the elements, including the rare earth metals. His representation has the advantage of being very compact, the drawback, however, being that many letters and figures are shown upside down. Twenty years later Scheele [50] published a number of systems, some of them based on the electronic theory (see 6.26). One of these is a circular system (Fig. 71) whose periods include the same elements as Janet's system, but, because of the basis chosen, do not form an uninterrupted whole. This system is therefore in conflict with the principle of increasing atomic weight.

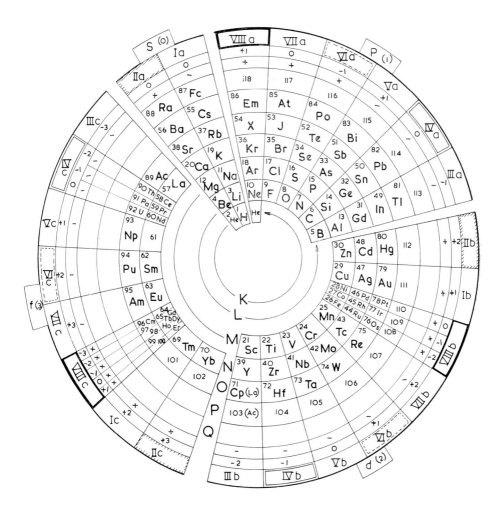

Fig.71   System of Scheele (1950).

6.12. Models related to the screw form

In 1914 Frederick Soddy *[51]* (1877–1956), a student of Rutherford's who became lecturer in physical chemistry at Glasgow and later professor of chemistry at Aberdeen and Oxford and won a Nobel prize (1921), constructed a three-dimensional system (Fig. 72) that can be visualized as being composed of a line winding around a cylinder with several changes of direction. These peculiar turns of the line arise from the application of a division into main and sub-group elements and rare earth metals.

Besides his circular system[15] (*6.11*), Janet devised an ingenious three-dimensional

[15] Another system by Janet, consisting of a spiral carrying three loops, offers no advantages.

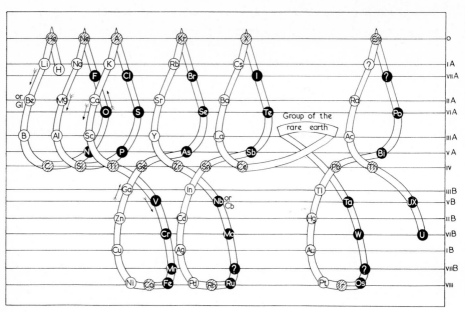

Fig.72   Three-dimensional system of Soddy (1911).

system[16] (Figs. 73 and 74) consisting of four cylinders of different sizes with a common tangent. Over these four cylinders the elements are distributed according to the periods to which they belong; thus, the first two elements of each period are found on the smallest cylinder, the next cylinder includes both the periods consisting of only eight elements from the other periods that are homologues of these elements of the *periods of eight*. The third and fourth cylinders carry the remaining ten elements of the large periods and the two series of rare earth metals, respectively.

6.13. The "pretzel" model

In his attempts to arrange the elements in a system, Sir William Crookes (1832–1919), who, as professor of chemistry at Chester, made many investigations into spectra and atomic structure, also sought a solution in a three-dimensional representation. After having done some preliminary research [52] in 1886 and 1887, in which he started with a graphic representation given by James Emerson Reynolds [53] (1843–1920), professor of chemistry at Trinity College, Dublin, he constructed [54, 55] a pretzel-shaped three-dimensional figure (Fig. 75). A graph provided by Crookes (Fig. 76) shows the origin of the elements on the basis of his so-called protyle theory, a refinement of Prout's hypothesis, in which protyle is the primordial substance from

[text continued p. 181]

[16] A more detailed discussion of these systems will be found in *14.4*.

Fig. 75   Crookes' "pretzel" model of the periodic system, made by himself. [Copyright Science Museum, London]

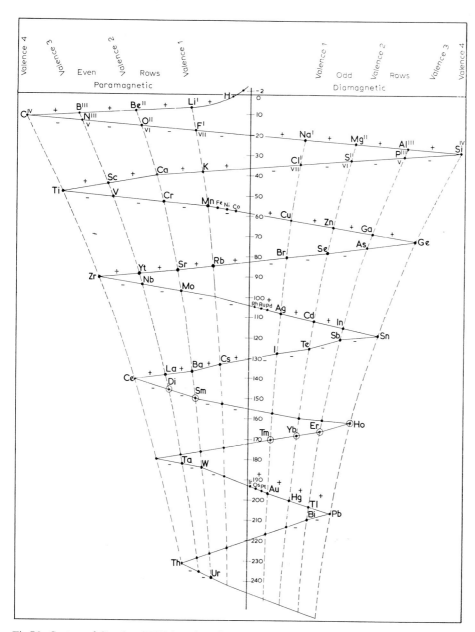

Fig.76   System of Crookes (1887) based on the system of Reynolds.

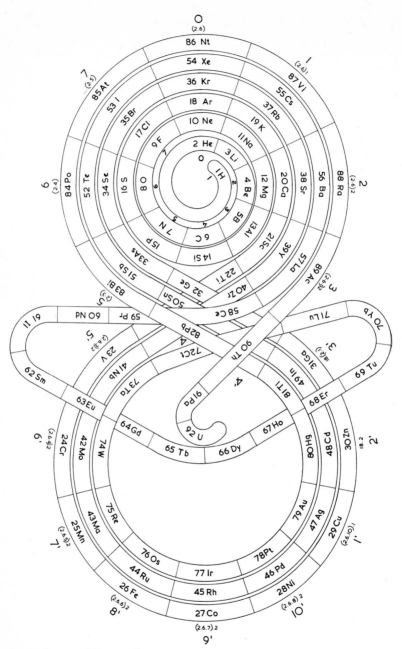

Fig.77   System of Romanoff.

which Crookes supposed the elements to have originated. This protyle was supposed to combine in all kinds of ways, thus liberating hydrogen or helium and energy, which explains the atomic weights deviating from a whole number. For instance, most calcium atoms would have an atomic weight of 40, but there would also be atoms having an atomic weight of 39.8, 39.9, 40.1 or 40.2. Crookes' hypothesis therefore anticipated the future isotopes and the mass defect.

Crookes named his spatial system *vis generatrix*. It is a lemniscate, actually a three-dimensional representation of Reynolds' curve (Fig. 112, p. 225), and indeed shows the analogies between the elements more clearly. In Reynolds' system Li, F, K, Mn, Rb, etc., for example, were in the same series; Crookes, however, made a division into Li, K, Rb, etc. and F, Mn. In this representation all the periods are of the same length, which is again a disadvantage as regards the arrangement of the rare earth metals and the elements of the short periods. We shall return to these problems in the relevant chapters (Chapters 9 and 11, and section *12.2*). In 1898 Crookes *[56]* also classified the noble gases.

In 1934 Romanoff *[15]* succeeded in avoiding both these disadvantages of Crookes' system. He gave the two short periods the form of a spiral (Fig. 77), starting the lemniscate[17] with argon only. The rare earth metals have been arranged on an extended line of the lemniscate. The resulting figure is no longer three-dimensional but a kind of projection (see also *6.7*).

6.14. The "arena" and related models

In the didactic literature we find systems related to the pretzel figure, for example those composed by Courtines and Clark. These authors appreciated the fact that the elements of both the short and the long periods have very definite interrelationships.

In 1925 Courtines *[58]* designed his system in the form of a cardboard structure which, when set up, gave a three-dimensional figure as shown in Fig. 78. As we see, this *stadium* or *arena model* was in the first place erected in two parts to distinguish the elements of the main and sub-groups. The small squares of the rare earth metals were folded like an accordion and so reduced to one place, i.e. the place belonging to lanthanum.

Clark *[59]*, after having considered the systems of twenty-two contemporary authors, presented a system of his own (Fig. 79) that may be seen as the projection of the arena model on a plane[18]. The resulting system somewhat resembles the ground-plan of an arena. The distances of the elements of the sub-group from the two elements of the main group are directly proportional to the increasing divergence in their properties, which Clark quite rightly claimed to be an advantage. Thus, the Cu-group

---

[17] Kipp *[57]* also designed a lemniscate, but did not include any rare earth metals, so that this system, although designed some years later, was retrograde.

[18] A system based on the same principle, but less representative, was designed by Payne *[60]*.

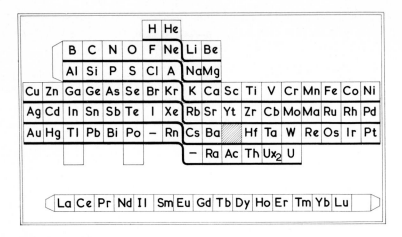

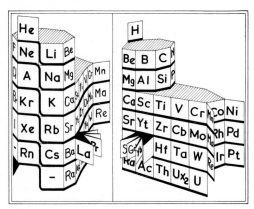

Fig.78   The periodic system as a cardboard construction.

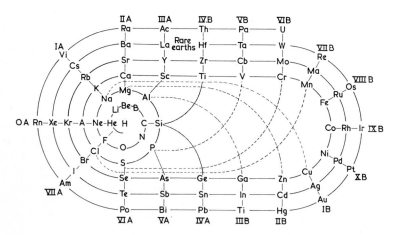

Fig.79   System of Clark (1921).

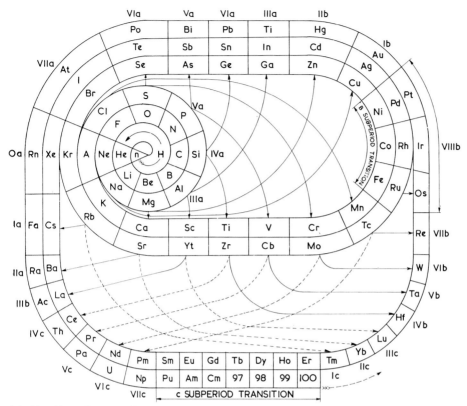

Fig.80 "Arena" ststem of Clark (1950).

is further away from Li and Na than the Ga-group is from B and Al. He regarded hydrogen as a homologue of the elements of both the alkali and the halogen groups. This system still does not provide any place for the rare earth metals. That was not accomplished until 1950 (Fig. 80), when he retained the outward form of his first system but incorporated the rare earth metals into the last two periods (see also p.232)[19].

6.15. The conic model

The cone was adopted as a representational form by Stedman [64]. His system (Fig. 82), which resembles Courtines', is constructed on a cone with a rounded top and two appendices, the larger one of which shows the elements of the groups Si→Th through Ge→Pb, the smaller one including both the elements of the series of the rare

[19] Ingo Waldemar Dagobert Hackh [61] (1890–1938), the future professor of chemistry at San Francisco, and noted for his research on radioactivity, found some points of resemblance between Clark's and his own systems dating from 1918 (Fig. 81) [62, 63]. His system may be called a true periodic system, however, because it shows the elements arranged according to increasing atomic number.

	4	5A	6A	7A		O		IA	2A	3A	4	
Vb	82 Pb	83 Bi	84 Po	85		86 Nt		87	88 Ra	89 Ac	90 Th	VI
IVb	50 Sn	51 Sb	52 Te	53 I		54 Xe		55 Cs	56 Ba	57 La	58 Ce	Va
IIIb	32 Ge	33 As	34 Se	35 Br		36 Kr		37 Rb	38 Sr	39 Y	40 Zr	IVa
IIb	14 Si	15 P	16 S	17 Cl		18 Ar		19 K	20 Ca	21 Sc	22 Ti	IIIa
Ib	6 C	7 N	8 O	9 F		10 Ne		11 Na	12 Mg	13 Al	14 Si	IIa
				1 H		2 He		3 Li	4 Be	5 B	6 C	Ia
III'	22 Ti	23 V	24 Cr	25 Mn	26 Fe	27 Co	28 Ni	29 Cu	30 Zn	31 Ga	32 Ge	III'
IV'	40 Zr	41 Cb	42 Mo	43	44 Ru	45 Rh	46 Pd	47 Ag	48 Cd	49 In	50 Sn	IV'
V'	58 Ce	59 Pr	60 Nd 61	62 Sa	63 Eu 64 Gd	65 Tb 66 Dy 67 Ho	68 Er	69 Ad	70 Cp 71 Yb	72 Lu		V'
V'	72 Lu	73 Ta	74 W	75	76 Os	77 Ir	78 Pt	79 Au	80 Hg	81 Tl	82 Pb	V'
VI	90 Th	91 Bv	92 U									
	4	5B	6B	7B		8		IB	2B	3B	4	

System of Hackh from 1918.

Fig.81   System of Hackh (1918).

earth metals and the actinides. A certain hesitation prevented Stedman from making a definite arrangement of this latter series, as shown by some double occurrences of these elements. Uranium through element 96 were classified in the larger appendix as a group of homologues of tungsten. The system's projection (Fig. 83) on the base adds to its comprehensibility and brings out the close relationship to Clark's system (see preceding section) clearly.

Stewart [65] adopted the true conical form to represent the elements. This three-dimensional figure enabled him to classify the isotopes as well, and in such a way that the isotopes of one element are in a vertical line one below the other, and each group of analogous isotope pleiads—thus the elements themselves—occupies a tangent of the imaginary cone. In the projection of this conical three-dimensional model on the base, the homologues are consequently situated on straight lines.

On the ground that the circumference of a cone increases proportionately to the distance from the apex, Hugo Stintzing [66] (b. 1888), the future professor of physical chemistry at Giessen and director of the Röntgen Institute at Darmstadt, who was noted for his many investigations into spectroscopy and papers on atomic theory, did not consider this shape to be suitable for the representation of the changes in length

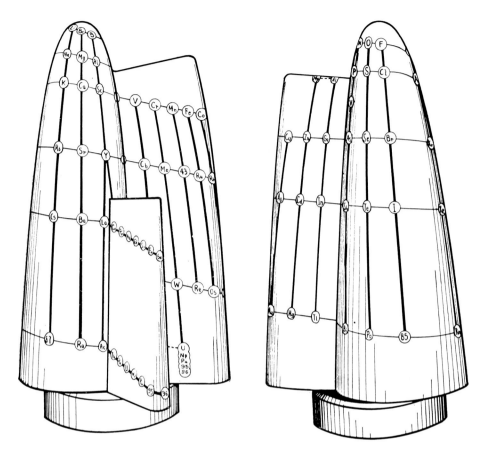

Fig.82 Conic system of Stedman (1947). Left: front view; Groups I-VIII. Right: back view; Cu, Ag, Au and electronegative groups.

of the periods of the elements. He therefore adopted a rotational body (Fig. 84)[20], whose shape he adjusted to suit these changes. A projection of this model on a plane creates a spiral system without any fundamental deviation from the systems already considered in detail (see *6.10*).

6.16. *A spherical system*

Another shape used to represent the periodic system, although a less effective one, was the sphere, adopted by John Albert Newton Friend *[68]* (b. 1881), head of the chemistry section of the Birmingham Technical College. This innovation offered

[20] Many years earlier Rantshev, in a paper read before the Russian Chemical Society, had already stated that he could depict the elements on a rotational surface at the points of intersection of other surfaces. This paper was not published, although Mendeleev *[67]* referred to it.

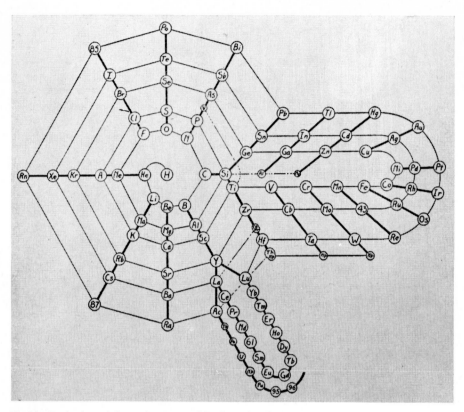

Fig.83   Projection of the conic system of Stedman (1947).

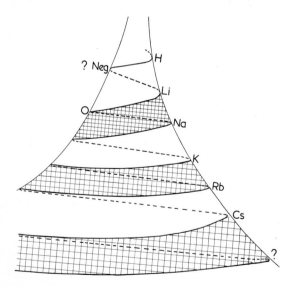

Fig.84   Rotational body of Stintzing (1916).

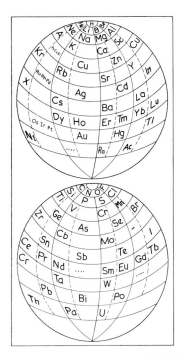

Fig.85   Spherical system of Friend (1934).

hardly any special advantages (Fig. 85). A sphere is too rigid to permit expression of all the small variations in the sizes of the periods. The rare earth metals, for example, cannot be classified separately.

6.17. A cubic model

When discussing Mendeleev's systems (*5.8*) we noted that he felt a three-dimensional figure, namely the cube, would be required to express all the analogies between the elements. This was demanded by the analogies in properties between the elements with small mutual differences in atomic weight, such as Fe, Co and Ni. Mendeleev did not think he was himself capable of drawing up a system of this kind, partly owing to the fact that not all the necessary data were available. Many years later, in 1911, Van den Broek *[69]* constructed a so-called cubiform system (Fig. 86). Van den Broek, although he was not a professional physicist but a lawyer, was nevertheless highly esteemed by such scientists as the student of quantum mechanics H. A. Kramers, professor at the University of Leyden. In 1913, he also introduced the atomic number as the basis of the periodic system (see also *10.8*, *13.2* and *13.5*). The price he had to pay for the composition of a cubical system was very high, however.

In order to bring it into line with his theory that the system of elements consists exclusively of triads, Van den Broek was obliged to assume many unknown elements, despite the fact that he had already made radioactive disintegration products fill many vacant spaces. Needless to say, these substances were not new elements[21] and should not have been given separate places in the periodic system. This had already been demonstrated but had not yet been generally accepted (see *13.2* and *13.3*).

		O			I			II			III			IV			V			VI			VII		
		1	2	3	1	2	3	1	2	3	1	2	3	1	2	3	1	2	3	1	2	3	1	2	3
A	1	He			Li			Be			B			C			N			O			F		
	2		Ne			Na			Mg			Al			Si			P			S			Cl	
	3			Ar			K			Ca			Sc			Ti			V			Cr			Mn
B	1	Fe			Co			Ni			Cu			-			-			-			Zn		
	2		-			-			-			Ga			Ge			As			Se			Br	
	3			Kr			Rb			Sr			Y			Zr			Nb			Mo			Ru
C	1	Rh			Pd			-			-			Ag			-			-			Cd		
	2		-			-			-			In			Sn			Sb			Te			J	
	3			Xe			Cs			Ba			La			Ce			Nd			Pr			Sm
D	1	Eu			AkC			Gd_1			Gd_2			Gd_3			Tb_1			Tb_2			AkB		
	2		Dy_1			Dy_2			Dy_3			Ho			Er			Tu_1			Tu_2			Tu_3	
	3			AkEm			AbX			Yb			RAk			Lu			Ta			W			Os
E	1	Ir			Pt			Au			Hg			ThC			Tl			ThB			Pb		
	2		Th_3			Bi			RaF			Th_2			RaC			Th_1			RaA			ThEm	
	3			RaEm			ThX			Ra			RTh			Th			Jo			U			-

Fig.86   Cubic system of Van den Broek (1911).

Van den Broek's three elements are not true triads, because in many cases their properties have little in common. This system can be visualized as so constructed that the elements of the triads are perpendicular to the plane of the diagram. The resulting three-dimensional figure is a rectangular block and not a cube, unless the distances of the elements in the three spatial directions—horizontal direction, 8 elements; vertical direction, 5 elements; and direction perpendicular to the plane, 3 elements—are taken unequally.

During his studies on the origin of the elements, to which we cannot refer here, Schmidt *[81, 82]* was confronted with the classification of elements. He considered the current periodic system to be only a projection of four independent three-dimensional systems on a plane. In this projection, several parts of these systems overlap.

The first of these partial systems designed by Schmidt, the primary system, is indeed of a very hypothetical nature. It includes a stellar gas, protometals, asterium and hydrogen. The secondary system is approximately the same as the main groups, and the ternary system as the sub-groups of the conventional system. The quaternary system includes the rare earth metals.

Emil Karl Kohlweiler *[83]* (b. 1896), in 1922 still a laboratory assistant but later to become assistant manager and manufacturer in a technical industry, conceived of

[21] Van den Broek *[70]* was not the only scientist to be too easily inclined to accept so many new elements simultaneously *[71–74]*. In this way the vacant spaces at the end of the system were easily filled *[75]*. It need hardly be said that these procedures gave rise to disputes *[71, 72, 76–80]*.

the elements as being composed of hydrogen and helium, as had Schmidt. His representation (Fig. 87) *[84, 85]* consisted of parallel planes connected by steps composed of the elements of the transition group and the rare earth metals. No distinction was made between the elements of the main and sub-groups.

Fig.87   System of Kohlweiler (1920).

Balarew [86] was also dissatisfied with a two-dimensional type of system. His main objection was that such systems do not show clearly the analogy between the elements of the same valence belonging to various columns. He wished to overcome this drawback by means of the three-dimensional system. For this purpose he used nine glass plates on which he arranged the elements so that each plate represented only one valence. Plate 1 carries the gases with zero valence, etc. The elements with more than one valence occupy different plates. When all nine slabs of glass are placed one above the other, a system results with the elements in their highest valence state. The places of elements with other valences are automatically covered. So Balarew's system provided not only for still undiscovered elements but also for valences not yet observed.

6.18. A wedge-shaped model

Loring's [87] step-wise wedge-shaped system (Fig. 88) differs from his previous system (Fig. 89) in that the rare earth metals, which had occupied such an unusual position, were now classified again as homologues of the other elements, whereas the elements Ta through RaF (polonium) appear as a separate part of the period (see also [88]).

6.19. Specific applications of the periodic system

Before going more deeply into the contributions of several scientists, we shall consider a few applications of the periodic system.

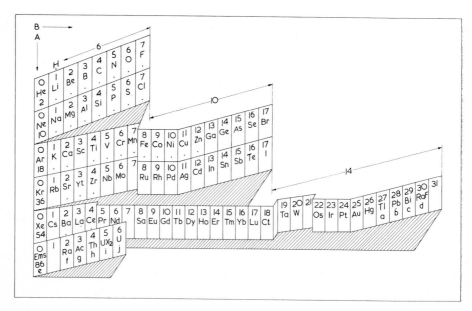

Fig.88   Wedge-shaped system of Loring (1920).

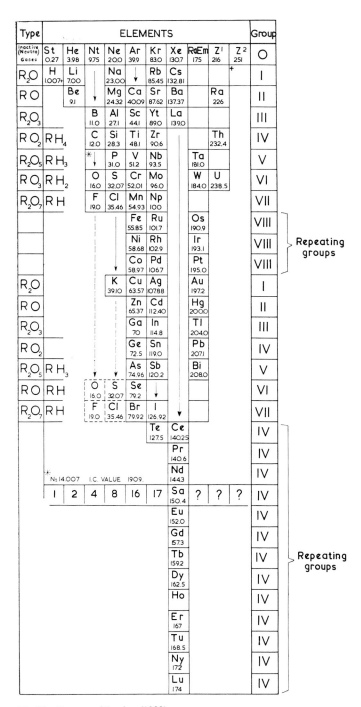

Fig.89   System of Loring (1909).

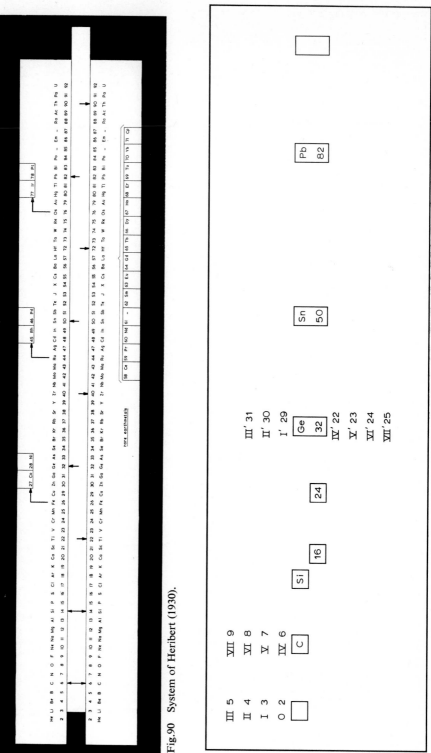

Fig.90   System of Heribert (1930).

Fig.91   Demonstration of the system of Heribert.

Very original indeed was the manner in which Heribert [89] drew up his system, namely in the form of a slide-rule (Fig. 90). On each of the two immovable parts, he arranged all the elements according to increasing atomic weight. On the movable part, i.e. between the two immovable parts, arrows are so arranged that when one arrow is placed next to any given element, the other arrows indicate the remaining elements of the homologous series. When, for example, the first arrow is shown at the element C, the other arrows will point to Si, Ti, Ge, Zr, Sn, Hf, Pb and Th, depending on whether the elements belong to the main or sub-groups. The rare earth metals and the transition metals are arranged beside the remaining elements, and so do not really have their share of the advantages of this system.

Another system devised by Heribert (Fig. 91), based on the same principle, consists of a plane surface on which all the elements with their atomic number are arranged on a strip of millimeter paper. Over this plane another strip of paper can be moved, the second strip carrying, at fixed distances, openings adjusted to the place taken by an element on the basic plane. If we take the same element as an example again, when the opening at the extreme left is placed over C, the other openings show the symbols Si, Ge, Sn and Pb of this fourth group.

Von Antropoff [90, 91, 92], too, was certain that the periodic system was the best means of expressing the elemental analogy of the elements (Fig. 59, p. 160). For that reason he published a *Chemiker-Block*, i.e. a pad of sheets with blank periodic system on which all kinds of properties could be filled in.

Over the years, several scientists have wished to use the periodic system to express special aspects of the analogies. As we have not yet considered these systems we shall now discuss a few examples to be regarded more as curiosities than as true scientific work. To keep in line with our chronological treatment of the subject, we must begin by mentioning Wilde [93], who in 1878 approached the regular relationships between the atomic weights from an unusual point of view.

6.20. Wilde's work

Henry Wilde (1833–1919), a physicist and astronomer whose research was concerned mainly with magnetism, observed a correspondence between the atomic weights of analogous elements and the distances of the planets from the sun, which he demonstrated by relationships like the one shown in Fig. 92. This work, however, was not original, because Hinrichs had already considered this phenomenon (see 5.6).

Wilde reserved the vacant spaces in both this series of relationships and other groups set up by him, for elements whose atomic weights were still too uncertain, principally the earth metals.

A system (Fig. 93) encompassing all the elements in the groups H $1n$ to H $7n$, but in which not all the elements belonging together are always in the same group, strikes us as rather peculiar, the alkali metals and the halogens being incorporated as one group. Many arrangements are incomprehensible. Various places are left vacant, e.g. 16, 32 and 36, while O = 16 and S = 32 are filled in group H $2n$; furthermore, C is placed above Al, B above P, N above Si, and there are two places with the atomic weight[22] of 177.

---

[22] Nine years later these views, on which he gave a lecture, were republished [94], and in 1881 an abstract of the former paper appeared [95]. In 1895 Wilde [96] still defended this system (see 9.3).

```
0 x 0    + 4  =    4 Mercury          0 x 0 +7 = Li  =    7
1 x 3    + 4  =    7 Venus            1 x 23 +0 = Na  =   23
2 x 3    + 4  =  10 Earth             2 x 23 −7 = K   =   39
4 x 3    + 4  =  16 Mars              3 x 23 −7 = Cu  =   62
8 x 3    + 4  =  28 Ceres, Pallas etc. 4 x 23 −7 = Rb  =   85
16 x 3   + 4  =  52 Jupiter           5 x 23 −7 = Ag  =  108
32 x 3   + 4  = 100 Saturn            6 x 23 −7 = Cs  =  131
64 x 3   + 4  = 196 Uranus            7 x 23 −7 = —   =  154
                                      8 x 23 −7 = —   =  177
                                      9 x 23 −7 = Hg  =  200
```

Fig.92   Relations between the atomic weights and the relative distances of the planets to the sun (Wilde; 1878).

1	+Hn−		+H2n−		H3n	H4n	H5n	H6n	H7n
2	Li = 7		Gl = 8		C = 12	− = 16	B = 10	− = 18	N = 14
3	Na= 23	F = 19	Mg= 24	O = 16	Al = 27	− = 32	P = 30	− = 36	Si = 35
4	K = 39	Cl= 35	Ca= 40	S = 32	− = 42	Ti = 48	V = 50	Cr = 54	Fe = 56
									Mn= 56
									Ni = 56
									Co= 56
5	Cu= 62		Zn = 64		− = 69	− = 72	As= 75		
6	Rb= 85	Br= 81	Sr = 88	Se= 80	Ga= 96	Zr = 92	Nb= 95	Mo= 96	
7	Ag=108		Cd=112		Y = 123	Sn = 116	Sb= 120		Pd = 105
									Rh = 105
									Ru = 105
									Da = 105
8	Cs =131	J =127	Ba=136	Te=128	In =150	La = 140	− = 140	− =144	
9	− =154		− =160		E =177	− = 165	− = 165		
10	− =177		− =184		Tl =204	D = 188	Ta = 185	W = 186	Au = 196
									Pt = 196
									Ir = 196
									Os= 196
11	Hg=200		Pb=208		Th =231	U = 240	Bi = 210		

Fig.93   System of Wilde (1878).

6.21. Preyer's work

Thierry William Preyer *[97]* (1841–1897), professor of physiology at Jena, a follower of Darwin, conceived of a revolutionary structure of both animate and inanimate nature and saw inorganic matter as inanimate organic matter *[98, 99]*. He thought that a system of elements must have a closed form and that the system of Mendeleev did not satisfy this condition. The so-called genetic system (Fig. 94), which Preyer based on the many relationships between the properties of the elements propounded by such men as Reynolds and Crookes and which were closely related to a theory based on the origin of the elements, consists of periods of equal length, i.e. each containing seven elements. The periods of the transition metals formed an exception, however. Preyer called the 14 periods set up in this way *Verdichtungsstufen*.

According to Preyer, there is a close relationship between the 14 elements of these *Verdichtungsstufen*. Five generations, arranged in the form of a family tree, show how one element originates from another. The elements of generation 2, for example, originate from the element of generation 1 (Fig. 95), whereas calcium, iron and magnesium are shown as originating from beryllium by *Verdichtung*. Preyer was struck by the fact that most of the *family* elements play a role in organic chemistry. He had already studied the position of these elements in the periodic system and their mutual relationships [100]. Three of the seven family trees include all fourteen elements supposed to have been generated one from the other as shown in Fig. 96: from element 1 → 3 → 6 → 9 → 14, but element 6 also gave rise to 11, etc.

When these family trees are more closely examined they are found to include not only the elements of the main and sub-groups but also other elements, even though

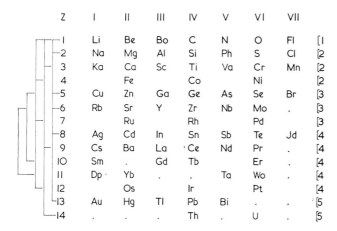

The brackets on the left indicate which elements of every pedigree are genetically connected.

Fig.94 The 14 *Verdichtungsstufen* of Preyer.

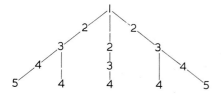

Fig.95 The 5 *Generationen* of Preyer.

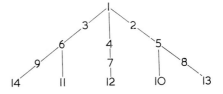

Fig. 96 The 14 *Stufenzahlen* of Preyer.

the properties of the latter, i.e. the transition metals and the elements of the rare earth series, do not support this. The closed system propounded by Preyer results in the latter elements' occupying a position equivalent to that of the remaining elements. Preyer accordingly placed elements yet to be discovered in the vacant spaces and even predicted some of their properties (see *7.8*).

His ideas are collected in a small volume entitled *Das genetische System der Elemente [101]* in which not only the differences in atomic weight but also physical properties, such as specific weight, specific heat and atomic volume, are discussed and compared as well as related to his ideas on the origin of the elements from each other.

The discovery of the noble gases (see *9.8*) induced Preyer some years later *[102]* to draw up another system (Fig. 97). Assuming the existence of noble gases, as yet unknown, and comparing these with the elements of the transition group and some rare earth metals, he left the vacant spaces (see *10.5*) which are the weak spot in his system. No elements have ever been discovered that would fill most of these vacancies.

±	+	+	+	±	–	–	–	±	±	±	+	+	+	±	–	–	–	±
He	Li	Be	Bo	C˙	N	O	Fl	Ar	21	22	Na	Mg	Al	Si	Ph	S	Cl	37
38	Ka	Ca	Sc	Ti	Va	Cr	Mn	Fe	Ni	Co	Cu	Zn	Ga	Ge	As	Se	Br	82
83	Rb	Sr	Yt	Zr	Nb	Mo	.	Ru	Rh	Pd	Ag	Cd	In	Sn	Sb	.	Jd	Te
130	Cs	Ba	La	Ce	Nd	Pr	.	.	Sm	.	.	Gd?	.	Tb	.	Er	.	Dp?
Yb	.	.	.	.	Ta	Wo	.	Os	Ir	Pt	Au	Hg	Tl	Pb	Bi	.	.	217
	.	.	.	Th	.	U	.											
	I	II	III	IV	III	II	I				I	II	III	IV	III	II	I	

Fig.97   System of Preyer (1896).

6.22. Rang's systems

One of the systems composed by Rang *[103]* (Fig. 98) was based on the electropositivity and electronegativity of the elements and differences in their melting-points, properties which indeed give good promise of a useful division. Thus, the first three columns (A) include highly electropositive elements, while group B contains the elements with a high melting-point. Group C, on the other hand, contains the elements with a low melting-point, whereas the elements of group D are the antipodes of those of group A, i.e. highly electronegative elements. The elements of the main and subgroups are separated, as had already been done by Otto (Fig. 116, p. 265) (see *12.5*). The rare earth metals, except for the assumed element didymium, were omitted. Rang was of the opinion that this system contained too many vacant spaces, but within two years [104], in 1895, most of them had been filled. A glance at the new system (Fig. 99) shows that a few spaces are occupied by elements which, though announced[23], proved not to be new elements. We may mention davium (Da) and uralium (Url) as respective homologues of manganese, mosandrium (Ms) as homologue of barium, neptunium (Np) as that of tantalum, and austriacum (Ast) as that of tellurium.

Norwegium (Ng), the assumed element considered to be a homologue of cadmium by both Lothar Meyer [113] in 1880 (Fig. 117, p. 266) and Friedrich Wilhelm Robert Otto [114] (1837–1907), professor of chemistry at the Brunswick Polytechnic in 1885 (Fig. 116, p. 265), was classified by Rang as a satellite of copper, with an atomic weight of 64 instead of 150, the value given by Meyer. Philippium (Pp) in Rang's system had to share its place with scandium. The atomic weight of philippium[23], determined by Delafontaine, was halved by Rang. He neglected mosandrium, separated from the same mineral. Rang actually classified only a few of the elements announced during the period 1877–1887 and for which an atomic weight had been determined, whether with the accepted value or with a value deviating by a certain factor. Rang

valence series	I	II	III	IV	V	VI	VII	VIII			I	II	III	IV	V	VI	VII
I	-	-	-	-	-	-	-	-	-	-	-	-	H	-	-	-	-
2	Li	Be	B	C	-	-	-	-	-	-	-	-	-	-	N	O	F
3	Na	Mg	Al	Si	-	-					-	-	-	-	P	S	Cl
4	K	Ca	Sc	Ti	V	Cr	Mn	Fe	Co	Ni	Cu	Zn	Ga	Ge	As	Se	Br
5	Rb	Sr	Y	Zr	Nb	Mo	-	Ru	Rh	Pd	Ag	Cd	In	Sn	Sb	Te	J
6	Cs	Ba	Di	-	Ta	W	-	Os	Ir	Pt	Au	Hg	Tl	Pb	Bi	-	-
7	-	Ms	-	Th	-	U	-	-	-	-	-	-	-	-	-	-	-
group	1	2	3			4							5			6	7
		A				B							C				D

Fig.98   System of Rang (1893).

[23] Davium was announced in 1877 by Kern [105], who thought he had found this element in platinum ore. It could then have been considered as a homologue of osmium, platinum and iridium. The atomic weight determined by Kern (approx. 154) was multiplied by Rang by $\frac{2}{3}$ so that it could pass for a homologue of manganese. In 1879 Dahl [106] thought he had discovered the "element" norwegium in cobalt and nickel ores, and determined its atomic weight at 145.9.

Brauner [107] believed that the "element" austriacum, announced by him in 1889, was an impurity of tellurium. This could well have been dvitellurium predicted by Mendeleev with an atomic weight of 212 (see 7.6).

Hermann [108] supposed in 1877 that he had discovered neptunium together with ilmenium. From the oxides of these "elements", which he saw as homologues of niobium, he determined their atomic weights as 118 and 104.5, respectively. Rang classified only neptunium, but with a doubled atomic weight. The atomic weight of mosandrium, which Smith [109] believed he had discovered in samarskite, a mineral containing calcium, cerium, thorium, magnesium, uranium and other elements, was determined as 109. Meyer and Seubert [110] converted this, in an unspecified way, into 139.5. In 1878 Delafontaine [111] also separated from the same mineral a substance he considered to be an element which he called philippium (Pp). Initially, he found an atomic weight in between 90 and 95, but later [112] he obtained a more exact value for this bivalent element, viz. 74, which was altered to 111 when tervalence was assumed for this metal.

gave no satisfactory justification for the changes he made in atomic weights, but he filled some vacant spaces, although the principle of analogy often suffered as a consequence.

The "elements" that Rang could not incorporate were ilmenium (Il), polymnestum (Pm), erebodium (Eb) and gadenium (Gd), despite the fact that he thought he knew their atomic weights and several other attributes. Rang believed that the many vacant spaces in his system between C and A and between Si and the place ^3IV_b were unreal: they could have appeared in the same way as the pole regions become enlarged when a globe is sectioned and projected on a plane. Exhaustive research has shown that, of the thirty assumed elements, there were only ten really new ones, i.e. Yb, Sc, Sm, Tu, Ho, Dy, Nd, Pr, Ge and Gd (see Table 4, p. 261).

VAL	I	II	III	IV	V	VI	VII	VIII			I	II	III	IV	V	VI	VII	VIII
1	..	..	..	..	..	..	..	..	..	..	..	..	H	He	*	(+)	(+)	(?)
2	Li	Be	B	C	..	..	..	..	..	..	..	..	..	A	N	O	F	(?)
3	Na	Mg	Aₗ	Si	..	..	..	..	..	..	..	..	..	(+)	P	S	Cₗ	(?)
4	K	Ca	Sc	Ti	V	Cr	Mn	Fe	Ni	Co	Cu	Zn	Ga	Ge	As	Sb	Br	(?)
5	Rb	Sr	Y	Zr	Nb	Mo	Da	Ru	Rh	Pd	Ag	Cd	In	Sn	Sb	Te	I	(?)
6	Cs	Ba	La	Yb	Ta	W	Ur	Os	Ir	Pt	Au	Hg	Tₗ	Pb	Bi	Ast	(+)	(?)
7	(+)	Ms	(+)	Th	Np	U	(?)	(?)	(?)	(?)								

₄IIIa=Sc,Pp ₄Ib=Cu,Ng ₆IIIa and ₆IV=La,Ce,Ny,Py,Sm,Gd,Tb,Ho,Er,Tm,Dc,Yb. ...

* Nameless element, a satellite of Helium.

+ Element as yet to be discovered.

Fig. 99   Completed system of Rang (1895).

6.23. Carey Lea's system

Carey Lea [115] is one of the few scientists who followed the completion and further development of the periodic system from the beginning and who never lost interest in it. Thirty-five years after his first publication and two years before his death in 1897, he published on the subject again. This scientist, who in the earliest period (4.15) had taken an active part in classifying elements according to several properties, did not approve of the then current system at all. Carey Lea did not think the difficulty of classifying the transition group had been removed and this weighed so heavily with him that he felt compelled to reject the 1895 system which he called "a singular

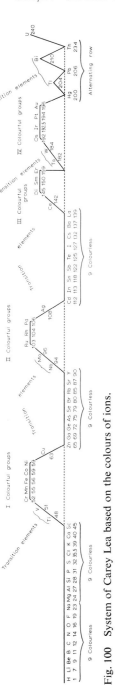

Fig. 100   System of Carey Lea based on the colours of ions.

			? η	Bi	Pb	Tl	
Ba	Cs	?ζ	J	Te	Sb	Sn	In
Sr	Rb	?ε	Br	Se	As	Ge	Ga
Ca	K	?δ	Cl	S	P	Si	Al	nodal points ("noeuds")
Mg	Na	?γ	F	O	N	C	Bo
Be	Li	?β	?α				
H	H	H	H	H	H	H	H

Fig.101   System of Lecoq de Boisbaudran (1895).

	Ba	137.0	Sr	87.5	Ca	40.0	Mg	24.7	Be	9.1
	Cs	133.0	Rb	85.4	K	39.1	Na	23.0	Li	7.0
"bascules"		2.0		1.05		0.45		0.85		1.05
	Te	127.8	Se	79.0	S	32.1	O	16.0	?η	214.0
	Sb	120.7	As	75.0	P	31.0	N	14.0	Bi	208.1
"bascules"		3.55		2.00		0.55		1.00		295(?)

Fig.102   Atomic weight relations of Lecoq de Boisbaudran (1897).

mixture of truth and error". Although Carey Lea's attempt to set up a system (Fig. 100) based on the ionic theory of Arrhenius was praiseworthy, it was highly unsuccessful. In fact, only the colour of the ions had been taken into account; no allowance had been made for their periodic properties.

Thomsen [4] found a correspondence between this system and his own (Fig. 53, p. 151), in which all ions of the elements of the middle parts of the long periods were indeed coloured, viz., those of Ti → Cu, Nb → Ag and Ce → Au. Carey Lea [116] accepted the conclusion arrived at by Thomsen, an authority on this point (see also 6.4), as evidence of the correctness of this system. He did not realize, however, that a collection of elements written out linearly, and therefore not including any elements as homologues of other ones, can hardly be called a periodic system, at least in the usual sense of the word. It is also very doubtful whether in such a collection the transition group shows to better advantage. Chromium, three rare earth metals and gold are in fact to be found in the regions of the metals of the transition group. It is therefore difficult to understand why Ackroyd [117] considered this system to be one of the finest ever constructed.

6.24. The work of Lecoq de Boisbaudran

Paul Emil Lecoq de Boisbaudran [118] (1838–1912), a private scientist who did research on spectroscopy, crystallography and rare earth metals, intended to show by his system (Fig. 101)—in which all the elements are included—that the incorporation of the then (1895) known noble gases rested on logical grounds (see also 9.4).

Every family of elements in this system contains the same number of members, i.e. six. Lecoq de Boisbaudran regarded hydrogen as a member of each group, although two years later [119] he discarded this view and showed this element accompanying only lithium. De Boisbaudran, whom history considers to have been one of the founders of the periodic system because of his discovery of gallium, which was predicted by Mendeleev, committed a strange error in his system. Like Carey Lea in his time, he accepted elements with a negative atomic weight; to these elements he assigned the first places of each group of elements. We can summarize De Boisbaudran's arguments as follows: in the series of atomic weight differences between the elements of two neighbouring groups (Fig. 102)—called *bascules* (balances)—a minimum is found for just those elements that in his system (Fig. 101) form the row of nodal points, the so-called *nœuds*. The lowest element in the group is dropped as soon as this nodal point lies above 34—the lowest *bascule* being occupied by the pair Cl–?δ. After this point, an element is added to the top of the group, as in the case of chlorine. In De Boisbaudran's opinion a system must therefore be based on the differences in atomic weight, which has the added advantage that, for instance, the Te–I reversal no longer constitutes an obstacle: only the difference in atomic weight is concerned and not the real value of the two atomic weights. This system is even more questionable as a periodic classification because by no means all the elements are included. Our only reason for describing it in such detail is because De Boisbaudran based his predictions of the noble gases on it (see *9.4*).

6.25. Müller's view

In 1944 R. M. Müller *[120]* compared the periodic system of elements with the natural system of living beings in which man is the undifferentiated middle, the germinal point, while the other organisms are to be considered as greater or lesser deviations (Fig. 103). He drew a parallel between this highest living being, man, who in organic nature stands above the animal kingdom, and the elements hydrogen and carbon as principal carriers of life, which stand above the inorganic kingdom. This analogy was carried even further. The carbon group, which includes hydrogen, comprises elements sometimes representing widely divergent modifications, just as there are also divergent animal types with as extremes wild and tame, comparable to graphite and diamond, respectively.

He compared the animal family tree, in which a distinction can be made between wild and tame species, with the family tree of the elements in which elements are alternately placed in the main and the sub-groups, for instance the copper group (tame) and the lithium group (wild). The "wild" elements are diverted to the twigs to shield the centre. He also correlated the origin of the elements with the development of life, which is an interesting comparison, although it is carried very far and is given insufficient philosophical consideration. Man, the crown of organic evolution, is compared to hydrogen, which has been seen as the origin of inorganic evolution although Müller (professor at the Technical College, Graz) gave this role to *prohydrogen*.

6.26. The periodic system and music

At first sight the title of this section may seem somewhat odd, but it should not be forgotten that one of the discoverers of the periodic system, Newlands, noticed points in common between the tones of the octave in music and the elements of a period. He did not go into this resemblance in great detail, but Alfred Partheil *[121]* (1861–1909), professor of pharmacy at Königsberg, did so in 1903. Partheil may have known that Pythagoras saw a relationship between the weight of the blacksmith's hammers and the tones produced by the blows on the anvil. This relationship explained harmonic progression and was related to the distances between the planets (see also *5.6*). Partheil thought he had made a most remarkable and interesting discovery, i.e. that of a connection between the harmonic relationships of the spectral lines of the elements and the wave numbers of the tones of the musical scale. If the spectra having something to do with the vibrations of the elements were in proportion to the atomic

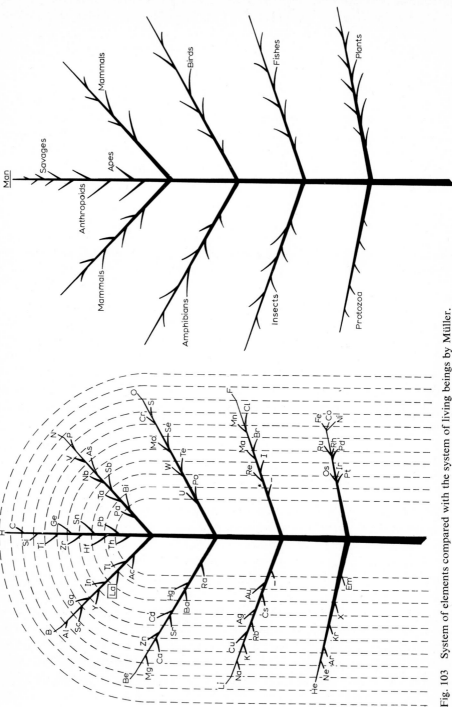

Fig. 103   System of elements compared with the system of living beings by Müller.

element	atomic weight	16 x atomic weight	pitch	vibration number
Li	7.03	112.48	# a^{-1}	112.5
B	11	176	# f^0	177.7
C	12.00	192	g^0	192
O	16.00	256	c^1	256
Na	23.05	368.8	b g^1	368.64
S	32.06	512.96	c^2	512
Ca	40.1	641.6	e^2	640
V	51.2	819.2	b a^2	819.2
As	75	1200	# d^3	1200
Br	79.96	1279.36	e^3	1280
Sb	120.2	1923.2	b^3	1920
Hg	200.0	3200	# g^4	3200

Fig. 104   Relations between the atomic weights of the elements and the vibration numbers of tones.

weights, a connection between the atomic weights of the elements and the vibration numbers of the tones would be demonstrable.

The atomic weights of the first and the last elements, hydrogen and uranium respectively, when multiplied by 16

$$1.008 \times 16 = 16.128$$
$$238.5 \times 16 = 3816$$

are indeed almost equal to the vibration numbers of the sub-contra $C = 16$ and the b'''' (b^4) = 3840, respectively. These numbers apply to pure tuning. In this, the b is the harmonic 15th. The divergence of the vibration numbers from the atomic weights multiplied by 16 are less than the differences between them and the corresponding vibration numbers in the equal temperament. In the twelve-tone system[24] the vibration number of $C_2 = 16.165$ and of $b^4 = 3906.168$. Fig. 104 contains a few other examples showing very striking points of a possible relationship. It must be borne in mind here that the elements have been chosen arbitrarily; hence there is no recognizable system. This is also true of the tones. Far from all the elements have been considered. Partheil also involved the thirty-one-tones system to illustrate his comparison[25].

[24] Using the present standard a = 440, $C_2 = 16.5$.
[25] This division of the octave into 31 tones was originated by Christiaan Huygens [122]. In 1661 and 1691, making use of the recently evolved method for calculating logarithms, he divided the octave into 31 equal intervals. The result is that the intervals sound purer than in the 12-tone system. An organ based on the 31-tone system has been built in the Teyler Foundation at Haarlem [123].

It was not until half a century later that Constantin G. Bedreag *[124]* (b. 1883), professor of theoretical physics at the University of Cernauti (Rumania), who did much research on the periodical physical properties of the elements, attempted to relate the relationship between the energy states of the electrons of the atoms to the harmonic intervals $1, \frac{5}{4}, \frac{3}{2}$ and 2 in music. These relations of the vibration numbers $1 : 2 : 3 : 4$ of the harmonics with the key-note compared with the relationship of the spectral lines of the elements were not worked out by Bedreag, who thereby created the impression that he meant only a formal relation.

In 1919 Schmiz *[125]* proposed somewhat similar numerical relationships between the atomic weights of homologous elements on the one hand and, on the other, the lengths of the sides (a, b, c) of the harmonic triangle (i.e. a right-angled isosceles triangle), the radius[26] of the harmonic circle (R) and the inner circle (r) of the triangle. To mention only one example: the relationship of the atomic weights of two successive homologous elements is in many cases equal to the relation $R : b$ (Fig. 105).

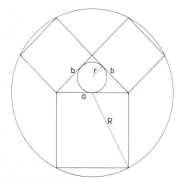

Y = 126.92	R = 126.92
Br = 79.92	b = 80.26
Te = 127.5	R = 127.5
Se = 79.2	b = 80.6
Sb = 120.2	R = 120.2
As = 74.96	b = 75.89
Ce = 140.25	R = 140.25
Zr = 90.6	b = 88.69
etc.	
etc.	

Fig. 105   Relations of Schmiz.

Schmiz observed similar relations between the vibration numbers of the tones. When, after he had written his treatise, Schmiz's attention was drawn to the relationships suggested by Partheil, he saw how to extend his own relationships. Nonetheless, his results cannot be called a system, because the picture is too fragmentary: "Überall in der Natur, wo sich Massen angeordnet, wo sich Systeme gebildet haben, finden wir die Einzelglieder dieser Systeme in den meisten Fällen nach Verhältnissen geordnet, die den Gröszenverhältnissen des harmonischen Dreiecks im harmonischen Kreis entsprechen oder sich direkt von diesen ableiten lassen. Sie spiegeln sich wider in den vom Künstler der Natur entnommenen Kunstformen, wie in der Baukunst, Bildhauerkunst und Malerei".

[26] I.e., of a circle drawn through the six free vertices of the three squares that can be set up on the sides of the harmonic triangle, whose centre coincides with the midpoint of the hypotenuse.

6.27. The periodic system and the modern atomic theory

Modern atomic theory, established by Rutherford and Bohr and developed further by several scientists, has contributed greatly to a better understanding of the periodic system of elements. The electronic structures of the elements provided an explanation of the properties of the atoms and thus reduced the known analogies of properties to an analogy in the structure of the outer atomic orbits. In Part II we shall deal with several parts and aspects of the periodic system of elements separately, and also discuss some systems based on the atomic and quantum theories (see Chapters 10,

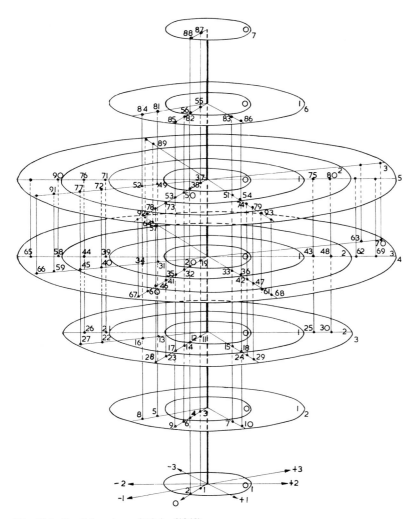

Fig. 106   Spatial system of Finke (1943).

12 and 13). However, a system of elements based on the regular sequence of the quantum numbers does not contain the elements in the order of increasing atomic number; thus in many cases inner orbits are filled up before the outer orbits have been completed.

As an illustration we have chosen Finke's system[27] (Figs. 106 and 107), in connection with which any comments on this type of classification [126] may be appropriate. This is a three-dimensional system in which the total quantum numbers $n = 1, 2 \ldots \ldots 7$ are placed vertically and determine the height of the levels on which concentric circles are found. The azimuthal quantum numbers $l = 0, 1, 2, 3$ are the radii of these circles. The magnetic quantum numbers $m_1 = 3, 2, \ldots 0, -1, \ldots -3$ determine the directions of the position of the elements on the circles. The spin quantum number $\pm \frac{1}{2}$ indicates whether the element is to be found inside or outside the standard circles.

Finke [127] considered a three-dimensional model to be less useful in practice. He therefore also made a two-dimensional model. We must not class this system among the periodic systems, however; the elements are arranged in pairs and not according to increasing atomic number.

A similar system was devised by Corbino[28] [128], but in his system, because of the premature completion of the d-orbits and the corresponding bivalence, Cu, Ag, and Au fall, not between Ni and Zn, and Pd and Cd, and Pt and Hg respectively, but above Zn, Cd and Hg; for example:

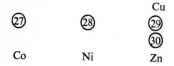

Finke gave his division of the 92 elements a popularized aspect: his elements occupy a house of 7 storeys (K, L, etc.). Each storey contains 4 apartments (s-, p-, d- and f-levels), each of which contains 7 rooms ($-3, -2, \ldots, +3$; magnetic quantum number). In these rooms the elements occur in pairs on the basis of their spin quantum number ($-\frac{1}{2}, +\frac{1}{2}$). Only 48 of the 196 rooms were occupied, however.

Finke [129] drew up a second system[29] (Fig. 107) in 1949; this one gives a better representation of the situation, because he took the order of increasing atomic number as his starting point. But he could not adhere to this rule consistently while at the same time maintaining the filling up of electronic structures in the order of the s-, p-, d- and f-orbits. Finke classified the elements up to 120, and therefore showed how he visualized the configuration of the as yet unknown elements beyond curium.

Wringley [131] and his co-workers constructed a system (Fig. 109) in the shape of a so-called *lamina*, also taking the atomic structure as the basis. The total quantum numbers determine a certain height (number of steps) in this system. The elements with $n = 1$ occupy a ground-floor position with respect to those with $n = 2$ on the first step, etc. Elements belonging to one period can therefore occur on several steps. For instance, of the sixth period, Cs and Ba, with their 6s-configuration, are found on the sixth step, La with its 5d-configuration together with Hf through Hg on the fifth step, Tl through Rn with their 6p-configuration again on the sixth step, while the rare earth metals, with their 4f-configuration, occur immediately on the fourth step.

[27] For the arrangement of the actinides and the elements with an atomic number larger than 96, see Chapter 14.

[28] Orso Mario Corbino (b. 1876), director of the Institute of Physics at Rome, did much research, some on spectral and electronic theories.

[29] Vallet [130] had already constructed a similar system a year earlier (Fig. 108).

K	H o 1	He • 2																														
L	Li o 3	Be • 4	B o 5	C o 6	N o 7	O • 8	F • 9	Ne • 10																								
M	Na o 11	Mg • 12	Al o 13	Si o 14	P o 15	S • 16	Cl • 17	Ar • 18	Sc o 21	Ti o 22	V o 23	Cr o 24	Mn o 25	Fe • 26	Co • 27	Ni • 28	Cu • 29	Zn • 30														
N	K o 19	Ca • 20	Ga o 31	Ge o 32	As o 33	Se • 34	Br • 35	Kr • 36	Y o 39	Zr o 40	Nb o 41	Mo o 42	Tc o 43	Ru • 44	Rh • 45	Pd • 46	Ag • 47	Cd • 48	La o 57	Ce o 58	Pr o 59	Nd o 60	Il ? o 61	Sm o 62	Eu o 63	Gd • 64	Tb • 65	Dy • 66	Ho • 67	Er • 68	Tu • 69	Yb • 70
O	Rb o 37	Sr • 38	In o 49	Sn o 50	Sb o 51	Te • 52	J • 53	X • 54	Cp o 71	Hf o 72	Ta o 73	W o 74	Re o 75	Os • 76	Ir • 77	Pt • 78	Au • 79	Hg • 80	Ac o 89	Th o 90	Pa o 91	U o 92	Np o 93	Pu o 94	Am o 95	Cm • 96	• 97	• 98	• 99	• 100	• 101	• 102
P	Cs o 55	Ba • 56	Tl o 81	Pb o 82	Bi o 83	Po • 84	At • 85	Em • 86	o 103	o 104	o 105	o 106	o 107	• 108	• 109	• 110	• 111	• 112														
Q	Fr o 87	Ra • 88	o 113	o 114	o 115	• 116	• 117	• 118																								
R	o 119	• 120																														
	0	0	-1	0	+1	-1	0	+1	-2	-1	0	+1	+2	-2	-1	0	+1	+2	-3	-2	-1	0	+1	+2	+3	-3	-2	-1	0	+1	+2	+3

o negative spin
• positive spin

Fig. 107 System of Finke (1949).

	l = 0		l = 1						l = 2										l = 3													
	s		p						d										f													
	J = 1/2		J = 1/2		J = 3/2				J = 3/2				J = 5/2						J = 5/2						J = 7/2							
K	1	2	–	–	–	–	–	–	–	–	–	–	–	–	–	–	–	–	–	–	–	–	–	–	–	–	–	–	–	–	–	–
n = 1.....	H	He	–	–	–	–	–	–	–	–	–	–	–	–	–	–	–	–	–	–	–	–	–	–	–	–	–	–	–	–	–	–
L	3	4	5	6	7	8	9	10	–	–	–	–	–	–	–	–	–	–	–	–	–	–	–	–	–	–	–	–	–	–	–	–
n = 2.....	Li	Be	B	C	N	O	F	Ne	–	–	–	–	–	–	–	–	–	–	–	–	–	–	–	–	–	–	–	–	–	–	–	–
M..........	11	12	13	14	15	16	17	18	21	22	23	24	25	26	27	28	29	30	–	–	–	–	–	–	–	–	–	–	–	–	–	–
n = 3.....	Na	Mg	Al	Si	P	S	Cl	A	Sc ″	Ti ″	V ″	Cr ′	Mn ″	Fe ″	Co ″	Ni ″	Cu ′	Zn ″	–	–	–	–	–	–	–	–	–	–	–	–	–	–
N	19	20	31	32	33	34	35	36	39	40	41	42	43	44	45	46	47	48	58	59	60	61	62	63	64	65	66	67	68	69	70	71
n = 4.....	K	Ca	Ga	Ge	As	Se	Br	Kr	Y ″	Zr ″	Nb ′	Mo ′	Ma ′	Ru ′	Rh ′	Pd	Ag ′	Cd ″	Ce ″	Pr ″	Nd ″	Il ″	Sm ″	Eu ″	Gd ″	Tb ″	Dy ″	Ho ″	Er ″	Tm ″	Yb ″	Lu ″
O..........	37	38	49	50	51	52	53	54	57	72	73	74	75	76	77	78	79	80	90	91	92	93	94	95	96	–	–	–	–	–	–	–
n = 5.....	Rb	Sr	In	Sn	Sb	Te	I	Xe	La ″	Hf ″	Ta ″	W ″	Re ″	Os ″	Ir	Pt ′	Au ′	Hg ″	Th ″	Pa ″	U ″	Np	Pu	Am	Cm	–	–	–	–	–	–	–
P	55	56	81	82	83	84	85	86	–	–	–	–	–	–	–	–	–	–	–	–	–	–	–	–	–	–	–	–	–	–	–	–
n = 6.....	Cs	Ba	Tl	Pb	Bi	Po	–	Rn	–	–	–	–	–	–	–	–	–	–	–	–	–	–	–	–	–	–	–	–	–	–	–	–
Q	87	88	–	–	–	–	–	–	–	–	–	–	–	–	–	–	–	–	–	–	–	–	–	–	–	–	–	–	–	–	–	–
n = 7.....	–	Ra	–	–	–	–	–	–	–	–	–	–	–	–	–	–	–	–	–	–	–	–	–	–	–	–	–	–	–	–	–	–

Fig. 108 System of Vallet (1948).

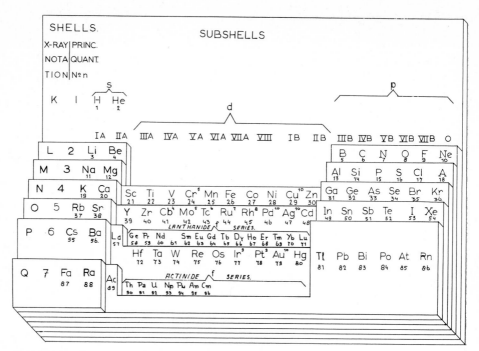

Fig. 109   Lamina system of Wringley c.s. (1949).

6.28. Conclusion

This chapter was intended as a survey of the development of the periodic system after its discovery. That this development was somewhat less spectacular than the genesis of the system, was to be expected. The principle of the general systematic structure was known as early as 1869, and was little influenced by subsequent theoretical and practical developments in the field of atomic research. With the acceleration of the latter after the first decade of this century, the difficulties which the system encountered in the matter of noble gases, rare earth metals, some pairs of elements inconsistent with the principle of increasing atomic weight, and radioactivity, were solved. The quantum theory, for instance, created more problems for the simple form of the periodic system than it solved. It cannot be denied that this theory confirmed the validity of the system on its old basis. It also forms a good guide for predicting properties of elements not yet discovered or synthetized.

We hope we have made it clear that, besides a number of scientists who contributed very positively to either the form or the theoretical derivation of the periodic system, there were authors who looked upon the periodic system mainly as a catalogue of elements, about which they provided only few interesting data. Many such systems have been published in the *Journal of Chemical Education*. They offer no new possibilities (see, e.g., *[132]*), or are obvious variations of the existing system *[133]*; there are also systems to be folded out *[134]* or put together like jig-saw puzzles *[135]* that have only limited educational value.

References

1 E. G. MAZURS, *Types of Graphic Representation of the Periodic System of Chemical Elements*, La Grange, Ill. 1957.

2 T. BAYLEY, *Phil. Mag.* [5], 13 (1882) 26; summary in *J. Chem. Soc.*, 42 (1882) 359.

3 H. BASSETT, *Chem. News*, 65 (1892) 3, 19.

4 J. THOMSEN, *Oversigt Kgl. Danske Videnskb. Selskab*, (1895) 132; *Z. Anorg. Chem.*, 9 (1895) 190.

5 A. WERNER, *Ber.*, 38 (1905) 914.

6 A. PICCINI, *Gazz. chim. ital.*, 35 II (1905) 417.

7 R. ABEGG, *Ber.* 38 (1905) 1386.

8 A. WERNER, *Ber.*, 38 (1905) 2022.

9 R. ABEGG, *Ber.*, 38 (1905) 2330.

10 G. RUDORF, *Chem. Ztg.*, 30 (1906) 595.

11 G. RUDORF, *Das periodische System, seine Geschichte und Bedeutung für eine chemische Systematik*, Hamburg–Berlin 1904.

12 N. BOHR, *Drei Aufsätze über Spektren und Atomen*, Brunswick 1922, p. 132.

13 N. BOHR, *The Theory of Spectra and Atomic Constitution* (three Essays), Cambridge 1922, p. 70.

14 N. BOHR, *Fysiks Tidsskrift*, 19 (1921) 152; transl. in *Z. Physik*, 9 (1922) 1.

15 V. ROMANOFF, *Rev. sci.*, 72 (1934) 661.

16 E. W. ZMACZYNSKI, *J. Chem. Educ.*, 14 (1937) 232.

17 E. I. ACHIMOV, *Zhur. Obshcheï Khim.*, 16 (1946) 961.

18 W. NERNST, *Theoretische Chemie*, 8th–10th ed., Stuttgart 1921, p. 192.

19 A. VON ANTROPOFF, *Z. Angew. Chem.*, 39 (1926) 722.

20 A. VON ANTROPOFF, *Z. Elektrochem.*, 32 (1926) 423.

21 F. M. SCHEELE, *Z. Naturforschung*, 4a (1949) 137.

22 E. HUTH, *Das periodische Gesetz der Atomgewichte und das natürliche System der Elementen*, Frankfurt-Oder 1884; 2nd ed., Berlin 1887.

23 T. CARNELLEY, *Chem. News*, 53 (1886) 183.

24 L. MEYER, *Die modernen Theorien der Chemie*, 2nd ed., Breslau (Wroclaw) 1872, p. 302.

25 J. H. VINCENT, *Phil. Mag.* [6], 4 (1902) 103.

26 G. J. STONEY, *Phil. Mag.* [6], 4 (1902) 411.

27 G. J. STONEY, *Proc. Roy. Soc. (London)*, 44 (1888) 115; *Chem. News*, 57 (1888) 163.

28 G. J. STONEY, *Phil. Mag.* [6], 4 (1902) 504.

29 H. ERDMANN, *Lehrbuch der Anorganischen Chemie*, 3rd ed., Brunswick 1902, Appendix III.

30 B. K. EMERSON, *Am. Chem. J.*, 45 (1911) 160.

31 B. K. EMERSON, *Chem. Revs.*, 5 (1928) 214.

32 W. D. HARKINS and R. E. HALL, *Z. Anorg. Chem.*, 97 (1916) 175, 336; revised edition of *J. Am. Chem. Soc.*, 38 (1916) 169.

33 E. I. EMERSON, *J. Chem. Educ.*, 21 (1944) 111.

34 H. SCHULTZE, *Naturwissenschaften*, 32 (1944) 58.

35 C. J. MONROE and W. B. TURNER, *J. Chem. Educ.*, 3 (1926) 1058.

36 See e.g. P. V. WELLS, *J. Wash. Acad. Sci.*, 8 (1918) 232.

37 W. D. HARKINS and E. D. WILSON, *Phil. Mag.* [6], 30 (1915) 723.

38 W. D. HARKINS and E. D. WILSON, *J. Am. Chem. Soc.*, 37 (1915) 1383.

39 W. D. HARKINS and E. D. WILSON, *Z. Anorg. Chem.*, 95 (1916) 20.

40 K. SCHERINGA, *Chem. Weekblad*, 14 (1917) 953.

41 W. D. HARKINS, *Proc. Nat. Acad. Sci., U.S.*, 2 (1916) 216.
42 W. D. HARKINS, *J. Am. Chem. Soc.*, 39 (1917) 856.
43 W. D. HARKINS, *Science*, 50 (1919) 577.
44 A. E. CASWELL, *Phys. Rev.*, 34 (1929) 543.
45 A. PIUTTI, *Gazz. chim. ital.*, 55 (1925) 754.
46 A. PIUTTI, *Rend. accad. sci. Napoli* [3], 31 (1924) 142.
47 C. R. NODDER, *Chem. News*, 121 (1920) 269.
48 F. J. WIIK, *Acta Soc. Sci. Fennicae*, 10 (1875) 413.
49 C. JANET, *Chem. News*, 138 (1929) 372, 388.
50 F. SCHEELE, *Z. Naturf.*, 5a (1950) 11.
51 F. SODDY, *The Chemistry of the Radio-elements*, Vol. II, *The Radio-elements and the Periodic Law*, London 1914.
52 W. CROOKES, *Chem. News*, 54 (1886) 115; 55 (1887) 83, 95.
53 J. E. REYNOLDS, *Chem. News*, 54 (1886) 1.
54 W. CROOKES, *J. Chem. Soc.*, 53 (1888) 487.
55 W. CROOKES, *Die Genesis der Elemente*, 2nd German ed., Brunswick 1895.
56 W. CROOKES, *Proc. Roy. Soc. (London)*, 63 (1898) 408; *Z. anorg. Chem.*, 18 (1898) 72; *Chem. News*, 78 (1898) 25.
57 F. KIPP, *Naturwissenschaften*, 30 (1942) 679.
58 M. COURTINES, *J. Chem. Educ.*, 2 (1925) 107.
59 J. D. CLARK, *J. Chem. Educ.*, 10 (1921) 675.
60 E. C. PAYNE, *J. Chem. Educ.*, 14 (1937) 593; 15 (1938) 180.
61 I. W. D. HACKH, *J. Chem. Educ.*, 11 (1934) 59.
62 I. W. D. HACKH, *Am. J. Sci.* [4], 46 (1918) 481.
63 I. W. D. HACKH, *J. Am. Chem. Soc.*, 40 (1918) 1023.
64 D. F. STEDMAN, *Can. J. Research*, 25B (1947) 199.
65 O. J. STEWART, *J. Chem. Educ.*, 5 (1928) 57.
66 H. STINTZING, *Z. Physik. Chem.*, 91 (1916) 500.
67 D. MENDELÉEFF, *Principles of Chemistry*, London 1891, Vol. II p. 20.
68 J. A. N. FRIEND, *Chem. News*, 130 (1925) 196.
69 A. V. D. BROEK, *Physik. Z.*, 12 (1911) 490.
70 A. V. D. BROEK, *Physik. Z.*, 14 (1913) 32.
71 A. S. RUSSELL, *Chem. News*, 107 (1913) 49.
72 A. SCHUSTER, *Nature*, 91 (1913) 30.
73 F. SODDY, *Chem. News*, 108 (1913) 168.
74 K. FAJANS, *Ber.*, 46 (1913) 422; *Le Radium*, 10 (1913) 171.
75 K. FAJANS, *Physik. Z.*, 14 (1913) 136.
76 F. SODDY, *Nature*, 91 (1913) 57.
77 N. R. CAMPBELL, *Nature*, 91 (1913) 85.
78 K. FAJANS, *Naturwissenschaften*, 2 (1914) 429, 463, 543.
79 A. V. D. BROEK, *Naturwissenschaften*, 2 (1914) 717.
80 A. FLECK, *Chem. News*, 107 (1913) 95.
81 C. SCHMIDT, *Z. Physik. Chem.*, 75 (1911) 651.
82 C. SCHMIDT, *Z. Anorg. Chem.*, 103 (1918) 79.
83 E. KOHLWEILER, *Physik. Z.*, 21 (1920) 203.
84 E. KOHLWEILER, *Physik. Z.*, 21 (1920) 311.
85 See also E. KOHLWEILER, *Z. physik. Chem.*, 94 (1920) 513.
86 D. BALAREW, *Z. Anorg. Chem.*, 121 (1922) 22.
87 F. H. LORING, *Chem. News*, 125 (1922) 386.
88 F. H. LORING, *Chem. News*, 144 (1932) 70.
89 H. HERIBERT, *Z. Elektrochem.*, 36 (1930) 687.
90 A. VON ANTROPOFF, *Z. Angew. Chem.*, 39 (1926) 725.
91 A. VON ANTROPOFF and M. VON STECKELBERG, *Atlas der physikalischen und anorganischen Chemie*, Berlin 1929.
92 A. VON ANTROPOFF, *Z. Elektrochem.*, 33 (1927) 475.

93 H. WILDE, *Chem. News*, 38 (1878) 66, 96, 107.

94 H. WILDE, *Mem. Proc. Manchester Lit. Phil. Soc.* [3], 10 (1887) 118.

95 H. WILDE, *A Century of Science in Manchester*, 1881, p. 359.

96 H. WILDE, *Chem. News*, 72 (1895) 291.

97 W. PREYER, *Ber. deut. pharm. Ges.*, 2 (1892) 144; *Verhandl. phys. Gesellsch. Berlin*, 10 (1891) 85; *Naturw. Wochenschr.*, 6 (1891) 523; 7 (1892) 4, 11.

98 W. PREYER, *Kosmos*, 1 (1877) 377.

99 See also R. HOOYKAAS, *Natural Law and Divine Miracle, a historical-critical Study of the Principle of Uniformity in Geology, Biology and Theology*, Leiden 1959, e.g., p. 176–177, 137–139.

100 W. PREYER, *Die organische Elemente und ihre Stellung im System*, Wiesbaden 1891.

101 W. PREYER, *Das genetische System der Elemente*, Berlin 1893.

102 W. PREYER, *Ber.*, 29 (1896) 1040; abstract in *Chem. News*, 73 (1896) 235.

103 P. J. F. RANG, *Chem. News*, 67 (1893) 178.

104 F. RANG, *Chem. News*, 72 (1895) 200.

105 S. KERN, *Chem. News*, 36 (1877) 114, 115; 37 (1878) 65.

106 T. DAHL, *Chem. News*, 40 (1879) 25; *Compt. Rend.*, 89 (1879) 47.

107 B. BRAUNER, *Chem. News*, 59 (1889) 295.

108 R. HERMANN, *J. Prakt. Chem.* [2], 15 (1877) 105; *Chem. News*, 35 (1877) 197.

109 J. L. SMITH, *Compt. Rend.*, 87 (1878) 148; *Chem. News*, 38 (1878) 100, 101.

110 L. MEYER and K. SEUBERT, *Die Atomgewichte der Elemente aus den Originalzahlen neuberechnet*, Leipzig 1883, p. 200.

111 M. DELAFONTAINE, *Compt. Rend.*, 87 (1878) 459.

112 M. DELAFONTAINE, *Compt. Rend.*, 87 (1878) 632.

113 L. MEYER, *Die modernen Theorien der Chemie*, 4th ed., Breslau (Wroclaw) 1880, p. 138.

114 T. GRAHAM and F. W. R. OTTO, *Lehrbuch der Chemie*, 3rd ed., Brunswick 1885, Vol. I Part II, p. 162.

115 M. CAREY LEA, *Am. J. Sci.* [3], 49 (1895) 357; transl. in *Z. Anorg. Chem.*, 9 (1895) 312.

116 M. CAREY LEA, *Z. Anorg. Chem.*, 12 (1896) 340.

117 W. ACKROYD, *Chem. News*, 73 (1896) 221.

118 P. E. LECOQ DE BOISBAUDRAN, *Compt. Rend.*, 120 (1895) 1097; abstract in *Chem. News*, 71 (1895) 271.

119 P. E. LECOQ DE BOISBAUDRAN, *Compt. Rend.*, 24 (1897) 127.

120 R. M. MÜLLER, *Über die Entfaltungsordnung und den Stammbaum der chemischen Grundstoffe; Eine morphologische Betrachtung*, Halle/Saale 1944.

121 A. PARTHEIL, *Ber. deut. pharm. Ges.*, 13 (1903) 466.

122 C. HUYGENS, *Oeuvres complètes*, The Hague 1940, Vol. XX p. 141.

123 See e.g. J. W. VAN SPRONSEN, *Het Orgel*, 48 (1952) 21, 42.

124 C. G. BEDREAG, *Bull. Inst. Polytech. Jassy*, 3 (1948) 317.

125 E. SCHMIZ, *Ber. deut. pharm. Ges.*, 29 (1919) 504.

126 W. FINKE, *Z. Physik.*, 122 (1943) 586.

127 W. FINKE, *Z. Physik.*, 122 (1943) 230.

128 O. M. CORBINO, *Nuovo Cimento*, 5 (1928) LVII.

129 W. FINKE, *Z. Physik.*, 126 (1949) 106.

130 P. VALLET, *Compt. Rend.*, 227 (1948) 58.

131 A. N. WRINGLEY, W. C. MAST and T. P. MC CUTCHEON, *J. Chem. Educ.*, 26 (1949) 216, 248.

132 K. E. ZIMMENS, *Festskrift till äghad J. Arvid Hedvall*, 1948, p. 635; *Trans. Chalmers Univ. Techn. Gothenburg* No. 78 (1948) 1–18.

133 L. S. FOSTER, *J. Chem. Educ.*, 26 (1949) 283; S. HECHT, *Explaining the Atom*, New York 1947.

134 G. A. SCHERER, *J. Chem. Educ.*, 26 (1949) 133.

135 F. J. HERRON, *J. Chem. Educ.*, 27 (1949) 540.

PART II

SPECIFIC ASPECTS

Chapter 7

Prediction of elements[1]

7.1. Introduction

When it first became necessary to assume the existence of an element that the available facts neither demonstrated conclusively nor denied, prediction became a useful scientific tool. For instance, the assumption of elements with an atomic weight greater than that of uranium or smaller than that of hydrogen, could by that time be called a prediction. In the field we are discussing, sufficient grounds for prediction were considered to be provided by (1) numerical relationships between atomic weights of (a) elements of a triad or (b) elements of a larger group of homologues; (2) gaps in a system of elements, chiefly the periodic system; and (3) data from a comparative study of the spectra of elements. It need hardly be said that the use of a periodic system with inconsistencies in the arrangement of periods and the division into main and sub-group elements led to erroneous conclusions.

When the spectra of the sun and other stars were investigated, many combinations of spectral lines could not be identified because they differed from those of the known elements. Many new elements were demonstrated from the spectra of terrestrial substances and solar spectra, but in addition, far-reaching predictions were made of elements whose existence has never been proven and has been excluded by later atomic theory. These predictions mainly concerned elements with a low atomic weight. The phenomena that gave rise to these predictions were eventually traced to certain atomic states of known elements.

Predictions concerning the existence of elements had already been made before the construction of the periodic system, but only by a few investigators, such as Döbereiner and Carey Lea. Among the constructors of the system, Newlands, Meyer and Mendeleev deduced the existence of new elements. De Chancourtois did not discuss undiscovered elements, whereas Odling and Hinrichs reserved places for them in their systems but ignored the consequences.

It is not surprising that only a few of these early workers were tempted to make predictions; the original triads and other series of elements showing some gaps were

---

[1] For the predictions of noble gases, rare earth metals, elements with an atomic weight above that of uranium or below that of hydrogen, elements between hydrogen and lithium, etc., see the relevant chapters (9, 10, and 15). The unexplained vacant spaces in the system are treated in the discussion of the individual systems (see Part I).

not always sufficiently prepossessing to convince the authors themselves immediately of their absolute value. Cooke (Fig. 6, p. 79) and Dumas (Fig. 12, p. 91) attempted to classify only the atomic weights of known elements. Döbereiner went a step further by assuming the existence of elements contingent on the validity of his law of triads. Others, like Pettenkofer and Lenssen, also used formulae to calculate atomic weights of elements which were difficult to determine. A very few, like Hinrichs, also calculated atomic weights of predicted elements never discovered later. These values were simply the result of a consistent application of the formulae used by these scientists, to which they assigned too much importance.

Two forms of prediction can be distinguished. In the first place, a suspicion of the existence of new elements could arise from closer examination of the atomic weights and the properties of elements already discovered. This occurred in the case of Mendeleev, who could even indicate various properties of predicted elements and some of their compounds. Secondly, the correct atomic weight or other properties of a known element inexactly examined can be predicted, as Newlands sometimes did. He was the first to make a proved correct prediction not purely based on chance.

7.2. Predictions made before the construction of the periodic system[2]

Of all the pioneers in this field, Döbereiner and Carey Lea were the only ones who predicted the existence of still undiscovered elements before 1862. When in 1829 Döbereiner could not place fluorine in the triad $Cl = (F + Br)/2$ because of its atomic weight, he concluded that it must be a member of another undiscovered triad. As already observed in the Introduction, this prediction depends on the validity of the law of triads. Döbereiner's faith in this was too tenuous to admit of any forecast as to the atomic weight and character of the missing elements. The relationships he found between the atomic weights of elements with analogous properties were sometimes applied by Pettenkofer and Lenssen to the determination of unknown or uncertain atomic weights of known elements, which was in fact also a matter of a prediction based on certain (theoretically unfounded) numerical relationships between the known atomic weights.

As we have already seen (4.15; Fig. 14, p. 93), Carey Lea assumed the existence of two new elements on the basis of the relations he had established. The difference of 44 or 45 between the atomic weights of homologous elements implies an atomic weight of 164. This proved to be half of the mean of double the atomic weights of antimony (240.6) and bismuth (416). With these double atomic weights Carey Lea had created a series with negative numbers. As was later proved, an element with an atom-

---

[2] Richter's assumption [1] of the existence of the element strontium in the series of the alkaline earths, because of relations between the equivalent weights of the oxides of these elements, in fact also comes within the province of these studies [2] (see 9.5).

ic weight of 164 does indeed exist, but as a member of the group of the rare earths, i.e. between elements having nothing in common with antimony and bismuth. The atomic weight of 152.5, which could be expected because of its relationship to the atomic weights of gold, silver and copper (Cu = 63.5; + 44.5 = 108 = Ag; + 44.5 = = 152.5 = ...; + 44.5 = 197 = Au) later proved to belong to a rare earth metal. Carey Lea, however, said nothing about the vacant places in his series where atomic weights of $\frac{1}{2}$ × 194, $\frac{1}{2}$ × 284 and $\frac{1}{2}$ × 372 were located.

These predictions are mentioned only to complete the picture; we have already stated that Carey Lea's contribution was of very little value.

7.3. Predictions made by Newlands

Newlands predicted the middle member of the triad of which Ir and Rh are the extremes in his very first publication *[3]*, in which he did not describe any system. He deduced this triad from the difference in atomic weight between iridium (99) and rhodium (52.2). This difference of 46.8 closely approximates the value of 48, which had already been observed several times as the difference in atomic weight between the extremes of a triad, e.g. Te (64) —S (16) = 48 and Ba (68.5) —Ca (20) = 48.5 (Fig. 18, p. 105). Subsequently, it was demonstrated that no element occurs between iridium and rhodium. The resemblance to other triads does not hold here because of a lack of analogy in electronic configuration, of which Newlands could not know any more than he could be aware of the resulting difference in length of the periods of elements. Moreover, the atomic weights employed by Newlands required doubling.

An element with an atomic weight of 163 should follow from the relationship Li + + 4K = 163 (see *5.3*).

A year later, in 1864, Newlands *[4]* thought an element had to be assumed between palladium (106.5) and platinum (197) (Fig. 22, p. 107). For the same reason as holds for the other transition metals, no element belongs here. Newlands' conclusion followed from the wrong determination of the atomic weight of vanadium, viz. 137 instead of 51. Newlands had arranged the triad Pd–?–Pt on the analogy of the triad Mo (96)–V (137)–W (184), to which vanadium belongs no more than any other element. Although Newlands had now collected the elements in a single system, that system did not clearly indicate the validity of a given triad because he did not arrange all the elements according to increasing atomic weight.

In the third group of this system Newlands assumed an element with an atomic weight of 73 as the middle term of a triad:

$$\frac{Si\ (28) + Sn\ (118)}{2} = 73.$$

He also worked out one consequence of this prediction. A triad with this unknown element gives an element with the atomic weight 50.5:

$$\frac{\text{Si} \ (28) + \ 73}{2} = 50.5.$$

This latter atomic weight belongs to titanium which, however, Newlands was just as unable to classify as other elements of this period, such as chromium, manganese and iron. These two predicted triads were found to be valid. The prediction of the element with an atomic weight of 73 could therefore be based on two grounds. Later, the element germanium with this atomic weight was discovered, so that Newlands [5] really deserved priority for a correct prediction, as he himself emphatically claimed. It should be noted, however, that he must have seen the consequences of the second triad, for just as titanium can be regarded as the middle member of the triad Si–Ti–Ge, so chromium and manganese, for instance, can be placed in the triads S–Cr–Se and Cl–Mn–Br, respectively. Newlands went rather far in considering Mn, Fe, Co, Ni and Cu as the middle elements of a purely speculative triad [6].

From the same system Newlands concluded the existence of a fifth halogen, but he did not predict its atomic weight. Furthermore, Newlands could certainly have said something about the atomic weight of the newly-discovered indium, had he known but a single property of this element. A few weeks after its announcement, Newlands [7], in reply to a certain Inquirer [8], said that he thought this element to be the last member of the group Li–Mg–Zn–Cd, not only because of its being found in zinc ores, but also because Roscoe thought he had demonstrated properties related to those of zinc. An atomic weight of 182 would then differ from the atomic weight of zinc by the same value as is found between the atomic weights of thallium and caesium:

In (182) − Cd (112) = 70.
Tl (203) − Cs (133) = 70.

The later determination of the atomic weight of indium as 74.14, which was incorrect because of a mistake concerning valence, caused Newlands [9] to have some doubts, however. Too little was known about this element. Indium could therefore be classified equally well between Si and Sn, i.e. in the place of the earlier predicted germanium to which Newlands gave the atomic weight 73, between Al and Y as the last member of the series Cr, Mn, Fe, Co, Ni, Cu, Zn and In, with Zn = (Fe + In)/2. Newlands furthermore realized that indium might possibly show an analogy with the elements of two groups. It was not until 1870 that indium, its atomic weight by then correctly determined at 113.4 by Bunsen using the law of Dulong and Petit, could be placed by Meyer [10] in the correct group (Fig. 38, p. 129). In 1866 Newlands [11] had also reserved a place between Ti and Zr, on the basis of his earlier considerations concerning his law of octaves.

The following may be said concerning Newlands' other predictions. On the analogy of the triad Pd–?–Pt, he established the triad Ag–?–Au. For the same reasons as in the case cited above, this triad could not exist. His interpretation of both groups (Cu, Ag, Au and Ni, Pd, Pt) as triads made Newlands' conclusions conflict with the atomic weights.

Two years later, in 1866, Newlands [11] no longer left vacant spaces in his system (Fig. 24, p. 108) for elements that might still be discovered. Gladstone raised objections to this [11] (see 5.3). As we have already said, Newlands [12] pointed out that the difference in ordinal number between phosphorus and arsenic was first 13 (Fig. 23, p. 107) and, to create a space for indium, became 14 (Fig. 24, p. 108), to which must be attached the consequence that the discovery of an element required the introduction of an entirely new group into the system.

7.4. Predictions made by "Studiosus"

Because a certain Studiosus [13] was convinced that all the atomic weights were multiples of 8, ye ascribed the missing multiples, i.e. the products 9, 19, 20, 21, etc., × 8, to elements still to be discovered. Studiosus was bothered neither by the fact that several multiples occured more than once, nor by the deviations from the multiple of 8. His reasoning lacked any basis and can hardly be called a prediction.

7.5. Hinrichs' equations for atomic weights

The relations between the atomic weights stated by Hinrichs [14] in 1869 resulted in a few vacant spaces in several groups (5.6). Of all the atomic weights belonging to these vacancies, only one proved to be valid: the homologue of iodine agreed with

	x	calculated at.w.	determined at.w.	difference
fluorine	1	18	19	+ 1
chlorine	2	36	35.5	- 0.5
bromine	3	81	80	- 1
iodine	4	126	127	+ 1
astatine [*]	5	216	isotope 216[*]	0[*]

[*]Entered by the present author

Formula used for $x = 1,2$: at.w. $= 9 \times 2^x$;
for $x = 3,4,5$: at.w. $= 9 (2^2 + 2.5 \times 2^{x-2})$

Fig. 110   Relations between atomic weights (Hinrichs; 1869).

the value later found for astatine (Fig. 110). This cannot be called true prediction, because Hinrichs did not comment on it and, furthermore, the equations for the atomic weight were not based on sufficient data.

References p. 234

7.6. Mendeleev's far-reaching predictions

From the very beginning, Mendeleev took the existence of still undiscovered elements into account and arranged his classifications accordingly. Not only did he reserve places for elements to be discovered, but also predicted the atomic weights and other properties of the elementary substances and of some of their compounds, all of which have since been found to possess virtually every one of the predicted attributes. In his first system *[15]* (Fig. 39, p. 129), constructed in 1869, he put question marks at the places of several known elements, as well as at the four places for which he had only reserved atomic weights, i.e. 45, 68, 70 and 180. Here, Mendeleev already foresaw the existence of the elements gallium (70) and germanium (73), for which he announced so many properties one year later. He soon put lanthanum in the place having the atomic weight 180, so that there was no space for hafnium, which was discovered later. An element with an atomic weight of 45 was to be placed between Ca and Er as a homologue of cerium. This mistaken prediction followed from an incorrect arrangement of the rare earth metals due to difficulties encountered in determining the exact atomic weight of these elements (see *9.3*). An atomic weight difference of about 47 units was evidently taken as the starting-point. That the element next to calcium later proved to be scandium, was fortuitous. Mendeleev cannot be said to have already foreseen this element in 1869; he did not do so until 1871, when he correctly classified yttrium between Sr and Zr.

Mendeleev actually postulated many more elements, as shown by his statement that only the series Li → Tl was complete, which implied the homologues of Cd, Sn, Te and I with a greater atomic weight as well as the homologues, with a smaller atomic weight, of the elements of the period Ti → Cu and of the first period Be → F.

The spiral system (Fig. 44, p. 135), given in the same publication, contained many more vacant spaces, because Mendeleev had not distinguished main and sub-group elements. Consequently, there are vacant spaces between S and Se, Se and Te, Cs and Tl, Cl and Br, and Br and I, for instance. However, from another system of 1869 (Fig. 131, p. 287) it is clear that Mendeleev meant these places for the elements Cr, Mo and Mn, among others .The atomic weight 58 he mentioned for an element between chlorine and bromine agrees reasonably well with that of manganese (55). At the place of the undiscovered element belonging between Br and I, he noted the atomic weight 100. The places next to iodine carry the atomic weights 160, 190 and 220 (Fig. 44, p. 135). Had Mendeleev not created a new series of elements between Ba and Ta in this system (as he indeed did not do in the system shown in Fig. 131, p. 287), thus avoiding a vacancy for an element with an atomic weight of 160, his atomic weights of 100, 190 and 220 would have constituted a prediction of the elements technetium (isotope 99), rhenium (186) and astatine (isotope 218), respectively. In 1871 his compact system with main and sub-groups of elements expressed the actual hiatus between elements much more effectively *[16]*; there were now only four of these, with atomic weights of 44, 68, 72 and 100. The extensive predictions associated by Mendeleev with ekaboron, eka-aluminium and ekasilicon are shown together with the demonstrated data in Fig. 48 (p. 139). After Lecoq de Boisbaudran [17] had discovered gallium in 1875, Mendeleev [18] rightly concluded that the validity of the periodic system of elements could no longer be questioned[3]. The confirmation of this prediction may certainly be called the culminating point in the history of the periodic system.

As we have already observed, the atomic weight 100 belongs to an isotope of technetium [20], which was finally synthetized in 1940. In 1871 Mendeleev had predicted only one property of technetium, its atomic weight. In the place next to zirconium, which Mendeleev kept vacant in 1869 with an atomic weight of 180, he now unfortunately placed lanthanum. This can be explained by the fact that cerium was here classified in the carbon group, and Mendeleev apparently regarded lanthanum as a homologue of this element. Cerium was indeed given the correct atomic weight, but for lanthanum Mendeleev used about $\frac{4}{3}$ of the exact value, i.e. less than twice the old value of 94.

The system of 1871 (Fig. 45, p. 137) contains a large number of vacant spaces. For three of these Mendeleev predicted elements with many of their attributes. For six spaces he predicted only the atomic weights of the elements. These six elements were ekacaesium (Ec = 175), dvicaesium[4] (Dc = 220), ekaniobium (En = 146), ekatantalum (Et = 235), ekamanganese (Em = 100), and trimanganese[5] (Tm = 190) (see Table 3). Mendeleev would have accepted ruthenium and osmium in place of ekamanganese and trimanganese, respectively, if the formula of the oxides of these elements had been R_2O_7 instead of RO_4. Actually, this exchange would not have helped Mendeleev to improve his system, because it would have caused the loss of the homologues of iron.

In the 8th, 9th and 10th periods, spaces became vacant because of the inclusion of several rare earth metals and the absence of elements with atomic weights between 140 and 178. That these intervals were not occupied by homologues of known elements, is obvious; the element ekacaesium would then be missing. Only dvicaesium would later prove to be the homologue of caesium. The nomenclature chosen for the homologues of manganese, i.e. eka- and tri-manganese, also played a role here. Dvimanganese, a homologue in the series of the rare earths, was neither named nor predicted by Mendeleev. Did Mendeleev feel intuitively that the vacant spaces in the 8th and the 9th periods had no real significance? Ekaniobium also belongs to this series. The remaining predicted elements were in fact discovered later (see Table 3). It is surprising that at a later date Mendeleev was no longer certain about all these predictions. In 1889 he predicted the homologue of tellurium [21]. Proceeding on the same grounds, he called this element dvitellurium. Once again an ekahomologue was

---

[3] Wurtz [19], however, saw a weakness in the analogy between eka-aluminium and gallium. But his criticism was based only on the relatively great difference in atomic weight (68 predicted, 69.9 found), which does not seem entirely justified.
[4] Dvi (Sanskrit) = two.
[5] Tri (Sanskrit) = three.

TABLE 3

PREDICTIONS OF ELEMENTS BY MENDELEEV, PREYER AND BRAUNER (EXCEPT THOSE OF EKABORON, EKA-ALUMINIUM, EKASILICON AND HAFNIUM)

name of the element [*]	symbol	name of the predictor	year of prediction	atomic weight	density	name of the element	year	atomic weight	dens
				predicted			discovered / determined		
eka-manganese	Em	Mendeleev	1871	100	± 11 (Brauner)	technetium	1939	isotope 99	
eka-niobium	En	Preyer	1892	98	10				
		Bassett	1892	100?					
		Mendeleev	1871	146					
dvi-manganese		Preyer	1892	147	13.5				
eka-caesium	Ec	Mendeleev	1871	175					
neighbour of tantalum		Preyer	1892	179	8	hafnium	1923	178	13.3
tri-manganese	Tm	Mendeleev	1871	190		rhenium	1925	186	20.5
neighbour of tungsten		Preyer	1892	187	21.2	rhenium	1925	186	20.5
		Bassett	1892	189?					
dvi-manganese		Brauner	1902	190	± 21	rhenium	1925	186	20.5
dvi-tellurium	Dt	Mendeleev	1889	212	± 9.3	polonium	1898	210	9.4
eka-tellurium		Brauner	1902	212	± 9	polonium	1898	210	9.4
		Preyer	1892	213	8.6				
eka-iodine [1]		Preyer	1892	217	7	astatine	1940	isotope 215	
		Bassett	1892	216?					
		Brauner	1902	214					
eka-xenon [2]		Brauner	1902	218		radon	1900	222	4.4 (li
		Preyer	1892	226	4.8				
dvi-caesium	Dc	Mendeleev	1871	220		francium	1939	223	—
eka-caesium [1]		Preyer	1892	223	2.8	francium	1939	223	—
		Bassett	1892	226?					
		Brauner	1902	220					
eka-barium [3]		Brauner	1902	225		radium	1898	226	5?
		Preyer	1892	229	7				
eka-lanthanum [4]		Brauner	1902	230		actinium	1899	227	—
eka-tantalum	Et	Mendeleev	1871	235		protactinium	1918	231	15.4
tri-manganese [*]		Preyer	1892	242	20	neptunium	1940	isotope 237	—

[1] called by Baur[80] supra-elements.

[2] may be, according to Brauner, the radioactive gas found by Rutherford (see Chapter 9).

[3] is, according to Brauner, very probably radium.

[4] is, according to Brauner, a trivalent radioactive rare earth.

[*] Preyer and Bassett did not name the predicted elements.

skipped. Not until the rare earths had obtained a separate place in the periodic system did trimanganese become dvimanganese and dvitellurium become ekatellurium, etc. This was accomplished by Brauner [22] in 1902.

Dvitellurium, which Mendeleev assumed to have an atomic weight of 212, was seen as a readily fusible, though not volatile, crystallizable grey metal with a specific weight of approximately 9.3. The oxide would be written as DtO_3, and the hydride H_2Dt would be less stable than H_2Te. Mendeleev thereby predicted polonium, and was proved correct in 1898: the atomic weight of polonium is 210 and the specific weight exactly 9.3. The oxide PoO_3 is also known. The metal did not, however, prove to have the volatility expected from the melting-points of the other elements of the oxygen group: its melting-point is 1800°C.

To solve the problem of the Te–I inversion, Brauner, also in 1902, started to investigate the purity of tellurium (see *8.2*). After learning that Mendeleev had predicted an atomic weight of 212 for the homologue of tellurium, Brauner *[23]* suggested that the element austriacum, which according to him caused the impurity of tellurium, might be identical to dvitellurium (see also *13.1*).

Carey Lea *[24, 25]* felt justified in stating in 1896 that, on the basis of his theoretical considerations (see also *4.15*), the atomic weight of the homologue of tellurium should differ from that of this element by 88 units and would therefore be 213.

In 1891 Mendeleev *[26]* predicted only one element, ekacadmium, with an atomic weight of 145.9 and did so only because, according to him, it would be identical to norwegium, which Dahl *[27]* thought he had found in nickel ore in Norway in 1879. This conclusion had also been drawn by Meyer in 1880 (see Fig. 117, p. 266) and Otto in 1885 (see Fig. 116. p. 265). The space for ekacadmium resulted from the creation of an interperiod (see *6.22*).

7.7. Predictions by Meyer in 1870

Before Mendeleev made his many predictions in 1871, Meyer *[10]* had in 1870 already stated his general views with regard to still undiscovered elements. In his system (Fig. 38, p. 129) he reserved a few vacant spaces for future elements. He also used his system and his graph of atomic volumes (Fig. 40, p. 130) to improve the values of poorly determined atomic weights. As examples, Meyer gave the elements Te, Pt, Ir, Os, In and U. He suggested the atomic weights 113.4 and 180 for indium and uranium, respectively. When used alone, this method of determining atomic weights resulted in less accurate values than those obtained by taking into account the values of vapour density and specific heat, as shown by the result for the atomic weight of uranium, which is too low by a factor of $\frac{3}{2}$.

7.8. Predictions made after the period of discovery[6]

In addition to Mendeleev's predictions from the years after the discovery of the periodic system, we may mention a few others. We are unable to agree with Haughton *[29]*, who stated in 1888 that the elements possessing a small atomic weight could include five new elements. He drew this conclusion from a graphic representation of the elements that he deliberately caused to differ from that of James Emerson Reynolds *[30]*. This professor of chemistry at Trinity College of the University of Dublin had placed the elements according to valence and atomic weight (Fig. 111) on a single

---

[6] For the reservation of places in the series of the rare earths, see Chapter 10. The elements with atomic weights of 48 and 64, suggested by Berthelot *[28]*, are derived from an incorrect interpretation of the periodic system, as Mendeleev *[21]* already observed in 1889.

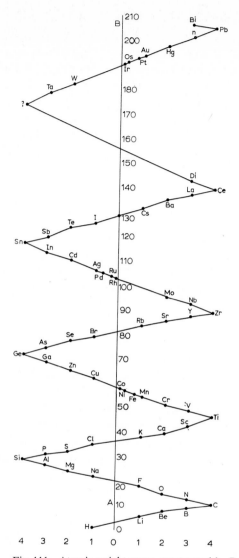

Fig. 111   Atomic weight curve constructed by Reynolds.

curve, but Haughton was of the opinion that the elements C, N, O, Mg and Si must be situated on a straight line (Fig. 112). This would leave spaces on the curve where the atomic weights 13, 16.4, 18, 25 and 26[7] belong, namely at the points of intersection of the curve with either the lines joining elements with the same valence or with other lines. In this interpretation Haughton lost sight of the fact that it implied either elements with a fraction for their valence or more than two elements with the same valence, which he vigorously denied.

[7] Haughton may have meant 28.

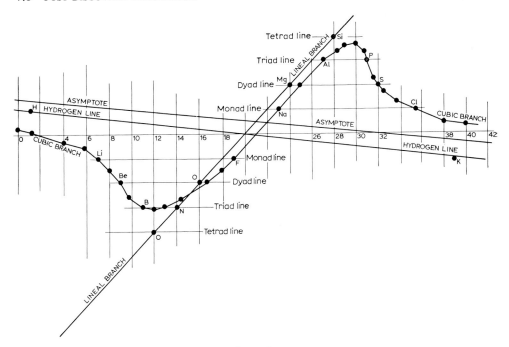

Fig. 112   Atomic weights, graph constructed by Haughton.

Preyer's system (Figs. 94–96, p. 195 and Fig.113, next page) led to 15 unknown elements, some of which fell between the correctly placed rare earth metals. In addition to their atomic weight, Preyer [31] also predicted their density, atomic volume and specific heat. The majority of his predictions have been reasonably well confirmed[8] (Table 3). They concerned the homologues of tellurium, iodine, manganese (three), caesium, barium, lanthanum and a few of the rare earths[9].

At the beginning of the present century, Brauner [22] predicted many elements[10] as we have seen. However, some of these had already been discovered (e.g. radium, radon and actinium) although they had not yet been definitely incorporated into the system (see Table 3). In addition to the atomic weight, Brauner indicated some densities. All of his specific predictions have been proved correct.

---

[8] The incorrect incorporation of dysprosium, due to an incorrect atomic weight, resulted in a few unjustified vacant spaces (10.5).

[9] In the same year Bassett [32] predicted not only the atomic weight of several elements of the rare earth series and the atomic weight of the transuranium metals (see 10.5 and 14.2), but also those of the homologues of iodine and caesium, and of the two homologues of manganese (see also 6.3; Fig. 52, p. 150).

[10] Mendeleev did not predict eka-iodine. Before Brauner, it was Preyer (1892), Bassett (1892) and Steele [33] (1901) who foresaw the discovery of francium. Steele, who later became professor of chemistry at the University of Brisbane, gave it an atomic weight of 218.

References p. 234

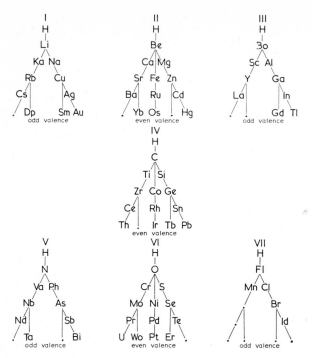

Fig. 113   Preyer's genealogical table with 5 *Generationen* and 14 *Verdichtungsstufen*.

7.9. Prediction of hafnium

Let us now concentrate on hafnium, an element that no-one predicted, at least not as a homologue of zirconium[11]. Its eventual place in the system had often been given to one of the rare earth metals, e.g. cerium, which also showed quadrivalence. Although in 1892 Preyer [31] gave the rare earth metals a place analogous to that of the remaining elements in the system, he nevertheless left a space vacant next to tantalum. The atomic weight that he gave to this hypothetical element, i.e. 179, agreed closely with the atomic weight of hafnium. Through a wrong determination of the specific weight of tantalum, namely 10.78 instead of 16.6, the value predicted of this property of hafnium (8) proved to have been far too low (correct value 13).

After the theoretical basis of the periodic system had been explained by Bohr's atomic theory, the zirconium minerals could be more effectively analysed, with the result that hafnium was discovered in 1923 by Von Hevesy [35] and Coster [36]. It

---

[11] Thorpe [34] reported in 1923, in connection with the discovery of hafnium, that in a conversation Mendeleev had told him more than once that a new element might be found in titanium minerals. According to Mendeleev, this could be expected on the basis of the excessively large atomic weight of titanium.

has been made clear why Mendeleev did not predict hafnium; Brauner also failed to expect it, and even in 1921 Nernst [37], who regarded cerium as a homologue of zirconium, reserved place number 72 for thulium II[12] (see 10.8). In 1895 Thomsen already regarded zirconium as a homologue of two elements, i.e. cerium and an element with an atomic weight of 181 (see Fig. 53, p. 151).

It is interesting to follow the general historical lines of the discovery of hafnium. As early as 1845 Svanberg [40] believed that he had found a new metal in zirconium ores, and called it norium. In 1866 Church [41, 42], the future professor of chemistry at the Royal Academy of Arts, London, named a similar element nigrium. In 1869 Sorby [43, 44], geologist and petrographer at Sheffield and originator of metallography, demonstrated an element, which he called jargonium, by means of spectral analysis; this element, like norium and nigrium, may have been hafnium [45]. Nearly fifty years later, in 1908, Ogawa [46, 47] announced a new element named nipponium (Np) which he had discovered as a silicate in thorianite and some other minerals. This silicate, according to Von Hevesy [48], really did contain 2% hafnium as well as zirconium. Ogawa wanted to place nipponium in the vacant space under manganese in the periodic system, and intended an atomic weight of 100 for it, although he thought that a higher value of 150 was also possible. Loring in 1909 included nipponium as a homologue of manganese in his system, which was published one year later (Fig. 89, p. 191). Celtium, which Urbain [49] announced in 1911 as element 72, was found to be lutetium, discovered by himself in 1907 [35].

7.10. Speculation about elements in celestial bodies

Very soon after its discovery, in 1859, spectrography was applied to the investigation of the chemical composition of celestial bodies. Helium, which had not yet been found on the Earth, was demonstrated in the solar spectrum as early as 1868 by the British astronomer Sir Norman Lockyer. After this success, Lockyer and several other astronomers were led to assume additional unknown elements, on the sole basis of the fact that the spectra of celestial bodies contained lines they could not identify. It was evidently difficult to believe that the same elements were present in both celestial bodies and Earth. In 1871 Lockyer [50] was of the opinion that he was dealing with a new gaseous element occurring in the corona of the sun. He called this element coronium, as did Gruenwald [51].

Other investigators thought coronium occurs in volcanic gases [52] and the northern lights [53], or attributed the spectral lines of coronium to fluorescence of a gas-

---

[12] Stephan Meyer [38], who, even before his appointments as professor of physics at the University of Vienna and director of the Institute for Radium Research of the Academy of Science, had done much research on the rare earths and radioactive elements, did the same in 1918. Kohlweiler [39], who was then studying at the Stuttgart Technical College and had published papers on atomic structure and radioactive elements, regarded thulium II as the 15th rare earth metal.

eous metal in the corona caused by solar radiation [54]. If coronium had to be a new element its atomic weight must have a value smaller than 1 [52, 55].

Coronium was not the only element assumed by Lockyer [50]. He found that the spectra of solar and astral elements showed a consistent deviation from those of the elementary substances cocurring on the Earth. He therefore called these elements protohydrogen and protometals, e.g. protomagnesium and protocalcium[13]. As early as 1896 Pickering [56], professor of applied astronomy at Harvard Observatory in Cambridge, Massachusetts, had mentioned protohydrogen as a new element on the basis of its spectral lines, which resembled those of hydrogen and were found in the spectra of the stars of the constellation Cygnus. Lockyer [50] also assumed an element named asterium among the elements found; for instance, on γ-Argus.

Fowler [57], who did a great deal of spectroscopic research and in 1918 became professor of astrophysics at London University, later established the formula of the spectral series for protohelium and introduced the name *Pickering series* after its discoverer.

Mendeleev was fascinated by the assumed discoveries of new elements [58], but not by the discovery of the electron, which interfered with his concept of the elementary essence of atoms. In 1904 he threw out suggestions as to the atomic weights of coronium and ether, calling the latter newtonium and regarding both as zerovalent elements. He assumed the ratio of the atomic weights of helium and coronium to be 10, so that coronium would have an atomic weight of 0.4. He extrapolated the atomic weight of newtonium from the series

$$\frac{Xe}{Kr} = 1.55; \qquad \frac{Kr}{Ar} = 2.15; \qquad \frac{Ar}{He} = 9.5; \qquad \frac{He}{newt.} = 23.6,$$

which gave 0.17 for this value. Mendeleev did not discuss the underlying principle of this series, so these results are of no value to us.

The astronomer, mathematician and physicist John William Nicholson [59] (1881–1955), who lectured at Cambridge and held afterwards professorships in mathematics at Oxford and London, must in a sense be regarded as one of Bohr's predecessors. In 1911 he was, like Ernest Rutherford [60], continuing the elaboration of the atomic hypothesis, put forward by Joseph John Thomson [61] in 1904. Taking e as the charge of an electron rotating in the positive sphere of an atom, Nicholson found the value for the inertia to be n^2e^2/a (n = number of electrons; a = atomic radius). But when he applied this result to the hypothetical elements coronium, nebulium and protofluorine, in order to compare his hypothesis with that of Mendeleev, he took the wrong direction. When he calculated the volume of atoms he arrived, with H = 1 for the standard, at atomic weights for the above-mentioned hypothetical elements. What Nicholson actually attempted to do was to put new life into the old protyle

---

[13] He assumed proto-Mg and -Ca on α-Cygnus and proto-Fe, -Ti, -Cu, -Mn, -Ni, -Cr, -V and -Sr on α-Canis major.

theory, on the one hand by means of spectroscopic data of the celestial bodies, and, on the other, by extending his atomic theory. He regarded all the elements as combinations of these protyles, to which he also added hydrogen and helium, and this took him a step further on this precarious course. That he continued to follow it, even after Bohr had published his atomic theory, is shown by the papers *[62, 63, 64]* he wrote in 1914 and 1919. Nicholson received support from Bourget *[65]*, director of the Observatory and professor of astronomy at the University of Marseilles.

Although Nicholson made a certain contribution to atomic physics, it is to be regretted that he applied his atomic theory primarily to unverified elements in which he was hampered by his conception of matter. According to him, a nebulium atom contains four and a protofluorine atom five electrons.

Benjamin Kendall Emerson *[66]* (1843–1932), professor of geology and zoology at Amherst College (Massachusetts), supported Nicholson's classification of these "new elements" in the periodic system as early as 1911. He tried to prove the existence and to deduce the atomic weight of protofluorine on the basis of the difference between the atomic weights of protofluorine and its homologue fluorine, and that between the atomic weights of sodium and lithium, being 16: Na $-$ Li $= 23 - 7 = 16$; F $-$ proto-F $= 19 - 3 = 16$. By this analogy the atomic weight of protofluorine becomes 3, which is one unit higher than Nicholson's value. The atomic weight of coronium was determined at 0.3 by Emerson on the basis of analogous atomic weight ratios. The ratio of the atomic weights of helium and coronium had to be 13.3 ($=1.9$ times that of Li : H, being 7).

Emerson postulated the atomic weight of nebulium at 2. It seemed probable to him that the atomic weights would be whole numbers, an assumption deriving from the symmetrical aspect of his helix-shaped system (Figs. 63 and 64, p. 165), which carries on its first spiral winding the elements H, Ne, Pf and He, with atomic weights of 1, 2, 3 and 4, respectively.

Coronium and ether (which he called electron) are also classified in his system: coronium as a noble gas, ether as a homologue of hydrogen. He placed nebulium halfway between hydrogen and protofluorine, which he classified as a homologue of fluorine. Of these elements, protofluorine and nebulium received atomic weights greater than 1. These suggestions date from 1911, long before Nicholson published his supplementary considerations.

It is clear that the gases discussed above, all of which had been demonstrated in the spectra of the sun and the stars, were not new elements. Two investigators, Bowen *[67]*, a student of Millikan's, who became associate professor of physics at Pasadena in 1928 and worked principally in spectral research, and Fowler *[68]*, professor of astrophysics at London University, proved in 1927 that the relevant lines of the nebulium spectrum came, for instance, from gases in an excited state or from completely ionized elements such as oxygen and nitrogen.

Much earlier, in 1919, Rutherford *[69]* had discovered artificial radioactivity. Clarke *[70]*, professor of mineralogical chemistry at George Washington University

and associated with the U.S. National Museum in Washington, D.C., who participated actively in the development of the atomic theory, considered nebulium to be the
reaction product of the bombardment of oxygen with α-particles. Protons could be
demonstrated in the majority of bombardment reactions of light elements with α-
particles. However, Rutherford could not identify the product of the reaction between oxygen and α-particles any more than that of the reaction between beryllium
and α-particles. Not until 1932 was a new elementary particle, the neutron, discovered
in the latter reaction. Nebulium had to be discarded.

7.11. Elements between hydrogen and helium[14]

For reasons of chemical analogy it was assumed that there must be elements to be
located between hydrogen and helium (or lithium, before the discovery of helium).
Before the discovery of the noble gases, all periods comprised seven elements if the
transition metals were not taken into account, and the system was not divided into
short and long periods. This periodicity was very well expressed in Newlands' system
and by its name, the law of octaves. Whether Newlands or others gave any consideration to the "sacred" number of seven or related this number of elements to the seven
planets and the seven metals of antiquity we do not know, but in none of the cases did
the number remain seven. On the other hand, the presumption resulted from the consistent application of either a formula relating number to atomic weight or a formula
indicating the lengths of the periods.

As early as 1869, Mendeleev [71, 15] was the first to arrange groups of homologous elements in his system in such a way as to create a series of six vacant spaces
between the elements hydrogen and lithium (Fig. 39, p. 129). Mendeleev did indeed
recognize the existence of the elements belonging to these vacancies, even after he had
removed hydrogen in 1871 from the copper group and placed it in the group of the
alkali metals. He remarked in 1869 that all places were occupied only in the series
Li $\rightarrow$ Tl.

Deeley's contention [72] that Mendeleev [73] assumed hydrogen compounds at
four of these six places between H and Li, is unfounded. Mendeleev only indicated in
his system that in the relevant columns the hydrogen compounds were written RH,
RH_2, RH_3 and RH_4, just as the oxide formulae R_2O, R_2O_2, etc., have an analogous
meaning. Deeley made this remark in connection with the assumption by the astro-
physicist Stoney [74] that the six places should be assigned to elements such as
infraberyllium, etc. (see 9.6).

The discovery of argon in 1894 apparently inspired the imagination of several
investigators. The above-mentioned creation of six infra-elements by Stoney was one
of its results. The inert argon was compared by Stoney to paraffin: like this hydro-

---

[14] Before the discovery of helium in 1895: elements between hydrogen and lithium.

carbon, argon might also be a hydrogen compound, i.e. a compound of supracarbon. Rang *[75]* found another reason to fill the six spaces between hydrogen and lithium in the classification of helium and argon as homologues of germanium (see Fig. 99, p. 198, and *6.12*). W. W. Andrews *[76]* also made a suggestion about the complexity of argon: this element, according to him, was built up of molecules consisting of 28 atoms of supraberyllium, a homologue of beryllium he had created himself and assigned an atomic weight 1.5. Andrews came to this conclusion because, on the one hand, he had no faith in the view that argon was not complex, but, on the other hand, because atoms with a small atomic volume, such as beryllium, attract each other and, in his opinion, can therefore yield large molecules. Gladstone *[77]*, in consequence of the discovery of helium, put elements with atomic weights of 2 and 4, respectively, in these vacant spaces (see also *9.6*).

We need hardly say that these investigators were all rather hasty and uncritical in their propositions, most of which were put forward at meetings of scientific societies. Stapley *[78]* even started from a hypothesis lacking any basis whatever. He put forward—although a few years earlier—the suggestion that all the elements might be oxygen compounds, either of one of the four undiscovered elements between H and Li, with the atomic weights 3, 4, 5 and 6, or of one of the elements belonging to the series lithium through oxygen, all these elements being oxides of the first member of the group in which they occur[15]. Stapley elucidated his views by means of a periodic system. Martin *[79]*, who later became a lecturer in chemistry at Birkbeck College of London University, encountered the problem of the vacant spaces between H and Li at a much later date, when he sought a place for hydrogen in the system, but not above fluorine because of the limited analogy between hydrogen and this halogen (see *12.7*). In his view, the placement of hydrogen above lithium would create places for six unknown elements.

Whereas from the shape of the periodic system six more elements could be assumed in the first period, the difference in atomic weight between hydrogen and lithium did not support this assumption. Yet several investigators could not accept that only helium (discovered in 1895) should be added to this period. Before the atomic number came to be regarded as the basis for the periodic system, Vincent *[80]* sought proof of the existence of elements between hydrogen and lithium from a formula he had deduced for the atomic weight as a function of the order number. This empirical formula was $W = N^{1.21}$, from which it would follow that elements would be found for $N = 2$ and $N = 4$. It is surprising that Vincent nevertheless gave more weight to this formula than to the principle of analogy resulting from the periodic system. Although he saw that the differences in atomic weight between two neighbouring elements deviated too greatly from a constant value, he could not accept the consequences; he too continued to think of six intermediate elements. The formulae

---

[15] To give some examples: Ni $= R_2^{III}O_3$ (in which R^{III} is the hypothetical element with an atomic weight of 6); Cu $= Li_2O_3$; S $= O_2$.

$y = 1.985\,x$ and $y = x^{1.23}$ used by Minet [81] for the first 20 elements and the remaining ones, respectively ($y =$ atomic weight, $x =$ order number), also produced the two elements mentioned above. But the substitution of two empirical formulae for one brings us no further. The fact that Vincent's formulae gave him $N = 92$ for uranium was an accident; he assumed only 12 rare earth metals instead of 14. Minet therefore arrived at an order number for uranium approaching 94. He was compelled to assume four unknown elements between bismuth and radium, as well as one between radium and thorium and one between thorium and uranium, some of which (polonium, radon and actinium) had been discovered in 1907, the year of his publication.

Rydberg's formula $4n^2$ (published [82, 83] in 1912) for the numbers of elements in the periods of the periodic system had a sounder basis even though, when applied consistently, the first period had to contain four elements or be split up into two periods of two elements. However, Rydberg could not yet draw on Bohr's atomic theory. He assigned to hydrogen and helium the order numbers 1 and 4. He created a new order number 0, which he assigned to the electron (E), and reserved the order numbers 2 and 3 for two new elements, homologues of the pairs electron–helium and hydrogen–lithium, respectively. Rydberg also concluded that element number 2 was a noble gas with an atomic weight of 2, and element number 3 a diatomic gas with a molecular weight of $2 \times 3 = 6$; for these he had coronium and nebulium in mind.

Van den Broek [84] pointed out, however, that Rydberg's formula [85] for the relationship between the wavelengths of the spectral lines and the order number of elements led to unreliable results, viz. the above stated existence of four elements in the first period of the periodic system. After isotopes of several elements had been demonstrated, Van den Broek (who in 1913 had given predominant importance to the atomic numbers of the elements) assumed the elements coronium and nebulium to be identical to isotopes of hydrogen and helium. He therefore became the first to indicate the correct order of the elements.

Yet, in 1931 Saz [86] could still suggest that the four elements of the first period concerned were electron, proton, hydrogen and helium. The subject "undiscovered elements" had not yet been disposed of.

Loring [87], who had already made several dubious assumptions, in 1913 replaced hydrogen in his system by two of its isotopes, protonium and deuterium, as homologues of lithium, in violation of the principle of periodicity.

7.12. Some later predictions

Richards [88] (1868-1928), professor of chemistry at Harvard University, who received the Nobel Prize in 1914 for his atomic weight determinations, stated in 1911 that the long-predicted elements eka-iodine and ekacaesium would prove to be unstable. These elements also received attention from Loring [89] and Druce [89, 90], who postulated an element after uranium *[91]* with the atomic number 93 (see *15.3*). Stintzing *[92]*, who taught chemistry at the University of Giessen and later became director of the Röntgen Institute of the Darmstadt Technical University where he did much theoretical research, in 1926 considered this element to be the heaviest. He deduced this from his hypothesis on the tetrahedral numbers in which the atoms are described as tetrahedrons, making use of the many symmetry properties of this three-dimensional figure. After 1926 more realistic attempts were made to explain the in-stability of several elements (see *15.4*).

Nevertheless, in 1926 yet more suggestions were put forward relating to missing elements. Washburn *[93]*, former professor of physical chemistry and allied to the U.S. National Bureau of Standards, made use of the graphic representation $A_x/A_0 -$ $- N_x/N_0$ as function of N (in which A and N are the respective atomic weight and atomic number of the element X or the next noble gas 0), to determine the atomic weights of missing elements. These predicted values do not differ much from the subse-quently determined values. In the same year Swinne *[94]*, an engineer employed by Siemens & Halske, Berlin, who devoted himself chiefly to the study of radioactivity and spectra, deduced from his classification of elements that, besides the known ele-ments cobalt, nickel, etc., the elements with atomic weights of 107 through 110 must belong to the siderophylic elements, i.e. elements found in the Earth's iron core and also in meteorites. He even thought he had observed element 108 by means of X-ray analysis in so-called polar dust, a ferrous dust from Greenland of cosmic origin, al-though his sample was too small to make sure. Just after neptunium and plutonium had been demonstrated and a few years before americium and curium had been syn-thetized, Turner *[95]*, professor of physics at Princeton University, who had worked with Compton and Russell, predicted that $^{237}_{93}$ Eka-Re, $^{244}_{94}$ Eka-Os, $^{243}_{95}$ Eka-Ir, and $^{250}_{96}$ Eka-Pt would be the heaviest β-stable isotopes. The hypothetical atomic weight of the last of these elements is certainly exceptional.

7.13. Conclusion

The predictions of elements made in the course of the last 150 years can be placed into several categories. Before the periodic system had been discovered, and even afterwards, elements could be anticipated either from triads or extensive numerical relationships between the atomic weights of elements with analogous properties. Whether these predictions were fulfilled or not, depended largely on the correctness of the relation, i.e. whether or not such a relation was confirmed by the final periodic system. Therefore, these first predictions should not be over-estimated; indeed, their theoretical basis could not be verified. When the existence of triads and several other relations between elements were confirmed by the periodic system—though not as mathematically exact relations—predicting became more successful. Newlands and Meyer, but especially Mendeleev, were the most important predictors of unknown elements, and they also used horizontal relationships between elements in the periodic system. That wrong predictions were also made was due to incorrect atomic weights or faulty groupings within the system (for example, the classification of the rare earths).

Just after the discovery of the periodic system, supplementary predictions were made. Most of these were confirmed, e.g. scandium, gallium, germanium, francium, astatine, rhenium and technetium. Incorrect incorporation of the rare earth metals sometimes led to incorrect prediction: the position of hafnium as a homologue of zirconium appeared to be practically unpredictable.

The predictions of elements with atomic weights less or a little more than that of hydrogen form a separate category. Some of these predictions were based on spectral lines of the sun and other celestial bodies, which also led to theories on the evolution of matter. On the other hand, some investigators concluded from the symmetry of the periodic system that the first period must also contain seven elements, and therefore assumed six more unknown elements between hydrogen and lithium. Both these methods of prediction were used by Mendeleev. It would appear that these ideas of Mendeleev's have not produced important contributions, just as his later speculations concerning world ether failed to do.

After the acceptance of the atomic number, incorrect predictions gradually ceased to appear. The discovery of artificial radioactivity (1919), however, as well as the synthesis of the first transuranium elements, did give rise to a number of unfounded speculations.

References

1 J. B. RICHTER, *Ueber die neuen Gegenstände der Chymie*, Breslau (Wroclaw)–Hirschberg–Lissa, Vol. VII (1796) p. 102 ff.; Vol. VIII (1797) p. 105 ff.; Vol. XI (1802).
2 J. W. VAN SPRONSEN, *Chem. Weekblad*, 60 (1964) 157.
3 J. A. R. NEWLANDS, *Chem. News*, 7 (1863) 70.
4 J. A. R. NEWLANDS, *ibid.*, 10 (1864) 59.
5 J. A. R. NEWLANDS, *ibid.*, 37 (1878) 255.
6 J. A. R. NEWLANDS, *ibid.*, 10 (1864) 94.
7 J. A. R. NEWLANDS, *ibid.*, 10 (1864) 95.
8 INQUIRER, *ibid.*, 10 (1864) 84.
9 J. A. R. NEWLANDS, *ibid.*, 10 (1864) 94.
10 L. MEYER, *Ann.*, Suppl. VII (1870) 354.
11 J. A. R. NEWLANDS., *Chem. News*, 13 (1866) 113.
12 J. A. R. NEWLANDS, *ibid.*, 13 (1866) 130.
13 STUDIOSUS, *ibid.*, 10 (1864) 11.
14 G. D. HINRICHS, *Proc. Am. Ass. for Advancement of Science*, 18 (1869) 112.
15 *Das natürliche System der chemischen Elemente, Abhandlungen von Lothar Meyer und D. Mendelejeff*, ed. by K. Seubert, Ostwald's Klassiker No. 68, Leipzig (1895).
16 D. MENDELEJEFF, *Ann.*, Suppl. VIII (1871) 133.
17 P. E. LECOQ DE BOISBAUDRAN, *Compt. Rend.*, 81 (1875) 493.
18 D. MENDELEJEFF, *Ber.*, 8 (1873) 1680.
19 A. WURTZ, *Chem. News*, 43 (1881) 15.
20 See e.g. J. W. VAN SPRONSEN, *Chem. Weekblad*, 47 (1951) 55; 48 (1952) 25.
21 D. MENDELEJEFF, *J. Chem. Soc.*, 55 (1889) 634.
22 B. BRAUNER, *Z. Anorg. Chem.*, 32 (1902) 1.

23 B. Brauner, *Chem. News*, 59 (1889) 295.

24 M. Carey Lea, *Chem. News*, 73 (1896) 203.

25 M. Carey Lea, *Z. Anorg. Chem.*, 12 (1896) 249.

26 D. Mendeléeff, *Principles of Chemistry*, London 1891, Vol. II p. 55.

    D. Mendelejeff, *Grundlagen der Chemie*, St. Petersburg–Leipzig 1892, p. 727.

27 T. Dahl, *Ber.*, 12 (1879) 1731; 13 (1880) 250, 1861.

28 M. Berthelot, *Les Origines de l'Alchimie*, Paris 1885, p. 309.

29 S. Haughton, *Chem. News*, 58 (1888) 93, 102.

30 J. E. Reynolds, *Chem. News*, 54 (1886) 1.

31 W. Preyer, *Ber. Deut. Pharm. Ges.*, 2 (1892) 144; *Das genetische System der Elemente*, Berlin 1893.

32 H. Bassett, *Chem. News*, 65 (1892) 3, 19.

33 B. D. Steele, *Chem. News*, 84 (1901) 245.

34 T. E. Thorpe, *Nature*, 111 (1923) 252.

35 G. v. Hevesy, *Ber.*, 56 (1923) 1503.

36 D. Coster, *Chem. Weekblad*, 20 (1923) 122.

37 W. Nernst, *Theoretische Chemie*, 8th–10th ed., Stuttgart 1921, p. 209.

38 S. Meyer, *Physik. Z.*, 19 (1918) 178.

39 E. Kohlweiler, *Z. Phys. Chem.*, 94 (1920) 513.

40 L. Svanberg, *Ann. Physik (Pogg.)*, 65 (1845) 317.

41 A. H. Church and H. C. Sorby, *Chem. News*, 19 (1869) 121.

42 A. H. Church, *ibid.*, 19 (1869) 142.

43 H. C. Sorby, *ibid.*, 19 (1869) 205.

44 H. C. Sorby, *ibid.*, 20 (1869) 7, 104.

45 F. D. Walker, *Nature*, 112 (1923) 831.

46 M. Ogawa, *Chem. News*, 98 (1908) 249.

47 M. Ogawa, *J. Coll. Sci. Imp. Univ. Tokyo (Japan)*, 25 (1908) art. 15.

48 G. v. Hevesy, *Kgl. Danske Videnskab. Selskab Mat.-Fys. Medd.*, 6 (1925) No. 7.

49 G. Urbain, *Compt. Rend.*, 152 (1911) 141.

50 N. Lockyer, *Inorganic Evolution as studied by Spectrum Analysis*, London 1900.

51 A. Gruenwald, *Chem. News*, 56 (1887) 232.

52 R. Nasini, F. Anderlini and R. Salvadori, *Atti Accad. Lincei* [5], 7, II (1898) 73; *Chem. Ztg.*, 22 (1898) 579.

53 L. Vegard, *Phil. Mag.* [6], 23 (1912) 211.

54 R. W. Wood, *Phil. Mag.* [6], 16 (1908) 184.

55 G. F. Becker, *Bull. Geol. Soc. Am.*, 19 (1908) 113.

56 E. C. Pickering, *Astrophys. J.*, 4 (1896) 369; 5 (1897) 92.

57 A. Fowler, *Proc. Roy. Soc. (London)* A, 90 (1914) 426.

58 D. Mendelejeff, *Prometheus*, 15 (1904) 97, 121, 129, 145.

    D. Mendelejeff, *An Attempt towards a Chemical Conception of Ether*, London 1904.

59 J. W. Nicholson, *Phil. Mag.* [6], 22 (1911) 864.

60 E. Rutherford, *ibid.*, [6], 21 (1901) 669.

61 J. J. Thomson, *ibid.*, [6], 7 (1904) 237.

62 J. W. Nicholson, *Compt. Rend.*, 158 (1914) 1322.

63 J. W. Nicholson, *Monthly Notices Roy. Astron. Soc.*, 74 (1914) 623.

64 J. W. Nicholson, *J. Chem. Soc.*, 115 (1919) 855.

65 H. Bourget, Ch. Fabry and H. Buisson, *Compt. Rend.*, 158 (1914) 1017.

66 B. K. Emerson, *Am. Chem. J.*, 45 (1911) 160.

67 I. S. Bowen, *Nature*, 120 (1927) 473.

68 A. Fowler, *ibid.*, 120 (1927) 582.

69 E. Rutherford, *ibid.*, 103 (1919) 1415.

70 F. W. Clarke, *J. Wash. Acad. Sci.*, 11 (1921) 289.

71 D. I. Mendeleev, *Zhur. Russ. Khim. Obshchestva*, 1 (1869) 60.

72 R. M. Deeley, *Chem. News*, 71 (1895) 87.

73 D. Mendeléeff, *Principles of Chemistry*, London 1891, Vol. II, p. 19.

74 G. J. Stoney, *Chem. News*, 71 (1895) 67.

75 F. Rang, *ibid.*, 72 (1895) 200.

76 W. W. Andrews, *ibid.*, 71 (1895) 235.

77 J. H. Gladstone, *ibid.*, 72 (1895) 267.

78 A. M. Stapley, *Nature*, 41 (1889) 56.

79 G. Martin, *Chem. News*, 84 (1901) 154.

80 J. H. Vincent, *Phil. Mag.* [6], 4 (1902) 103.

81 A. Minet, *Compt. Rend.*, 144 (1907) 428.

82 J. R. Rydberg, *Acta Regia. Soc. Physiograph. Lundensis, Kgl. Fysiograf. Sällskap. Lund Handl.* 24 (1912) No. 18.

83 J. R. Rydberg, *Lunds Univ. Arsskr.* [2], 9 (1913) No. 18.

84 A. van den Broek, *Phil. Mag.* [6], (1914) 630.

85 J. R. Rydberg, *Phil. Mag.* [6], 28 (1914) 144.

86 E. Saz, *Iberica*, 35 (1931) 186.

87 F. H. Loring, *Chem. Products*, (1943) 29.

88 T. W. Richards, *J. Chem. Soc.*, 99 (1911) 1201.

89 F. H. Loring and J. G. T. Druce, *Chem. News*, 131 (1925) 289, 305, 321, 337.

90 J. G. T. Druce, *ibid.*, 130 (1925) 322.

91 J. G. T. Druce, *ibid.*, 131 (1925) 273.

92 H. Stintzing, *Z. Physik*, 40 (1926) 92.

93 E. W. Washburn, *J. Am. Chem. Soc.*, 48 (1926) 2351.

94 R. Swinne, *Z. Techn. Physik*, 7 (1926) 166, 205.

95 L. A. Turner, *Phys. Rev.*, 57 (1940) 950; 58 (1940) 181.

Chapter 8

Deviation from the order of increase in atomic weight

8.1. Introduction

The periodic system includes four pairs of elements that cannot be classified according to increasing atomic weight without disturbing the arrangement by analogous properties. During the years in which the system was in course of discovery, it was observed that tellurium and iodine, and to a lesser degree cobalt and nickel, did not satisfy the requirement of increasing atomic weight. The atomic weights of the latter pair of elements differed so slightly that at first the extent of the problem was not appreciated, the more so because their properties differed so little.

As some time elapsed before argon and protactinium were discovered, the other two exceptions were, of course, not then realized. Once, however, argon was discovered (in 1894), the problem became extremely disturbing. The difficulty of even determining the atomic weight of this noble gas added greatly to the confusion. No-one knew that argon was mono-atomic; furthermore, before the discovery of neon, an atomic weight of $40:2 = 20$ seemed acceptable for argon. Three problems had to be solved simultaneously if argon was to be correctly placed among the other elements.

When protactinium was discovered in 1918, the assignment of its exact place in the periodic system no longer presented a problem, in spite of the fact that its atomic weight was less than that of the preceding element, thorium. Since 1913, when Moseley, Rydberg, and Van den Broek had demonstrated that the atomic number was the essential factor, systematization had proceeded according to atomic number rather than atomic weight, and the theoretical background had also been clarified (see Chapter 13). Much later, it was demonstrated that in all cases one of the elements of the pairs was radioactive with a long half-life. It is assumed [1] that in prehistoric times the element with the highest atomic number was the heaviest one[1]. From the relevant literature published before 1913, it is evident that this problem was a crucial matter even leading to some doubts in regard to the validity of the periodic system or, alternatively, to the correctness of the atomic weights of the elements concerned.

---

[1] The radio-active isotopes would then have been β-disintegrators, so that a proton changed into a neutron: $^{40}K \rightarrow {}^{40}Ar$; $^{59}Ni \rightarrow {}^{59}Co$; $^{129}I \rightarrow {}^{129}Te$.

References p. 244

8.2. The pair tellurium–iodine

It need hardly be said that the *Te–I problem* did not arise before the discovery of the periodic system. Until then, only series of homologous elements were considered; their mutual relationships did not enter in. Dumas, considering the latter in 1858, did not localize the problem since he compared only the elements of the fluorine and nitrogen groups and those of the magnesium and oxygen groups with each other.

Although the problem became urgent after 1862, De Chancourtois did not deal with it yet because it did not emerge from his graphical method of classification. In his *Vis tellurique* (Fig. 16, p. 99), tellurium was classified as a homologue of selenium, and iodine as a homologue of bromine. Iodine and bromine were not placed below fluorine and chlorine, however, because the atomic weights of both these elements were a few units more than, and those of iodine and bromine one unit less than, a multiple of 16.

Odling, Newlands, Hinrichs, and Meyer soon overcame the obstacle inherent in the fact that a smaller value had been determined for the atomic weight of iodine than for that of tellurium, probably because no absolute value was placed upon the practical atomic weights. Experimentation might still yield other results, as had already occurred many times before. These investigators depended upon the chemical properties of the elements concerned, which showed very distinct differences. Their belief in the periodic system was too strong to admit such an exception. Moreover, an isolated exception of this kind was not acceptable. This was also Mendeleev's opinion [2] when he discussed the point in 1869. In 1871 he considered 125 rather than 128 to be the best value for the atomic weight of tellurium [3].

Soon there were investigators who supported Mendeleev's view and did their best to confirm the assumed atomic weight experimentally. Some scientists did not doubt the first value, like Wills [4] who consequently classified tellurium after iodine in the periodic system. Brauner, who worked on the periodic system during a great part of his life, also attempted to solve the tellurium–iodine problem in 1883, after he had come to a final decision on the incorporation of beryllium (see *12.6*) and had started working on the problem of the insertion of the rare earth metals (see *10.4* and *10.6*). For tellurium he determined [5, 6] an atomic weight between 124.94 and 125.40, and considered the problem solved. A great many atomic weight determinations were necessary, however, before Brauner and others became convinced that the atomic weight of tellurium was greater than that of iodine. Brauner [7, 8] first found values between 125 and 140. With apparently very pure tellurium he determined an atomic weight of 127.64 for this element[2], after which it did not fit in the system. Brauner could only conclude that he had worked with impure tellurium. Since he could not detect a known element in his sample, he thought the impurity must have been

---

[2] At present, the atomic weight of tellurium is accepted as 127.61 on the basis of determinations made in 1932 and 1933. Brauner's value corresponds very well with this. He had really used pure tellurium and could have trusted his own measurements.

austriacum or austrium[3], which might be identical to the dvitellurium predicted by Mendeleev (see *7.6*). A few years later it was suggested that tellurium contained more than one unknown element. Even in the years around 1889, Newlands persisted in his view that tellurium had to be placed before iodine, if need be with an atomic weight of 125. He mentioned this without any supporting arguments during the discussion that took place after Brauner's lecture before the Chemical Society [10].

Immediately after the discovery of argon in 1894, Brauner [11] made an even more radical assumption. He suggested that tellurium, whose atomic weight he had himself determined at 127.71 (which was almost the correct value), was a combination of the pure element with an atomic weight of 123.4 and an element named tri-argon or argon No. 4 (Ar^4). The formula of the compound was then indicated by Brauner to be $Te_2Ar_2{}^4$, and the atomic weight determined for the impure tellurium, i.e. the molecular weight of the compound, was written as $(125.4 + 130) : 2 = 127.7$. The reason for Brauner's assumption of just this formula is not clear from his publication.

Some years earlier, in 1891, Retgers had considered the tellurium problem more closely. In one of his extensive treatises [12] on isomorphism, he argued that tellurium ought to be classified after iodine not only because of its atomic weight (128) but also because of its properties that placed this element in the eighth group between ruthenium and osmium as a homologue of these two elements. Solely on the ground that potassium tellurate ($K_2TeO_4 \cdot 5$ aq.) was isomorphous with the osmiate and not with the selenate, sulphate and chromate, Retgers drew the conclusion that tellurium should be considered as a homologue of ruthenium and osmium. This conclusion—very premature for a serious investigator—also meant that a new series of three elements was added to the transition group, of which two elements, the homologues of Rh and Pd, had not yet been discovered. An additional complication was that the former place of tellurium below selenium was left vacant, requiring another element. It may be taken as a mitigating circumstance here that the octavalence of osmium had not yet been definitely proved (see *11.3*). The very next year Retgers [13] partially retracted his conclusion by stating that tellurium might belong below selenium.

Ludwig Staudenmaier [14] (1865–1933), who later taught chemistry at the Royal Lyceum in Freising, combined the suggestions made by Brauner and Retgers by suggesting that tellurium consisted of three elements which could then fit into the transition group well; Retgers [15] was unable to agree with him.

In 1895 Retgers [16] thought he had found the solution of the tellurium–iodine problem by gathering the 6th, 7th and 8th groups of elements together in a single column (Fig. 132, p. 309). This column then contained bivalent elements including thorium. He also placed other elements together in his system without making much distinction between them, such as the rare earths, the halogens, the alkali metals, and the elements of the Cu-group, with which he also classed Tl and Hg as univalent metals. The advantage of Retgers' objective, i.e. to leave as few vacant spaces in the

---

[3] In 1886 Linnemann [9] believed that he had found this element in orthite.

system as possible, was completely lost by substituting what was only an arrangement of elements according to valence. In this new system tellurium was still classified with respect to its atomic weight, after iodine.

On the eve of its solution, the Te–I problem had reached its culminating point. The investigators were divided into two camps. One group held firmly to the definition of the periodic system by which arrangement according to increasing atomic weight is the essential criterion and was therefore compelled to doubt either the correctness of the atomic weight of tellurium or its elementary character. The other group still adhered to the correctness of the experimental data, which implied a reconsideration of the definition or perhaps of the whole periodic system of elements.

We may now examine some opinions held at that time, to elucidate the problems that could arise from only one exception to the periodic law.

Wetherell [17] was of the opinion that tellurium contained a satellite element. Tellurium itself would then have an atomic weight of 124. Wilde [18] and Seubert [19] became convinced, partly as a result of the latter's accurate atomic weight determinations which gave a value of 127.6 for tellurium and 126.85 for iodine, that the irregularity in the order of atomic weight had to be accepted; tellurium was beyond question a homologue of selenium, and iodine a homologue of bromine. Felix Alexander Gutbier (1876–1926) and Ferdinand Flury [20] (1877–1947), a pharmacist who had performed many atomic weight determinations, could not accept this and suspended the incorporation of tellurium.

Partheil [21] took a completely new approach. His study of the connection between the atomic weights of the elements and the tones in music had suggested to him that, although the value of the magnitude at.w./64 for iodine, i.e. 1.98, was smaller than that of tellurium, i.e. 1.995, the order of these elements in the periodic system had to be reversed. He noted an analogous problem by the intervals for C flat and B sharp: the order of tones C and B had to be reversed too. In spite of the striking result, little value can be put upon Partheil's theory because of its unstable foundation.

Loring [22] tried to combine his prediction about the noble gas nitron (9.5) with the excessive atomic weight of tellurium by considering the former element, together with helium, as components of tellurium.

An explanation of the occurrence of elements with an excessive atomic weight in the system was sought by some scientists in the origin of the elements from elementary matter. Curt Schmidt [23] made two suggestions. First, tellurium could have originated by polymerization after iodine, which could explain its larger mass. The anomaly persisted in the equivalent volume, as is clearly shown by a graph of the values of this property of all the elements. Secondly, the heat generated during polymerization could have been responsible for a partial decomposition of matter, leading to the lighter element iodine. However, he reserved this hypothesis for the explanation of the greater atomic weight of argon as compared to that of potassium.

Even after the introduction of the atomic number as criterion, Szymanowitz [24] put forward a similar explanation. Partly on the basis of Prout's hypothesis, he assumed the elements to be composed of hydrogen and nebulium, and irregularities in condensation accompanying the formation of the elements to cause the atomic weight of tellurium to become greater than that of iodine.

8.3. The pair cobalt–nickel

Although cobalt and nickel had been known for a long time and their atomic weights had been accurately determined, their relative position had received little attention. The small difference in atomic weight between them, which is only 0.26 units (Co = 58.95 and Ni = 58.69), probably accounts for this. Although the determinations were fairly accurate, this small difference was not considered very important. Furthermore in many cases atomic weights were rounded off, which in this instance gave both elements the same atomic weight of 59.

We shall see (11.3) that in 1870 Mendeleev was the first to place cobalt and nickel in the system in their correct order with respect to the analogy with the other elements of the eighth group. Although he did examine the obviously incorrect atomic weights of Ir, Pt and Os—elements which in this order prevented a logical arrangement of the transition group on the basis of analogy—he did not refer to the Co–Ni inversion. We must ascribe this to the above-mentioned reason, a conclusion which is justified by consideration of the atomic weights of the elements in several systems [e.g. Mendeleev, Figs. 45 and 46 (pp. 137 and 138): Co = 59, Ni = 59; Meyer, Fig. 41 (p. 132): Co = 58.6, Ni = 58.6; Wilde, Fig. 93 (p. 194): Co = 56, Ni = 56; and Bayley, Fig. 51 (p. 149): Co = 58.6, Ni = 58.6].

In 1884 Newlands [25] suggested that the resemblance between the atomic weights of cobalt and nickel might be ascribed to the configuration of the physical atoms contained in the chemical atoms, thus attaching a kind of isomerical phenomenon to the elementary particles. The same number of physical atoms, differently arranged, might form two or more distinct elements. This did little to solve the problem, especially since Newlands did not agree with the conclusion that the atomic weight of cobalt was greater than that of nickel.

In 1892 Bassett [26] broke with the tradition to assign the same atomic weight to cobalt and nickel, as a consequence of which nickel, with an atomic weight of 58.7, was placed in his system (Fig. 52, p. 150) before cobalt, which he gave an atomic weight of 59. The atomic weights of these elements had been determined in 1886 by Zimmermann [27], whose results (Co = 58.74, Ni = 58.56) were considered sufficiently reliable to permit the conclusion that the atomic weight of cobalt was really greater than that of nickel.

Krüss and Schmidt even attempted to separate other elements from elementary nickel and thought they had succeeded. They called one of their products gnomium but this "element" was later found to be a mixture [28].

Using Zimmermann's values for the atomic weights of both elements, Richards (Fig. 119, p. 269) and Werner (Fig. 54, p. 153) incorporated cobalt and nickel in 1898 and 1905, respectively. But these investigators put so much emphasis on the analogies in properties of these elements with their homologues, e.g. the isomorphism of the alumns of Co, Rh and Ir, that they placed these elements in reverse order. Loring

(Fig. 89, p. 191) and Emerson (Figs. 63 and 64, p. 165) were not convinced by this argument.

Finally, this problem too was solved in 1913.

8.4. The pair argon–potassium

No element has had such an eventful scientific history as the noble gas argon. Although the attempts at classification occupied only a relatively short period of time, this only serves to emphasize the number and farfetchedness of the suggestions and counter-suggestions (see 9.9).

Ramsay [29, 30], who had discovered argon, still adhered to his conclusion that the problem was the same as that of tellurium and iodine. Howe [31] was also convinced of this. This conclusion could not have been very difficult to arrive at, because argon clearly belonged to the series of noble gases whose members had been all discovered within the space of a few years. Furthermore, the position of potassium as an alkali metal could not be doubted.

As in the other instances we have discussed, several investigators sought for an explanation of this reversal. Wetherell [17], who also tried to solve the problems associated with tellurium (see 8.2) and beryllium (see 12.6), believed in 1904 that the atomic weight of argon could be reduced to 36 by assuming that it concluded a satellite element.

Partheil [21] made an attempt to attack the problem of the Ar–K reversal by the same approach he had applied to the Te–I problem. He pointed out that the frequency of F flat, although located higher in the octave, makes a smaller interval with C than E sharp. We have already pointed to the unscientific character of Partheil's explanation.

Schmidt's attempt [23] to explain the reversal a few years before the solution was found in 1913 has already been discussed (8.3).

8.5. The pair thorium–protactinium

This inversion was solved without great difficulty. Although some investigators had assigned a place between thorium and uranium to UX_2, called brevium (Figs. 68 and 81, pp. 170 and 184 respectively), which was separated in 1913 from the radioactive disintegration products of uranium, Hahn and Lise Meitner [32] immediately replaced brevium by protactinium, which they had discovered. In 1919 they determined its atomic weight at about 230. The reversed position of protactinium with respect to thorium, for which an atomic weight of 232.2 had been accepted, was immediately pointed out, but it was concluded at the same time that the atomic weight was no longer significant for the placement of an element in the periodic system. The accurate atomic weight of protactinium (230.6) was determined only many years later by Aristid Victor Grosse (b. 1905) [33, 34], professor of chemistry at the University of Chicago.

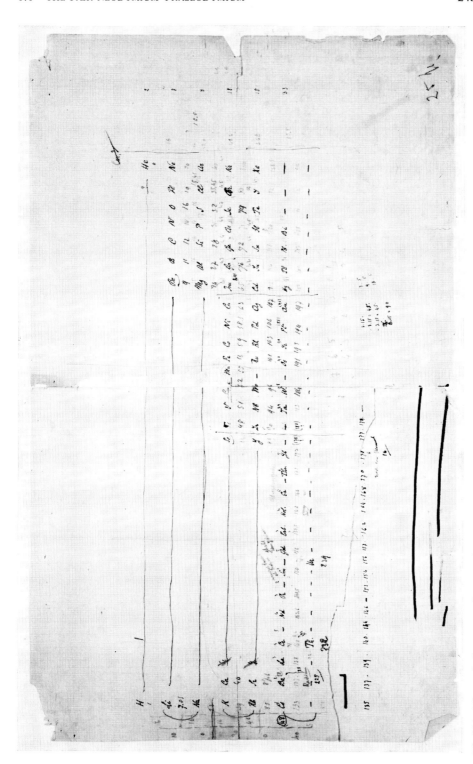

Fig.114   The periodic system of Werner handwritten by himself.

8.6. The pair neodymium–praseodymium

In Werner's system (1905; Figs. 54. p. 153, and 114, respectively) neodymium and praseodymium were arranged according to decreasing atomic weight [35]. The only reason given by Werner for this proposal consisted in the resemblance in colour of hydrated cobalt and neodymium salts, which are red and that of nickel and praseodymium, which are green. In the book [36] he published in 1913 he gave as additional arguments the regular trends of the melting points and the heats of formation of the oxides (R_2O_3) in the order La, Nd, Pr, Sm. Other regularities would demonstrate just the opposite. Although Werner did not say so explicitly, the acceptance of incorrect atomic weights may have influenced his conclusions. As we shall see (Table 4, p. 261), before 1898 the atomic weights of Pr and Nd were set at 144 and 141, respectively. In that year more accurate determinations showed that the sequence of these elements had to be reversed. There was no reason, therefore, to consider the situation with the pair Nd–Pr to be analogous with the pairs Te–I, Ar–K, and Co–Ni.

References

1 L. H. ACERA, *Ion*, 14 (1954) 78; *Chem. Abstr.*, 48 (1954) 11855.
2 *Das natürliche System der chemischen Elemente. Abhandlungen von Lothar Meyer und D. Mendelejeff*, ed. by K. SEUBERT, Ostwald's Klassiker No. 68, Leipzig 1895, p. 39.
3 *Ibid.*, p. 102.
4 W. L. WILLS, *J. Chem. Soc.*, 35 (1879) 704.
5 B. BRAUNER, *Ber.*, 16 (1883) 3055.
6 B. BRAUNER, *Zhur. Russ. Khim. Obshchestva*, 15 (1883) 433.
7 B. BRAUNER, *Monatsh.*, 10 (1889) 411.
8 B. BRAUNER, *Proc. Roy. Soc. (London)*, Abstr. 5 (1889) 94.
9 E. LINNEMANN, *Nature*, 34 (1886) 59; *Monatsh.*, 7 (1886) 121.
10 B. BRAUNER, *J. Chem. Soc.*, 55 (1889) 382; *Z. physik. Chem.*, 4 (1889) 344.
11 B. BRAUNER, *Chem. News*, 71 (1895) 196.
12 J. W. RETGERS, *Z. Physik. Chem.*, 8 (1891) 6.
13 J. W. RETGERS, *Z. Physik. Chem.*, 10 (1892) 385.
14 L. STAUDENMAIER, *Z. Anorg. Chem.*, 12 (1896) 98.
15 J. W. RETGERS, *Z. Anorg. Chem.*, 12 (1896) 98.
16 J. W. RETGERS, *Z. Physik. Chem.*, 16 (1895) 577.
17 E. W. WETHERELL, *Chem. News*, 90 (1904) 260.
18 H. WILDE, *Compt. Rend.*, 127 (1898) 613.
19 K. SEUBERT, *Z. Anorg. Chem.*, 33 (1903) 246.
20 F. A. GUTBIER, AND F. FLURY, *J. Prakt. Chem.* [2], 75 (1907) 99.
21 A. PARTHEIL, *Ber. Deut. Pharm. Ges.*, 13 (1903) 466.
22 F. H. LORING, *Chem. News*, 100 (1909) 281.
23 C. SCHMIDT, *Z. Physik. Chem.*, 75 (1911) 651.
24 R. SZYMANOWITZ, *Chem. News*, 119 (1919) 58.
25 J. A. R. NEWLANDS, *On the Discovery of the Periodic Law*, London 1884, p. 32.
26 H. BASSETT, *Chem. News*, 65 (1892) 3, 19.
27 C. ZIMMERMANN, *Ann.*, 232 (1886) 324.
28 See T. W. RICHARDS and A. S. CUSHMAN, *Chem. News*, 79 (1899) 163.
29 W. RAMSAY, *Ber.*, 31 (1898) 3111.
30 W. RAMSAY, *Rev. gén. Chim.*, 1 (1899) 49.

31 J. L. Howe, *Chem. News*, 80 (1899) 74.
32 O. Hahn and L. Meitner, *Ber.*, 52 (1919) 1812.
33 A. V. Grosse, *J. Am. Chem. Soc.*, 56 (1934) 2501.
34 A. V. Grosse, *Proc. Roy. Soc. (London)*, A 150 (1935) 363.
35 A. Werner, *Ber.*, 38 (1905) 3022.
36 A. Werner, *Neuere Anschauungen auf dem Gebiete der Anorganischen Chemie*, Brunswick (1913), p. 4.

Chapter 9

The noble gases

9.1. Introduction

Once discovered, the periodic system of elements received much scientific attention. As many as 40 publications appeared on the subject during 1895. This increased activity arose mainly from the discovery and isolation of argon and helium. The discovery of one or two new elements alone might not have caused such a stir, but these were exceptional because of a peculiar chemical inertness. There were no vacant spaces for them in Mendeleev's system or in most of the others. Moreover, it proved impossible to determine their atomic weights with certainty, this being the first encounter with monatomic elementary gases. Furthermore, as we now know, argon's atomic weight is greater than that of potassium. No-one wanted to place argon between potassium and calcium; and, because it did not fit better before potassium, several investigators thought its atomic weight should be halved. It seemed as though conjecture about the correct place in the periodic system of this and other noble gases would never end, and the elementary nature of these substances was therefore not easily accepted.

9.2. Discovery and classification of the noble gases [1]

The first element of the series of noble gases to be discovered was argon. Although Rayleigh already knew of the existence of this element in 1893, it was not until a year later that he and Ramsay were able to isolate this gas.

Helium was separated in 1895, by Ramsay alone, many years after it had been demonstrated in the solar spectrum by Lockyer *[2]* in 1868. The discoveries of neon, krypton and xenon were made by Ramsay and Travers *[3]* in 1898. At the same time, another gas was discovered and named meta-argon[1], but this gas was later found to be a mixture of argon and a carbon compound.

The last element of this group—for the time being at least—was not demonstrated until just after the turn of the century, in various modifications such as radium-, thorium-, and actinium-emanation. It was first indicated as radon, niton and emanon,

---

[1] Several investigators were confused by the announcement of this element. Howe *[4]* placed it in his system even before he could have found out how to place krypton and xenon (see *9.4*).

until officially called radon. From the very beginning, Ramsay himself [5] assumed that this gas, together with argon, had to be classified within a new group of the periodic system between the halogens and the alkali metals. He could not yet see whether the atomic weights of helium and argon should be 4 and 20 or 8 and 40, respectively. In either case the difference in atomic weight between the two elements, i.e. 16 and 32 respectively, occurred frequently in the periodic system[2]. Ramsay first wondered whether, on the analogy of the triads of the transition group, there might also be a triad at the end of the first period, of which only one element, argon, just discovered, was known [7].

Raleigh and Ramsay [8], however, believed that it would be incorrect to place argon between potassium and calcium, although that was the most likely place according to its atomic weight, 40. Taking the latter value to be correct, they were faced with the choice between doubting the validity of the periodic system and regarding argon as a mixture of two elements, e.g. with atomic weights of 37 and 82. In the latter case the element in the mixture with an atomic weight of 37 would be placed in the periodic system after chlorine, the other, with an atomic weight of 82, after bromine. With this assumption Rayleigh and Ramsay unwittingly anticipated the discovery of krypton. These elements would then have to occur in a ratio which would make their mean atomic weight 40.

Thus, the atomic weight of argon remained a controversial point. Only a few years later Ramsay [9, 10] was the first to advance the idea of relating the incorporation of argon to the tellurium–iodine inversion (see 9.9), which supplied the key to the solution of the problem. The atomic weights of neon, krypton and xenon also remained uncertain for some years after their discovery. Ramsay and Travers [11] did not arrive at a definite placement until 1901[3], in a new group between the halogens and the alkali metals. The series then appears as:

$$He(4) - Ne(20) - Ar(40) - Kr(82) - Xe(128).$$

The discoverers were not entirely certain of the elementary character of xenon; it might well contain a very small amount of another element, as an impurity.

Ramsay [1] arranged these noble gases in a system in which, after the second period of elements, each of the two parts of the system with eight groups can cover the places of the other. The incorporation of the heaviest noble gas into the system involved many difficulties. Instead of a single element there appeared to be three equally heavy gases with the character of a noble gas, viz. radium-, thorium- and actinium-emanation, all discovered within a period of a few years at the beginning of this century. Nevertheless, Ramsay [12, 13, 14] believed that each of them should be placed in the

---

[2] Ramsay in fact accepted a quite new group of elements; it cannot, therefore, be said of him, as Wightman did [6], that he only sought triads. On this point Wightman followed Travers [7], Ramsay's biographer.

[3] Not in 1908, as stated by Wightman [6].

system below xenon. Difficulties with the determination of their atomic weights were still to come, however. The values determined by several investigators for the atomic weight of radium-emanation varied from 170 to 180, with an average of about 175.

In 1908 Ramsay [15] placed this element in the system immediately below xenon where, according to him, an element with an atomic weight of 172 belonged. For reasons of analogy and atomic weight differences (about 45 units), this investigator assumed, between the period fragments Sb → Ba and Bi → Ra, a series of elements with atomic weights of 164, 169, 171, 172, 177 and 182, but these elements did not exist. Ramsay did not take the rare earth metals into account. He admitted to the period fragment Bi → Ra some four unknown elements of which he thought the one with an atomic weight of 216 to be thorium-emanation, which he proposed to call niton because of its phosphorescent properties (L. *nitens*, shimmering). Two years later [16] he proposed a detailed period fragment (atomic weights 252 → 271) on the basis of similar atomic weight differences analogous to the series Bi → Ra. In 1908 he needed this period to be able to place actinium-emanation, which he thought to have an atomic weight of 260 (in 1910: 263), as a homologue of thorium-emanation. In 1910 he believed, on the basis of Debierne's research, that the atomic weight of radium-emanation was 220. Actinium- and thorium-emanation no longer elicited much discussion.

9.3. Predictions prior to the discovery of the noble gases

Sedgwick[4] deserves the credit for having predicted the noble gases as far back as 1890. In his book *Force as an Entity* [26] he came to the conclusion, partly because of differences in atomic weight between the elements in the periods, that there were elements without valence. His reasoning was that, as the atomic weights of F (19), O (16), N (14) and C (12) differ by about two units, the element without valence would have an atomic weight of 20 in this series. In the ideal case the atomic weights of the other zero-valent elements would be 40, 80 and 120, i.e. each a multiple of 20.

Sedgwick had a curious view of atoms. He imagined the atoms of the zero-valent elements as being small balls, those of the univalent elements as flattened on one side,

---

[4] Wilde [17] contended in 1895 that the noble gases could be fitted directly into his system, too (Fig. 93, p.194). Although he explained his views (partly as the result of a discussion with Gladstone on priority; see 16.2), it does not follow, in our opinion, that one of his groups, H_n to H_{7n}, left a place appropriate for a noble gas, particularly as Wilde [18, 19] doubted the simplicity of these substances and also because the gases krypton and xenon, discovered later, did not fit into a proper place in his periodic system [20].

Deeley [21, 22, 23], too, proves to have had more than one string to his bow. He suggested different places successively for helium and argon, sometimes implying the existence of an unknown element. His systems are not periodic classifications. Wilde [24], who was not easily taken in, commented on the doubtful character of Deeley's work, saying: "A first examination of these tables left the impression that Mr. Deeley was a humorist adopting an ironical method of bringing the periodic system into ridicule". Fortunately, Wilde [25] also knew Deeley from better publications.

those of the divalent elements having two flat surfaces, and so on. Concerning the existence of entirely spherical atoms he said: "......it is quite possible that there are such atoms, and possible indeed that they are thickly distributed throughout space in the intervals between the heavenly bodies large and small, and that we are unable to perceive them because they have no angles or flat places on which force can lay a sufficiently firm hold to be able to collect them and bring them within our reach, or to enable them to offer a sensible resistance to the motion of the heavenly bodies in their orbits", which proves that Sedgwick did not expect the discovery of the zero-valent elements to take place soon. Although Sedgwick's hypothesis of zero-valent elements proved to be correct, his atomic hypothesis is not acceptable. In an article published in 1895, Sedgwick [27] referred to his prediction and showed once again that the existence of zero-valent elements could be deduced from the atomic weights of the other elements. The zero-valent, inactive elements would have the atomic weights 20, 37, 82, 130, and 5, 62, 105, 201, i.e., the atomic weights of elements of the zero and eighth group. In the years following the appearance of his book he continued to occupy himself with this problem, though he never published on the subject again.

Thomsen [28, 29] also claimed that he had predicted the inactive elements, his only reason for not publishing his prediction being that he did not want to advance another unverifiable hypothesis. In 1895, however, he considered the time had come to make his theories known. For the atomic weights of the elements concerned Thomsen had assumed 4, 20, 36, 84, 132, 212 and 292, the last of which would fall far outside the system. Thomsen represented the properties of these elements located in the system between the electropositive and electronegative elements by the $\pm \infty$ sign deriving from his theory. He arranged the elements with the atomic weights $4 \rightarrow 36$ around a

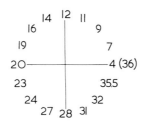

Fig. 115 Thomsen's arrangement of atomic weights.

circle (Fig. 115) and used the following formula to indicate the electrochemical nature[5] of the elements with the above-mentioned atomic weights:

$$e = \cotg \frac{a-4}{16} \pi$$

in which a represents the atomic weight ($\pi = 180°$). Thomsen thus expressed the electrochemical nature by a number that, he said, did not have an absolute meaning.

---

[5] For the valency he formulated the function:

$$v = 4f \, (\sin \frac{a-4}{16} \pi)^2 \, \text{(in which f = function sign).}$$

For the other elements he deduced similar relations. The properties of the inactive elements always represent, in the above formula, the value $\pm \infty$, which is clarified by substitution of the atomic weights. Flavitzky [30] had set up a similar relationship eight years earlier, but he did not predict the noble gases from it. Reed [31] also mentioned the fact that he had already foreseen the existence of inert elements. He contended that, in a graph dated 1885, by which he expressed the connection between valence and atomic weight [32], six spaces between the univalent metals and the univalent metalloids with the atomic weights 4, 20, 36, 84, 132 and 196 are perceptible. These are the points of intersection of the broken line indicating the connection between valence and atomic weight and the line corresponding to the valence 0. On further consideration this broken line proves to intersect the 0-axis at more then six points. It is evident that this selection made by Reed, which resulted in six atomic weights, is arbitrary.

For that matter, Reed had no plan to claim priority. He only wished to point out that the prediction made by Lecoq de Boisbaudran [33] in 1895, which was after the discovery of argon, was not original (see below), a conclusion with which we cannot agree[6].

9.4. Predictions made after the discovery of argon and helium

Although it may at first seem paradoxical, the real difficulties started only after argon had been discovered and the atomic weight of this noble gas had been determined. No space was available for an element with an atomic weight of 40 between potassium and calcium, i.e., between a univalent and a divalent element. Moreover, the zero-valence of this new element accounted for the disagreement among chemists over the exact atomic weight (20 or 40) as well as the degree of complexity of the element argon.

After argon had been discovered, Lecoq de Boisbaudran [34] predicted the atomic weights of the homologues of this element, which he believed to have an atomic weight of 20.0945. The successive atomic weights of the other elements would be 36.40 ± 0.08, 84.01 ± 0.20, and 132.71 ± 0.15. De Boisbaudran did not mention how he had calculated these values. His predicted atomic weights of krypton and xenon correspond fairly well with the present ones, which are 83.7 and 131.3, respectively. De Boisbaudran remarked that the first two elements would occur in nature, but not the others. Where the latter would be found, he did not say.

---

[6] In his turn, Lecoq de Boisbaudran argued that he, too, had foreseen the existence of some noble gases. He obtained a statement from Friedel and Moissan—in an addendum to his publication [33]—that the atomic weights 3, 19 and 20 had been mentioned by him as belonging to elements still to be discovered.

De Boisbaudran [33] thought he could predict not only perfectly accurate atomic weights for unknown elements, but could also determine atomic weights with even greater accuracy. By unspecified methods he computed the atomic weights of the eight basic elements of the periodic system accurately to seven decimal places as follows:

Be	9.0156250	O	15.9375000
Li	6.9921875	N	13.9843750
$?\beta^7$	3.8906250	C	11.9453125
$?\alpha^8$	2.9375000	B	10.9218750

Even before helium had been discovered, Lecoq de Boisbaudran had deduced its atomic weight, viz. 3.9. The atomic weight 20 for argon continued to worry him. But when he used 7.8 and 40 respectively, i.e., doubled values, for He and Ar, they fitted no better into his system (Fig. 102, p. 199). This system contains a range of so-called *nœuds* (nodal points): Ca, K, ?δ, Cl, S, P, Si and Al. The unknown element represented as ?δ might be, according to De Boisbaudran, an unknown member of the argon family. This place was not reserved for argon for the reasons mentioned at the beginning of this section. For the same reasons De Boisbaudran did not have to double the atomic weight of helium as a consequence of the doubling of the atomic weight of argon. His readiness to double the atomic weights of these newly discovered elements is explained by the doubtful atomicity (mono-, di- or tri-atomicity) of their molecules.

These elements also revived interest in Prout's hypothesis. According to De Boisbaudran, because of the differences in atomic weight, all elements must be composed of particles with a weight not exceeding 1/128th of the atomic weight of hydrogen (compare these with the precise atomic weights). An unexplained splitting into two unequal parts would give rise to an electropositive and an electronegative part, i.e. the basic substances of the metals and non-metals, respectively. De Boisbaudran also believed in the transmutation of elements in nature.

According to J. L. Howe [4] (1859–1955), professor of chemistry in the Washington and Lee University, Lexington, Virginia (U.S.A.), who had made a thorough study of inorganic chemistry in terms of the periodic system, it was quite possible that several more noble gases would be classified in the transition group, as homologues of the elements of the transition triads. He mentioned as an example the elements with atomic weights 82 and 84, one of which might correspond to that of meta-argon announced by Ramsay and Travers (see 9.2). It is curious that, although he was acquainted with the discoveries of krypton and xenon, these elements obtained no place in his system, while on the other hand meta-argon, a substance about which nothing was known, was given the place mentioned above. Howe [65] indeed referred, a few years later in 1900, to Ramsay's suggestion that krypton be included in the eighth group, and Howe himself proposed that elements with the atomic weights 130 (xenon) and 150 be placed in this group. The latter atomic weight proved to belong to the series of the rare earth metals. Howe's failure to give a definite judgment on krypton was caused by the incorrect atomic weight of 59 determined by Ladenburg and Krügl.

---

[7] Unknown member of the family of argon.
[8] Unknown member of the halogen family.

9.5. Predictions made after the discovery of the other noble gases

Uncertainties about atomic weights led to the assumption of noble gases outside the zero group; consideration was even given to an extension of this group. A splitting of a few noble gases into other elements, whether or not already discovered, was considered by several scientists to be a consequence of what they believed to be correct incorporation into the system.

We have already seen *(9.2)* that one of the discoverers, Ramsay *[16]*, assumed as late as 1910 that noble gases would still be discovered, viz., those with atomic weights of 173 and 263. Loring *[35]* was also of the opinion that it would be possible to find the former element, whose place, however, ultimately had to be given to the rare earth metal lutetium. He justified this on the basis of an atomic weight relationship which strikes us as rather obscure. In addition *[36]*, he predicted the noble gases satellite (St) and nitron (Nt), with exact atomic weights of 0.27 and 9.75, respectively. To make room for nitron in the zero group, he created two vacant spaces, one between Li and Na and one between Be and Mg; and, to classify radium-emanation with the incorrectly determined atomic weight 175, he created four vacant spaces (Fig. 89, p. 191). Moreover, Loring did not explain why his postulated atomic weight was less than 1. Three noble gases, satellite, nitron and radium-emanation were even taken as components of nitrogen, thereby exceeding the limits of Dennstedt's assumption *(9.7)* that this element in the tri-atomic state must be considered to be argon, which in itself was far-fetched enough!

Many years later, when the periodic system had already been provided with a theoretical basis and the lengths of the periods could be described by a formula, William Symes Andrews *[37]* (1847–1929), an American electrical engineer who worked briefly with Edison, predicted a noble gas as the last element of the seventh period, with an atomic number of 118 $[= 2 \times (1^2 + 2^2 + 2^2 + 3^2 + 3^2 + 4^2 + 4^2)]$. He suggested that this element, which he called hypon[9], might occur on the sun and other stars, where, as first member of a radioactive series, it might supply the great energy of these bodies by its disintegration. Andrews developed this idea in some detail in 1928, eight years before Bethe suggested the hydrogen–helium cycle.

A decade earlier, Silbermann *[39]* had postulated the existence of a parent substance of all elements. In his book *Der Weltanfang und die Bildung von Energie und Stoffen* *[40]* he tried to demonstrate that most noble gases consist of a mixture of other noble gases which must be considered to be parent substances. Thus niton (radon), for instance, would be split into helium and one or more other noble gases. Argon would be split into two elements, having atomic weights of 36.4 and 153, the latter of which in-

---

[9] A few years later, in 1931, Swinne *[38]* called this element eka-emanation. The final elements of the eighth and ninth periods, each with 50 (2×5^2) places, he called dvi-and tri-emanation, respectively (see also *7.12*). These noble gases, which would then have the atomic numbers 168 and 218, respectively, could only exist if the same irregularities occurred in the structure of the electron shells as in those of the preceding elements.

deed occurred as a member of the rare earth metals but, by virtue of its arithmetical relation (Nt + Kr) : 2, might be a noble gas. Hence, Silbermann, instead of solving it, made the problem of the noble gases even more complicated.

The hypothetical element coronium (Cn) was apparently regarded by Emerson [41] as a noble gas (Fig. 63, p. 165; see also p. 229 and 6.10), as Achimov [42] had done for the neutron (see 13.2) and Rydberg [43, 44] for the electron to which he assigned the atomic number 0.

9.6. Elements between hydrogen and helium

As we have already seen (7.11), the discovery of argon aroused speculations about the existence of elements between hydrogen and lithium. According to Stoney [45], argon might possibly be a combination of infracarbon (the hypothetical element above carbon belonging to the first period of the periodic system with an atomic weight of 2.5 to 3) and hydrogen, this combination being a compound similar to paraffin C_7H_{16}, for example, a substance which is also rather inert. In asserting at the same time that the periodic system did not need to undergo any alteration, Stoney forgot that he had introduced no less than six new elements, namely infraberyllium through infrafluorine.

W. W. Andrews' hypothesis [46] is already known to us (see p. 231). His assumption that argon was a 28-atom molecule of supraberyllium—Stoney's infraberyllium— could be deduced, in his opinion, from an extrapolation of the curves indicating the connection between the atomic weights of the elements and their atomic volumes, melting-points, and so on. By such extrapolation, the atomic weight of supraberyllium would be 1.5.

Henry Edward Armstrong [47] (1848–1937), professor of chemistry at the City and Guilds Technical College, London, encountered difficulties in fitting the newly-discovered noble gases into the system of elements for his lectures. To begin with, he started by assuming the di-atomicity of all these elements. Thus, argon could indeed be incorporated tolerably well between fluorine and sodium, but neon, krypton and xenon were left with a column inserted between beryllium and boron. Moreover, Armstrong based his system on the idea that, in principle, for every whole number there was a corresponding atomic weight of an element, which followed from his hypothesis of the complexity of the elements, which, in turn, he considered as following from the periodic system. We are more inclined to believe that he tried to adapt the periodic system to his views, to its detriment. One consequence of his views was that he had to take into account two undiscovered noble gases, which he incorporated into the system between helium and lithium, an interval that he believed had room for four elements. For a homologue of helium (atomic weight 18) he thought of a place between oxygen and fluorine.

References p. 258

9.7. Argon and helium considered as molecules of other elements

Immediately after the discovery of argon[10] and helium[11], doubts arose concerning their lack of complexity, particularly with respect to the former. In this context two different groups of investigators become apparent, namely those who saw these elements as molecules of still unknown elements, and those who thought that either argon consists of molecules of nitrogen, or helium of molecules of hydrogen.

As early as 1895, Rang *[50]* tried to solve the difficulty that argon did not fit into the eighth group because of its atomic weight, by regarding the molecules of this noble gas as tri-atomic, i.e., as containing three atoms of a still unknown element, to which he assigned the symbol A. To argon Rang assigned the formula A_3, with the

structure .

This must be considered a risky hypothesis because it created an element with an atomic weight of 13 that had to be placed between C and N, according to Rang with a valence of 4. The relation of this element A to carbon would then be similar, for example, to that of Ge to Ti, and of Cu to K (Fig. 99, p. 198). That this involved no consequences for the remaining places between the elements of the second period is due to the double filling of the fourth group. The fact that helium as a homologue of the element A was thus considered to be quadrivalent and yet another element was created in this group, presented the negative aspect which weighed all the more heavily now that in both the first and eighth groups so many elements remained to be discovered (see also *7.11*). Furthermore, it is inconsistent that only the second and third periods should have two pairs of elements with the same valence and very small differences in atomic weight, which runs counter to the basic divisions of the periodic system. Another characteristic of the system, the arrangement of elements with analogous properties in columns, is entirely lost with the insertion of helium and argon in the germanium group, and the classification according to the highest valence is also disturbed. One of Rang's undiscussed assumptions was a satellite of helium in the fifth column. The third column did include hydrogen (see *6.22*).

Now that an element with nearly the same atomic weight as that of nitrogen was required to express the tri-atomicity of argon, it was almost inevitable that a tri-atomicity of nitrogen itself should be considered to explain the existence of argon, the more so because the real atomic weight of the latter was not exactly $3 \times 13 = 39$. On the other hand, $42 = 3 \times 14$ was on the high side. Dennstedt *[51]*, director of the

---

[10] Gladstone's objection *[48]* to the (correct) atomic weight of 39.9 was that it differed by not even one unit from the atomic weights of its neighbouring elements potassium and calcium, which would constitute a precedent in the periodic system.

[11] Gladstone *[49]* examined the gas liberated from clevite, but believed that it could be demonstrated to contain two elementary substances. He wanted to place the latter, with atomic weights of 2 and 4, between H and Li.

Chemical State Laboratory at Hamburg, was the first to propose this tri-atomicity.

He compared N_3 with O_3 and gave argon the structure in which a central bond

appears to account for the stability of the noble gas. He believed he had found an analogy in organic chemistry, i.e. with pyridine. Dennstedt, whose broad practical investigations dealt especially with pyridine and pyrrole, assigned the following formula to pyridine:

This compound is comparable to one of the structural formulae considered for benzene. Brauner [52] took this view and added the molecule H_3 to represent helium[12], which in his opinion could not be incorporated into the periodic system, since its atomic weight was 2.

It is interesting to read what one of the discoverers of both argon and the periodic system of elements had to say on this point. In the years just after the discovery of argon, Ramsay [53] likewise failed to find a definite place for this element, but he apparently placed so much reliance upon his experiments that he could not accept a non-elementary or a polyatomic interpretation. We have seen (9.2) that Ramsay was the first to break through the problem by assuming a third irregularity in the increase of the atomic weight in the periodic classification.

Mendeleev [54, 55] thought it might well be that argon was polyatomic. Should this new gas be diatomic, it would as a member of the eighth group receive a place between fluorine and sodium; this group, however, included the transition metals, which have no properties in common with argon. Mendeleev might also have gone along with Dennstedt in his view of the tri-atomicity of nitrogen. Piccini [56], professor of pharmacy in the Department of Pharmacy of the University of Florence, later came to support Mendeleev's idea of the eighth group, and remarked that the eighth group with the noble gases was quite different from that of the transition metals, which began only after the third period. In 1899, when most of the noble gases had been discovered, Piccini could be more positive about his ideas than Mendeleev in 1895, when only two noble gases were known and the problems raised by argon were such as to cause Piccini even to doubt that the noble gases were elements. The colligative (i.e. binding) function, which he regarded as a very essential feature of the transition metals between the main and sub-group elements, could not be disturbed by placing other elements, such as noble gases, in the eighth group.

---

[12] Wilde [19] also advanced this idea.

9.8. Incorporation of the noble gases in the periodic system leading to the assumption of unknown elements

Soon after the discovery of argon and helium, even scientists who had done much work on the system of elements could find no way of incorporating these new gases. One of these scientists was Preyer *[57]*, who had recourse to a system with periods of the same length (Fig. 97, p. 196). As we shall see (*10.5*), Preyer included the rare earth metals as homologues of other elements, and now wished to do the same with the noble gases. The remarkable sequel to this was the acceptance of an entirely new group of elements headed by helium, as well as of other elements with atomic weights of 38, 83 and 130, and the element ytterbium. Preyer could not regard argon as the element with an atomic weight of 38, because this atomic weight infringed the progression. He included argon with an atomic weight of 20 in the transition group, which implied the existence of elements with atomic weights[13] of 21 and 22. This transition group also included samarium and two undiscovered rare earth metals. In this classification Preyer somewhat neglected the analogy of properties.

9.9. The atomic weight of argon considered in the light of the Te–I reversal[14]

One of the main difficulties associated with the incorporation of argon into the periodic system was the fact that its atomic weight was greater than that of potassium. Although it took him four years, Ramsay's clear thinking led him to associate argon with the Te–I inversion *[9, 10]*. In 1898 he was almost certain that the atomic weights of helium and argon were 4 and 40, respectively, in spite of the difficulties of the determination caused by the inert behaviour of these gases. He was therefore forced to conclude that the problem of classifying argon was similar to that of classifying tellurium and iodine. None the less, Ramsay still left open the possibility that helium and argon had atomic weights of 2 and 20. With these atomic weights both elements would fit into the system without complication, but the atomic weights of 4 and 40 would leave a space for an element with an atomic weight of about 20. It is remarkable that Ramsay did not think of neon, which he and Travers *[3]* had just discovered together with xenon and krypton. The fact that the atomic weight of neon had not yet been correctly determined was presumably the main cause of this hesitation. This seems the more likely in view of the difficulties encountered in determining the atomic weight of xenon, which was first given a value of 81 to 83 rather than the exact value of 131. Remarkably enough, krypton has an atomic weight of 83.

Howe *[4]* also saw a connection with the difficulties in placing the pairs Te–I and Co–Ni (see p. 243).

---

[13] Reynolds *[58]* had already come to this conclusion in 1895 on the basis of analogy with the periodicity of a vibrating string.

[14] See also *8.4*.

9.10. The place of the noble gases in the periodic system according to the electronic theory[15]

The formulation of the electronic theory and its experimental confirmation put an end to the difficulties with this part of the periodic system. The electronic theory gave each element its proper place in the periodic system. It appeared, however, that the noble gases whose atoms possess either a complete *s*-orbit (helium) or complete *s*- and *p*-orbits, could be arranged in two ways. In a system arranged according to electronic structure, the noble gases Ne, Ar, Kr, Xe and Rn would become members of the same column, but He would fall outside it and head the second group because of its complete *s*-orbit, which it has in common with the alkaline earths. This classification was favoured by Janet *[59]* (Fig. 128, p. 281). A few years ago, a system on the same basis was also published by the present writer *[60]* (Figure on the cover). The agreement in chemical properties provides a better basis, even though it involves a departure from the criterion that elements in the same column of the periodic system must fill up or complete the orbit of the same kind in the electron shell of their atoms.

9.11. Doubts as to the validity of the periodic system

In spite of all the difficulties caused by the insertion of the noble gases in the periodic system, few scientists lost their belief in its validity. Whenever possible, an attempt was made to preserve the law of periodicity, even by forced solutions such as the representation of argon by N_3, and so on. Nevertheless, there were also scientists who rejected the periodic system as a classification of elements. One of these was Grégoire Wyrouboff *[61]* (1843–1913), professor of chemistry in the Collège de France, who investigated the rare earths as well as isomerism and problems mainly in the field of analytical chemistry. His reason for rejecting the periodic system was the difficulty of incorporating tellurium, although his error was pointed out by Piccini *[56]*, who, however, himself believed that the noble gases could not be classified.

Nasini *[62]*, professor of chemistry at Padua, was also of the opinion that the periodic system had to be rejected, but he was not consistent because he stated in 1895 that the noble gases belonged between the halogens and the alkali metals. His compatriots Checchi and Tarugi *[63]*, the future professor of pharmaceutical chemistry at Pisa, had difficulty in placing argon even in 1901. But on the whole, the investigators who acquired much experience with the periodic system were not so pessimistic. Sooner or later they found a satisfactory solution, as did, for instance, Crookes *[64]*, who, as early as 1898, assigned a definite place to the noble gases in his three-dimensional system, even though this involved some modification of it (Fig. 75, p. 178).

---

[15] See also Chapter 12.

9.12. Conclusion

The discovery of the noble gases in the last decade of the preceding century created many problems, particularly for those with as firm a belief in the periodic system of elements as the discoverers of the noble gases, Ramsay and Rayleigh. Before it had become clear that this discovery meant the addition of a new group of elements, of which only helium and argon were known, it could still be doubted that both these gases were elements and they could be considered, for instance, as molecules (polymers) of other well-known elements, such as hydrogen and nitrogen. The discovery of argon also revived assumptions of the existence of elements with atomic weights between those of hydrogen and lithium. This left the periodic system with six still unoccupied places in its first period, a view finally supported by Mendeleev. The difficulties occasioned by argon have already been indicated (see 9.1).

Nonetheless, there were some investigators who had actually foreseen a group of elements between the most strongly electronegative elements, i.e., the halogens in group VII, and the most electropositive elements, i.e., the alkali metals in group I. Understandably enough, the other noble gases were predicted after the discovery of helium and argon. The classification of the rare earths as homologues of the other elements resulted in the incorrect assumption of a noble gas with an atomic weight of about 175. After the discovery of the noble gases, which came to many as a complete surprise, there were even those who began to doubt the periodic system—the validity of which had just been conclusively proved by this very discovery. Both the electronic theory, introduced a little more than 10 years later, and the hypothesis of the 8-electron structures, could only reinforce the demonstration of its validity.

References

1 See e.g. W. RAMSAY, *The Gases of the Atmosphere*, London 1902.
2 See e.g. N. LOCKYER, *Inorganic Evolution as studied by Spectrum Analysis*, London 1900.
3 W. RAMSAY and M. W. TRAVERS, *Proc. Roy. Soc. (London)*, 63 (1898) 405, 437.
4 J. L. HOWE, *Chem. News*, 80 (1899) 74.
5 W. RAMSAY, *Proc. Roy. Soc. (London)*, 58 (1895) 81.
6 W. P. D. WIGHTMAN, *J. Roy. Inst. Chem.*, 82 (1958) 688.
7 M. W. TRAVERS, *A Life of Sir William Ramsay*, London 1956, pp. 110, 166.
8 LORD RALEIGH and W. RAMSAY, *Z. Physik. Chem.*, 16 (1895) 344.
9 W. RAMSAY, *Ber.*, 31 (1898) 3111.
10 W. RAMSAY, *Rev. Gén. Chim.*, 1 (1899) 49.
11 W. RAMSAY and M. W. TRAVERS, *Phil. Trans. Roy. Soc.*, A 197 (1901) 47.
12 W. RAMSAY, *Einige Betrachtungen über das periodische Gesetz der Elemente*, Leipzig 1903.
13 W. RAMSAY, *Verhandl. Ges. Deut. Naturf. Aerzte*, 1 (1904) 62.
14 W. RAMSAY, *Rev. Gén. Chim.*, 6 (1903) 449.
15 W. RAMSAY, *Proc. Roy. Soc. (London)*, A 81 (1908) 178.
16 W. RAMSAY and R. W. GRAY, *Compt. Rend.*, 151 (1910) 126.
17 H. WILDE, *Chem. News*, 72 (1895) 291.
18 H. WILDE, *Compt. Rend.*, 125 (1897) 649.
19 H. WILDE, *Phil. Mag.* [5], 40 (1895) 466.
20 H. WILDE, *Compt. Rend.*, 134 (1902) 770.

21 R. M. DEELEY, *Chem. News*, 72 (1895) 297.

22 R. M. DEELEY, *ibid.*, 73 (1896) 13.

23 R. M. DEELEY, *ibid.*, 74 (1896) 278.

24 H. WILDE, *ibid.*, 73 (1896) 35.

25 R. M. DEELEY, *J. Chem. Soc.*, 63 (1893) 852; summary in *Proc. Chem. Soc.*, 9 (1903) 70.

26 W. SEDGWICK, *Force as an Entity, with Stream, Pool and Wave Forms*, London 1890.

27 W. SEDGWICK, *Chem. News*, 71 (1895) 139.

28 J. THOMSEN, *Z. Anorg. Chem.*, 9 (1895) 282.

29 J. THOMSEN, *Oversigt Kgl. Danske Videnskab, Selskab Forh.*, (1895) 137.

30 F. FLAWITZKI, *Z. Physik. Chem.*, 2 (1888) 102, transl. from *Trans. Sci. Soc. Univ. Kasan* (Russ.) 1887.

31 C. J. REED, *Chem. News*, 71 (1895) 213.

32 C. J. REED, *Trans. Acad. Sci. St. Louis*, 4 (1885) 4649.

33 P. E. LECOQ DE BOISBAUDRAN, *Compt. Rend.*, 120 (1895) 1097; summary in *Chem. News*, 71 (1895) 271.

34 P. E. LECOQ DE BOISBAUDRAN, *Compt. Rend.*, 120 (1895) 361; *Chem. News*, 71 (1895) 116.

35 F. H. LORING, *Chem. News*, 99 (1909) 148, 167.

36 F. H. LORING, *Chem. News*, 100 (1909) 120, 218.

37 W. S. ANDREWS, *Sci. Monthly*, 27 (1928) 535.

38 R. SWINNE, *Wiss. Veröffentl. Siemens-Konzern*, 10 (1931) 137.

39 T. SILBERMANN, *Ber.*, 49 (1916) 2219.

40 T. SILBERMANN, *Der Weltanfang und die Bildung von Energie und Stoffen*, Halle/Saale 1917, see e.g. p. 37.

41 B. K. EMERSON, *Am. Chem. J.*, 45 (1911) 160.

42 E. I. ACHIMOV, *Zhur. Obshcheï Khim.*, 16 (1946) 961.

43 J. R. RYDBERG, *Acta Regio Soc. Physiograph. Lundensis, Kgl. Fysiograf. Sällskap. Lund Handl.*, 24 (1913) No. 18.

44 J. R. RYDBERG, *Lunds Univ. Arsskr.* [2], 9 (1913) No. 18.

45 G. J. STONEY, *Chem. News*, 71 (1895) 67.

46 W. W. ANDREWS, *Chem. News*, 71 (1895) 235.

47 H. E. ARMSTRONG, *Proc. Roy. Soc. (London)*, 70 (1902) 86; *Chem. News*, 86 (1902) 86, 103.

48 J. H. GLADSTONE, *Nature*, 51 (1895) 389.

49 J. H. GLADSTONE, *Chem. News*, 72 (1895) 200.

50 L. T. RANG, *Chem. News*, 72 (1895) 200.

51 M. DENNSTEDT, *Chem. Ztg.*, 19 (1895) 2164.

52 B. BRAUNER, *Chem. News*, 71 (1895) 271.

53 W. RAMSAY, *Chem. News*, 73 (1896) 283.

54 See also D. MENDELEEV, *Osnovy Khimii*, 6th ed., St. Petersburg 1895, p. 754; D. MENDELÉEFF, *Principes de Chimie*, Paris 1899.

55 D. MENDELEEV, *Zhur. Russ. Khim. Obshchestva*, 27 (1895) 69.

56 A. PICCINI, *Z. Anorg. Chem.*, 19 (1899) 295.

57 W. PREYER, *Ber.*, 29 (1896) 1040; summary in *Chem. News*, 73 (1896) 235.

58 J. E. REYNOLDS, *Nature*, 51 (1895) 486.

59 See e.g. CH. JANET, *La Classification hélicoïdale des Eléments chimiques*, Beauvais 1928.

60 J. W. VAN SPRONSEN, *Chem. Weekblad*, 47 (1951) 55.

61 G. WYROUBOFF, *Actualités Chim.*, 1 (1896) 18; *Chem. News*, 74 (1896) 31.

62 R. NASINI, *Gazz. Chim. Ital.*, 25 (1895) 37; summary in *Chem. News*, 72 (1895) 247.

63 N. TARUGI and Q. CHECCHI, *Gazz. Chim. Ital.*, 31 II (1901) 417.

64 W. CROOKES, *Proc. Roy. Soc. (London)*, 63 (1898) 408; *Z. Anorg. Chem.*, 18 (1898) 72; *Chem. News*, 78 (1898) 25.

65 J. L. HOWE, *Chem. News*, 82 (1900) 15, 30, 37, 52.

Chapter 10

Incorporation of the rare earths into the periodic system[1]

10.1. Introduction

Before, and even long after the period in which the periodic system had been discovered, the rare earths were not assigned the exceptional position they are now known to possess. Before this period, when attempts were made to find series of elements with analogous chemical and/or physical properties and whose atomic weights form a simple arithmetical relation, the few rare earths then known were included as ordinary elements in homologous series. Actually, cerium was the only representative of this series of elements known with certainty. Lanthanum was discovered in 1839, and the oxides of terbium and erbium were isolated in 1843. Not before 1878 was the series expanded by the addition of the elements ytterbium and holmium. Thereafter, however, the discoveries followed each other in rapid succession (see Table 4) until lutetium, the last element in numerical order, was isolated in 1907.

As long as only a few rare earth elements were known, the classification presented no insurmountable difficulties. It may even be said that the periodic system of elements could have been set up comparatively soon after the Congress of Karlsruhe, precisely because most of the rare earths were missing. As these exhibited marked mutual analogies and only slight differences in atomic weight—the atomic weights had been by no means always exactly determined—and since the assumptions made concerning the valence were not always correct, the investigators who sought to arrive at a periodic classification would not have known what to do with some 14 of these elements, had they known of their existence. This is evident from the difficulties encountered even with the small number of known elements and from the large number of spaces kept open between the other elements for elements with atomic weights between those of cerium and tantalum. The rare earths cannot, of course, be fitted between the other elements, and later it was seen that there was no need to place them in the columns of the known elements, but we can understand that places were left open for them. At that time there were great gaps in the sequence of atomic weights.

The exceptional position occupied by the rare earths is due not only to the small differences in their characteristics, which also complicated their isolation, but also to their scarcity, as their name indicates. This factor also impeded chemical analysis, as a

---

[1] We include lanthanum in this series. Unlike the nineteenth century investigators, however, we do not include scandium and yttrium among the rare earths.

TABLE 4

DISCOVERY AND CLASSIFICATION OF THE RARE EARTH METALS

Legend:

⊕ discovered

× classified without atomic weight

□ atomic weight determination; no classification

140 (underlined) first correct classification

= after this period of no importance

○ predicted element

(148) $\ldots$

M proved to be a mixture

!) not (exactly) the correct value

2) N,Yb

result of which the correct placing of these elements was delayed to a time well beyond the discovery of the periodic system. Even after the theoretical background had been solved, most scientists continued to treat the rare earths as a separate group rather than assign them a place of their own in the system.

10.2. Classification of the rare earths before the discovery of the periodic system

Gladstone was one of the first chemists to attempt placing *all* the known elements in one system, taking as his basis the classification data of Gmelin (*4.3*) who, however, did not include any rare earths in his system. But only four elements of this series—if lanthanum is included—required placing, notably lanthanum (La), cerium (Ce), erbium (Er)[2] and terbium (Tr)[2]. Yttrium was often added to this series. The "element" didymium, to which Gladstone gave an atomic weight of 50, was later found to be a mixture. Didymium was given a place in many systems, and Mendeleev included it in his system several times, each time with a different atomic weight.

Gladstone's system (Fig. 3, p. 71) is characterized by its series of seven elements: G, Er, Y, Tr, Ce, Di, La, which, on the analogy of other series of elements, must be considered as a group of homologues. Needless to say, beryllium (G) and yttrium do not belong to the rare earths. Nevertheless, the inclusion of yttrium is understandable, because, like erbium and terbium, this element had been separated from yttria and the atomic weights of these latter elements had not yet been determined. Lanthanum which, as later theoretical considerations have shown, should not be included among the rare earths, was placed here by Gladstone and later by other investigators because of chemical properties analogous to those of other elements. Some years later, in 1857, Lenssen (*4.10*) grouped three of these elements in a triad: (La + Di) : 2 = Ce (Fig. 7, p. 81). He, too, treated yttrium as a homologue of erbium and terbium in terms of the triad he had composed with these three elements. Lenssen's belief in the triad form was so great that he used it as a basis for the prediction of atomic weights. His law of enneads (Fig. 9, p. 83), with the atomic weight of yttrium taken as 32.2, supplied the atomic weights 37 and 42 for erbium and terbium, respectively.

10.3. Incorporation of the rare earths into the periodic systems of the discoverers

Consideration of the systems developed during the era of discovery leads to the following conclusions. Lanthanum, cerium, didymium and yttrium are indeed included in the *Vis tellurique* of De Chancourtois (Fig. 16, p. 99), but this can hardly be called a systematic classification. Yttrium even occurs twice, depending on its valence (2 or 3), i.e., with atomic weights of 64 and 100; the correct atomic weight is 89, but this was

---

[2] The symbols used here are according to Gladstone.

not known until about 1870. Whenever an atomic weight is not a term of a simple arithmetical progression, this causes considerable distortion of the *Vis tellurique*.

In 1864 Odling assigned a place to cerium only, i.e., in the 3rd period between zirconium and molybdenum (Fig. 27, p. 114). The homologous elements with a smaller and a greater atomic weight are both missing. The former should have been vanadium, which, with an excessive atomic weight, now shared its place with tungsten. The second gap existed because hafnium had not yet been discovered. Cerium had also disappeared from the 1868 system (Fig. 28, p. 115).

In 1865 Newlands tried to solve the problem by placing cerium and lanthanum together, at number 33, between Sr and Zr in the series of the tervalent elements Bo, Al, Cr, Y, Ce, and La, U, Ta, Th, in which yttrium is also found (Fig. 24, p. 108). Didymium appears, together with molybdenum, a few places further in the nitrogen group.

Neither Meyer, in his first publication, nor Hinrichs classified any rare earths. Apparently, they considered the properties of these metals to be too divergent to permit of their inclusion among other elements. These two investigators were convinced that each element had its homologues[3].

Mendeleev could at most classify the rare earths, including yttrium, in pairs according to mutual analogous properties (see e.g. Fig. 39, p. 129). These elements, together with indium and thorium and an undiscovered element with an atomic weight of 45, form as it were the two "tails" of the third and fourth period, respectively. Mendeleev thus attempted to be consistent in classifying the rare earths. These elements, with the exception of indium and thorium (see 5.8), actually form a separate series. In 1869 the values taken as atomic weights were not yet correct, but just one year later Mendeleev [1] was able to classify cerium with its exact atomic weight.

In his detailed treatise of 1871 Mendeleev [2] devoted much attention to the rare earths (Fig. 45, p. 137). The old atomic weight of cerium (92) had been calculated from the formula of the oxide Ce_3O_4. A valence of 4, however, yields virtually the correct atomic weight, i.e. 138, and this also gives cerium its proper place in the carbon group. Here, it is again clear how much confidence Mendeleev had in his system. Deviating from current opinions on valence and the atomic weights derived from it[4], he gave cerium both the correct atomic weight and its proper place on the basis of its characteristics. Lanthanum was unfortunately placed erroneously. Had Mendeleev assigned it a valence of 3 instead of 4, it would have occupied the place of didymium. Mendeleev had doubled the old atomic weight of about 90. Meyer [5] was the first, in 1876, to give the correct atomic weight, viz., 139. For erbium, Mendeleev gave 3 as the possible valence and thus, with an atomic weight of 178, it was placed in the third column, ahead of cerium. Only in 1880 did erbium receive virtually its correct atomic weight (166), given by Meyer. The way in which Mendeleev arrived at the atomic weight value

---

[3] Meyer also assigned an exceptional position to nickel.

[4] In 1873 Rammelsberg [3] disagreed with Mendeleev because of the results he had obtained from the analysis of cerium salts, but Mendeleev [4] refuted his arguments on the basis of isomorphic phenomena.

References p. 282

of this element still seems strange. Multiplied by 3, the value he used in 1869 should have given $3 \times 56 = 168$, which is almost the correct value. He wanted to consider this element as a member of the group of the tervalent homologues yttrium and didymium, and therefore doubted the determined atomic weight.

Mendeleev was the first to give yttrium its correct atomic weight (88)[5], two years before Cleve and Höglund found a way to approximate the exact value (89.5); previously the wrong atomic weight of 60 had been used. Here, Mendeleev's faith in his periodic law prevailed once again. A striking fact is that, although their atomic weights were changed more or less radically, in 1871 as in 1869 the rare earths were left close together in the system, thus, however, creating 15 spaces between Ce and Er. This classification is, therefore, far superior to those constructed by Kremers and Baumhauer during the same years (see 5.9). Kremers classified only erbium and yttrium (Fig. 49, p. 140), the former with Mendeleev's old doubled atomic weight value (112.6) and the latter with the old value unchanged (61.7). These values do indeed bring erbium and yttrium into the column of the tervalent elements, but, like so many elements in this system, they are not arranged in the order of increasing atomic weight. Baumhauer, who could find no solution to the problem of the rare earths, simply used the same atomic weights for these two elements and gave both lanthanum and cerium the atomic weight of 92 (Fig. 50, p. 141).

10.4. The rare earths considered as homologues of other elements

In 1872, when he published the second edition of his textbook, even Meyer [6] did not know how to classify any of the rare earths. In his opinion, their atomic weights had not been determined with sufficient accuracy and their characteristics alone did not provide an adequate basis. To strengthen his argument, he pointed to Mendeleev's system with its many vacant places and the changes the latter had to make after 1869 in the atomic weights of 8 elements to include them in his system. The fact that most of them (including Y, Ce, In and Th) had meanwhile been given their correct atomic weights, had apparently escaped Meyer's notice. As we have already said, in 1876 Meyer [5] placed the rare earths as follows: erbium and cerium, like yttrium, in the boron group as tervalent elements, and lanthanum, now for the first time with its correct atomic weight of 139, but in the column of quadrivalent elements, in which he also expected there would come an element with an atomic weight of about 173. At first glance, the position assigned to lanthanum is incomprehensible, because Mendeleev had given as its atomic weight 94 and 180 in 1869 and 1871, respectively, the latter for quadrivalent lanthanum. The atomic weight 139 would then fit with tervalence ($\frac{3}{4} \times 180 = 135$ and $\frac{3}{2} \times 94 = 141$). This placement was based on Hillebrand's correct determination of the atomic weight of lanthanum [7] in 1876, combined with

---

[5] The exact atomic weight is 88.92.

an incorrect determination for cerium (138). Although Meyer knew that the oxides of these elements were La_2O_3 and CeO_2, respectively, he assigned more importance to the new atomic weights, of which that of lanthanum was correct[6].

The classification of these rare earths as homologues of the other elements led to several vacant places, e.g., between lanthanum and didymium, between didymium and erbium, and below lanthanum between erbium and tantalum. For some of these open spaces Meyer had specific elements in mind, these being provided with the atomic weights 165(?), 169(?) and 173(?). Other systems dating from the same period also have vacant spaces for elements still to be discovered, e.g., the system given in Graham and Otto's textbook [9] (Fig. 116).

Groups	I	2	3	4	5a	6a	7a		8		Ia	2a	3a	4a	5	6	7
Row I	Li	Be	Bo	C											N	O	Fl
Row II	Na	Mg	Al	Si											P	S	Cl
Row III	K	Ca	Sc	Ti	V	Cr	Mn	Fe	Ni	Co	Cu	Zn	Ga	–	As	Se	Br
Row IV	Rb	Sr	Yt	Zr	Nb	Mo	–	Rt	Rh	Pd	Ag	Cd	In	Sn	Sb	Te	J
Row V	Cs	Ba	La	Ce	Di	–	–	–	–	–	–	Ng	–	–	–	–	–
Row VI	–	–	Yb	–	Ta	Wo	–	Os	Ir	Pt	Au	Hg	Tl	Pb	Bi	–	–
Row VII	–	–	–	Th	–	Ur	–	–	–	–	–	–	–	–	–	–	–

Fig. 116   System in Graham and Otto's textbook (1885).

In 1878 Meyer [10] suggested that the elements La, Ce and Di might form a new group with boron as first term. This was strange, because it led to a group of analogous elements of which three had nearly the same atomic weights. This suggestion was consistent with the shifting of beryllium by Nilson and Pettersson [11] to the third group on the basis of its specific heat value, as a result of which boron became available as first element of the group (see 12.6).

In the same year Delafontaine, while carrying out chemical experiments with the rare earths, came to the conclusion that didymium was not an elementary substance; nevertheless, for many years after 1878 some investigators continued to class didymium as one of the elements. Crookes [12], for example, assumed as late as 1888, after Lecoq de Boisbaudran's separation of samarium in 1879 and Auer von Welsbach's separation of praseodymium and neodymium from didymium in 1885, that thix "element" was a homologue of niobium, even though Crookes knew of these experiments [13, 14] (Fig. 76, p. 179).

Experiments like these mark a period of discoveries of new rare earths and of atomic weight determinations for these elements (Table 4). In 1878 Marignac discovered ytterbium and in 1879 holmium was discovered by Cleve. For the former element, the correct atomic weight (172) was determined in the same year as its discovery. Meyer [15] classified ytterbium in 1880 as first term of a triad in the third group (Fig. 117; see also below). Otto [9] considered this element to be only a homologue of lanthanum (Fig. 116).

[6] Other investigators who, like Wilde [8], used the new atomic weights, also left lanthanum quadrivalent (Fig. 93, p. 194).

I	II	III			IV	V	VI	VII	VIII		
Li 7.01 _15.98_	?Be 9.3 _11.6_	B 11.0 _16.3_			C 11.97 _16_	N 14.01	O 15.96 _16.95_	F 19.1 _16.3_			
Na 22.99 _16.05_	Mg 23.94 _15.96_	Al 27.3 _17.1_			Si 28 _20_	P 30.96 _20.2_	S 31.98 _20.4_	Cl 35.37 _19.4_			
K 39.04 _24.3_	Ca 39.90 _25.0_	?Sc 45 _25_			Ti 48 _23_	V 51.2 _23.7_	Cr 52.4 _26.5_	Ma 54.8 _25.0_	Fe 55.9	Co 58.6	Ni 58.6
Cu 63.3 _21.9_	Zn 64.9 _22.3_	Ga 69.9 _19_			? 72 _18_	As 74.9 _19_	Se 78.9 _16.9_	Br 79.75 _19_			
Rb 85.2 _22.5_	Sr 87.2 _24.4_	?Y 88	X 88.5 _25_	X 89	Zr 90 _27.8_	Nb 94 _28_	Mo 95.8 _30.5_	? 99 _28_	Ru 103.9	Rh 104.1	Pd 106.2
Ag 107.66 _24.3_	Cd 111.6 _25.2_	In 113.4 _26_			Sn 117.8	Sb 122	Te 126.3	J 126.53			
Cs 132.0	Ba 136.8	La 139	Di 140	Ce 141	? 142	? 143	? 144	? 146	? 147	? 148	? 149
? 150	?Ng 150?	? 158 ?			? 160	? 162	? 164	? 165			
? 172	? 173	Yb 174	X 176	X 177	? 180	Ta 182	W 184	? 187	Os 198?	Ir 192?	Pt 195
Au 196.2	Hg 199.8	Tl 203.6			Pb 206.4	Bi 210	? 212	? 213			
? 222	? 227	X 229	X 231	X 232	?Th 233.9	? 238	?U 240				

Fig. 117. System of Meyer (1880).

Nb 94	Zr 90	Y 89
Di 146	Ce 142	La 138
Ta 182	—	Yb 173
—	Th 231	—

Fig. 118 Arrangement of the rare earth metals in Mendeleev's 1881 system.

Notwithstanding the fact that the atomic weight of holmium had been determined correctly as early as 1880, it was not placed in a system until eight years later, by Crookes [12] (Fig. 76, p. 179). This long hesitation was probably due to the fact that the element had not been obtained in a pure state. In 1886 Lecoq de Boisbaudran isolated from it a new element: dysprosium. The atomic weight of terbium was determined in 1878, but not very accurately (about 150 instead of 159). Although this element had been discovered as early as 1843, it was seldom placed in the system[7], unlike erbium, which was discovered in the same year. An atomic weight for terbium was determined, albeit incorrectly, as early as 1865 by Delafontaine. Brauner classified it in his system with another, likewise incorrect, atomic weight in 1882 (see below). Not until 1891, its correct atomic weight (159) having been determined a year earlier, was terbium given its proper place by Preyer [16]. Samarium and thulium were discovered in 1879. In the following year the atomic weight of the latter element was determined. Brauner [17] was the first to classify these elements, including gadolinium, with their correct atomic weights. Gadolinium was discovered in 1880 by Marignac, who gave it

[7] Lenssen made terbium part of a triad.

the symbol Yα. It received its name only after its rediscovery by Lecoq de Boisbaudran in 1886.

The problem of the classification of the rare earths became increasingly important, partly because chemists were no longer so preoccupied with the "beryllium affair" (see *12.6*). This does not alter the fact that some investigators, including Nilson and Pettersson [18], who were working on the determination of atomic weights, began to lose confidence in the validity of the periodic system of elements, as a result of the difficulties with both beryllium and the rare earths. Not, however, Mendeleev [19]! He continued, as he had begun in 1871, to classify the rare earths as homologues of the other elements, which led to long series of vacant spaces. Ten years later, he classified lanthanum, cerium and didymium[8] (138, 142 and 146, respectively) and ytterbium (173) as homologues of lanthanum and yttrium, elements of the third column (Fig. 118).

Meyer also followed closely the developments resulting from the discovery of the rare earths. In 1880 he removed lanthanum from the fourth group, partly on the grounds of the erroneous determination of the atomic weight of cerium [15]. Now that a better value was available for this element, he classified La (139), Di (140) and Ce (141) in the third group, which he expected to contain three more triads. The first terms of two of these triads were yttrium (88) and ytterbium (171). The other elements of these triads would have atomic weights of 88.5 and 89 and of 176 and 179 respectively. The third triad would have elements with atomic weights of 229, 231 and 232. The system suggested by Meyer (Fig 117) thus showed a rather large number of vacant spaces.

Brauner (1855–1935), in those years a young chemist inspired by Mendeleev's considerations on the periodic classification, carried out detailed investigations in the field of the rare earths, while he worked under Roscoe in Manchester, having finished his studies in Prague and Heidelberg. Brauner, unlike Nilson and Pettersson [20, 21], was convinced of the validity of the periodic system and tried to refute their arguments by new atomic weight determinations. Like Mendeleev, Brauner [17, 22] favoured the quadrivalence of cerium, which removed the difficulties connected with the sequence La (139)–Ce (141.6)–Di (146.7). These elements were classified according to their highest valence, as were terbium (148.8) and erbium (166), which brought terbium into Group III and erbium into Group V. Brauner believed that the remaining rare earths could not yet be classified because exhaustive analyses had not been performed [23]. Nevertheless, he classified these elements—Sm (150), Yα (159)[9], Tm (169) and Yb (173)—as homologues of other elements[10]. In fact, he seems to have been the first to classify almost all the known rare earths after their discovery. For this purpose he had to determine some atomic weights himself. It is also remarkable that Brauner [17, 22,

---

[8] Mendeleev was apparently not yet aware of the complexity of this element.

[9] This is gadolinium, discovered by Marignac.

[10] Long after it was known that the rare earths formed a separate group, investigators continued to treat them as homologues of the other elements, e.g. cerium as that of zirconium [24, 25].

26] should still continue to consider didymium as an elementary substance. Because of its atomic volume, he thought it had two valences: 3 and 5. Brauner determined its atomic weight as 146.58 by analysing the sulphate, which forms an instructive example of how "exact" measurements can be made on a non-existent element!

10.5. The rare earths considered as a special group

The first to become convinced that the rare earths should not be treated as homologues of other elements was Bayley *[27]*, but he did not give further consideration to the special places of the rare earths in his system (Fig. 51, p. 149).

The few elements he classified in 1882, i.e. Ce, La, Di and Er (still with their old atomic weights) do not even form an independent series, but as a precursor of Bassett, Thomsen, Werner and Bohr, he did reject insertion of the rare earths between the other elements. As a result, his system left a rather large number of spaces between bismuth and thorium, on account of his respect for symmetry, and admitted no more than 9 rare earths.

The first to classify the rare earths as a separate group and also to regard the known actinides as homologues of the rare earths, was Bassett *[28]*. When he published his system (Fig. 52, p. 150) in 1892, 12 elements of the first series and their atomic weights were known. Bassett classified 10 of these, Gd and Dy being missing. Unfortunately, it is not quite clear why he assumed 18 to be the total number of rare earths. By doing so he unnecessarily increased the number of vacant spaces in both the lanthanides and actinides series. If this implies a prediction, an element with an atomic weight of about 148 (later synthetized as promethium) would be indicated.

A few years later, Thomsen *[29]*, who was unaware of Bassett's publication, included 14 rare earths in his system. Among them was cerium, which he considered to be a homologue of zirconium. One of the two unknown elements supposed to follow ytterbium would, in Thomsen's opinion, also belong to the series of the rare earths; yet he considered this element and ytterbium itself in a certain sense to be homologues of cadmium and indium, respectively, as indicated by the dotted line between the elements in his system (Fig. 53, p. 151). Thomsen, like Bassett, left a vacant space between Pr and Sm and also between Sm and Gd for europium, to be discovered later. Unlike Bassett, however, Thomsen made no mention of a second series of rare earths. His is, nevertheless, one of the most convenient systems suggested. The fact itself that he assumed the correct number of rare earths bears out his judicious use of the known atomic weight values and their mutual differences, leaving just enough vacant spaces. Thomsen's omission of holmium and dysprosium can be attributed to his use of Clarke's atomic weights list *[30]*, in which they were not mentioned.

A few years earlier, in 1891, Preyer *[16, 31]* had suggested a systematization (Figs. 94 and 113, pp. 195 and 226) which also included precisely 14 rare earths. For the first time neodymium, praseodymium and dysprosium were assigned places. The atomic weight of the last of these was not even known; it was not determined until 1915. Dysprosium was consequently placed erroneously. The incorrect determination of the atomic weights of neodymium and praseodymium resulted in the reversal of the actual sequence of the neighbouring elements. For the first time[11] since 1857, Preyer classified terbium, which had been discovered in 1843, with its proper atomic weight[12], after gadolinium. In 1896, Preyer [32] rejected this system for a new one (Fig. 97, p. 196), but his new system contained more vacant spaces, some of which had no real significance. These spaces arose partly from Preyer's assumption of a triad with the newly-discovered argon as first term and samarium[13] as middle term, on the analogy of the transition elements triad in the iron group. The other rare earths were considered to be homologues of the remaining elements, such as niobium, molybdenum, cadmium, etc.

10.6. Systems with collective places for, or omitting, the rare earths

A few examples since the turn of the century may serve to illustrate the persistent difficulty of incorporating the rare earths into the system of elements. Richards' field of study [34] included the determination of atomic weights of many elements (Fig. 119), for which he received the Nobel Pirze for Chemistry in 1914. How-

			K =3940	Rb =8544	Cs =1329	. . . . .	. . . . .
			Ca =400	Sr =8786	Ba =13743	. . . . .	. . . . .
			Sc =440	Y =890	La =1385	. . . . .	. . . . .
			Ti =4817	Zr =905	Ce =1400	. . . . .	Th=2330
	Li = 703	Na =23050	V =514	Cb (Nb)=94		Ta=1830	. . . . .
	Gl= 91	Mg=2436	Cr =5214	Mo=960	. . . . .	W =1844	U =2400
H=10075	B =1095	Al =271	Mn=5502	. . . . .		. . . . .	
	C =12001	Si =284	Fe =560	Ru =1017	. . . . .	Os=1908	
	N =14045	P =310	Co =5900	Rh =1030		Ir =1930	
	O =16000	S =32065	Ni =5870	Pd =1065		Pt =1952	
	F =1905	Cl =35455	Cu =6360	Ag =10793	. . . . .	Au=1973	
			Zn =6540	Cd =1123	. . . . .	Hg=2000	
			Ga =700	In =1140		Tl =20415	
			Ge =725	Sn =1190	. . . . .	Pb=20692	
			As =750	Sb =1200	. . . . .	Bi =2080	
			Se =790	Te =1275?	. . . . .		
			Br =79995	I =12685	. . . . .		

Fig. 119 System of Richards (1898).

ever, he was no more successful in classifying the rare earths than Rang[14] [35] (Fig. 99, p. 198; see also 6.22) and the physico-chemist Walker [36] (1863–1935) (Fig. 120). Stoney [37] was able to classify only lanthanum, cerium and, as late as 1902, didymium, (Fig. 62, p. 164).

[11] Lenssen made this element part of one of his triads.
[12] Determined as early as 1890.
[13] Benedicks [33], who performed many experiments with gadolinium, also considered this element to be a member of the eighth group.
[14] Rang took into account the possible existence of other rare earths still to be discovered.

In 1899 Brauner [38] thought he had found a long-sought solution by creating *pleiads*, which gave the same place to Ce (140.8), Pr (140.8), and Nd (143.6). Brauner also thought this grouping explained the unpredictability of the existence of other rare earths. But he was not satisfied with this single possibility. After Steele [39] had point-

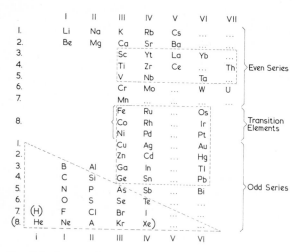

Fig. 120   System of Walker (1901).

Ce	140	Pr	141	Nd	144	-	145
-	147	Sm	148	Eu	151	-	152
-	155	Gd	156	-	159	-	160
Tb	163	Ho	165	Er	166	-	167
Tm	171	Yb	173	-	176		

Fig. 121   Arrangement of the rare earth elements by Brauner in 1902.

Groups

row	O	I	II	III	IV	V	VI	VII	VIII		
8	Xe	Cs	Ba	La	Ce	Pr	Nd	Sm	Eu	-	-
	128	132.9	137.4	139	140	141	144	150	152	-	-
9	-	-	-	Gd	Tb	Dy	Ho	Er	Tm	N Yb	-
	-	-	-	157	159	162	165	167	169	171.7	-
10	-	-	-	Lu	?	Ta	W	-	Os	Ir	Pt
	-	-	-	173.8	177	181	184	-	191	193	195

Fig. 122   Arrangement of the rare earth elements by Brauner in 1908.

ed out several years later that, on the basis of atomic volumes and magnetic susceptibility, 15 places could be occupied between cerium and tantalum, Brauner [40, 41] reexamined the problem.

A first possible solution agreed with Steele's considerations: all tervalent elements assigned to Group III. In Brauner's system this meant 19 elements, of which 11 are mentioned and the others indicated by atomic weights, distributed over 5 horizontal series (Fig. 121). Dysprosium, which had been discovered in 1886 but whose atomic weight had not yet been determined, was not included. In 1908 it was incorporated [42] with the atomic weight 162 as determined in 1905 (Fig. 122). Brauner then also classified lutetium, the last term of the series and discovered in 1907, with the atomic weight 177. The element neo-ytterbium, which Urbain thought he had separated together with lutetium from ytterbium, was classified by Brauner, with an atomic weight of 171.7 (symbol Nyb), in the eighth group as part of a small sub-system within the larger system, the sub-system being distributed over Groups III–VIII. Neo-ytterbium, however, proved to be not a new element, but pure ytterbium. One year later we also find this element (with symbol Ny) included in Loring's system [43] (Fig. 89, p. 191).

The second possibility considered by Brauner was the normal classification of these elements in the eight columns. The valence changes of these elements in their oxides do in fact suggest analogies. Naturally, the tervalence of copper implied by this arrangement must be regarded as an exception (Fig. 123).

I	II	III	IV	V	VI	VII		VIII	
Cs	Ba	La	Ce	Pr	Nd	Sm	Eu	Gd	Tb	Ho

analogous to e.g.

K	Ca	Sc	Ti	V	Cr	Mn	Fe	Co	Ni	Cu

Fig. 123   One of the classifications of the rare earth metals according to Brauner.

In a third hypothesis put forward by Brauner all the rare earths are considered to be homologues of cerium to be placed in a third dimension relative to the two-dimensional periodic system. The latter view had already been put forward by Beketov [44]. Obviously, however, this way out is very similar to Brauner's first hypothesis.

As compared to Brauner's system of 1902, that of 1908 left fewer spaces open; actually none between Nd and Sm, where promethium would later require placement. The most notable difference is that Brauner saw vertical analogies in his 1908 system (Fig. 122). For instance, he considered La, Gd and Lu as each other's homologues because of their main valence[15] of 3.

Since Brauner kept no place for element 61, the analogies between the elements of the two series into which he split the rare earths were distorted; but he maintained that lanthanum and gadolinium showed analogous properties and was therefore forced to retain vacant spaces.

Many other attempts at systematization were made during those years. Biltz [45], who later became professor of chemistry at Breslau (Wroclaw), treated lanthanum, cerium, praseodymium and neodymium as a single element, at least with respect to their place in the periodic system. These elements, indicated by the symbol Σ Ce, were

---

[15] This grouping has since been proved valid. The electron configurations $4f^0$, $4f^7$ (half-filled orbit) and $4f^{14}$ (full orbit) explain the analogies in properties (see 10.9).

placed below yttrium. Biltz assumed that other rare earths could also be classified in this manner. Marc [46] who also concentrated on the investigation of the rare earths, concurred with Biltz's views and classified Tb (157?), Er, Ho (about 164) and Yb (173), with the symbol Σ Er, as homologues of Σ Ce in the group of tervalent elements, adding the comment that he was not quite certain about the tervalence of cerium: as a quadrivalent element it would fit better into the carbon group. Muthmann and Weiss [47] also argued in favour of the inclusion of Sc, Y and La in the Ce-pleiad.

Benedicks [48] thought a modification of Biltz's classification was required. He distributed the rare earths over the two places in the 3rd and 4th columns. This approach to the grouping of the rare earths was in fact due to an underestimation of the utility of the periodic system, which underestimation was justifiably criticized by Rudorf [49]. Other elements could be similarly classified, but this introduced some arbitrary factors in the arrangement of elements.

10.7. The number of elements in the rare earths series, according to Thomsen, Brauner and Werner

Before the introduction in 1913 of the atomic numbers, which were based on the theories and experiments of Rydberg, Moseley and Bohr, it was not known that there are 14 rare earths. Around the turn of the century, some of the elements of this series had not yet been discovered. The last element, promethium (atomic number 61), was not synthetized until as recently as 1947, long after Hopkins, Harris and Yntema had thought, in 1926, that they had discovered illinium in the X-ray spectrum of monazite sands as the element missing between neodymium (60) and samarium (62).

Having discussed the positional assignment of the rare earths by the leading investigators of the last century, i.e. Bayley, Bassett, Thomsen, Crookes and Preyer (*10.5*), we can now turn to the results achieved by several investigators of the twentieth century. As we have seen, Thomsen took the number of the rare earths to be 14 (Fig. 53, p. 151), assuming that five elements remained to be discovered[16]. To provide an impression of Brauner's, Thomsen's and Werner's conception [50] of the series of the rare earths, Fig. 124 shows their arrangements together with the present evaluation. All three investigators took into account some undiscovered elements. As we have seen, Brauner's view in 1908 differed from that held in 1902. After first assuming 19 elements, in 1908 he reduced this number to 18, three of which were still undiscovered.

One outstanding difference is that Thomsen and Brauner considered cerium to be the first member, whereas Werner began the series of the rare earths with lanthanum (Fig. 54, p. 153). The latter conception implies that an assumed element coming directly

---

[16] Closer consideration leads to the conclusion that Thomsen in fact assumed only 11 elements: he took the element cerium to be a homologue of zirconium, and regarded ytterbium and the next unknown element as homologous to cadmium and indium, respectively.

	Ce	Pr	Nd	Pm (1947)[*]	Sm	Eu (1901)[*]	Gd	Tb	Dy	Ho	Er	Tu	Yb	Lu (1907)[*]
Thomsen 1895	Ce	Nd	Pr	-	Sm	-	Gd	Tb	-	Er	-	Tu	Yb	-
Brauner 1902	Ce	Pr	Nd	- -	Sm	Eu --	Gd --	Tb	Ho	Er	-	Tm	Yb	-
Brauner 1908	Ce	Pr	Nd		Sm	Eu --	Gd	Tb	Dy	Ho	Er	Tm	NYb -	Lu
Werner 1905	La Ce	Nd	Pr	- -	Sa	Eu	Gd	Tb	Ho	Er	-	Tu	Yb	

[*] here only the years of discovery of the elements discovered after 1895 have been mentioned

Fig. 124   The number of rare earth metals in the systems of Thomsen, Brauner and Werner.

after ytterbium would have to be considered a homologue of yttrium. Werner accepted this view. The reversal of the order of neodymium and praseodymium, inexplicably favoured by Werner, appears to have been adopted by Gutbier and Flury [51]. Werner seems to have expected two more members between praseodymium and samarium and one between erbium and thulium. In his handwritten system (Fig. 114, p. 243) the two members were put between cerium and neodymium and between praseodymium and samarium, respectively.

R. J. Meyer [52], who investigated the rare earths, concluded that "die Gruppe der seltenen Erde ein kleines periodisches System für sich bildet, in dem alle Beziehungen des Hauptsystems im kleinem nachgebildet sind". This most striking statement was of course generally confirmed by spectral studies (see 10.9).

10.8. Systematization between 1913 and 1922

In 1913, after Bohr had formulated his atomic theory and Rydberg, Van den Broek and Moseley[17] had established the atomic number as the primary systemic factor[18], the problem of the rare earth metals had been essentially solved. That some investigators continued to deny the importance of these elements, and not only out of mere conservatism, requires explanation. Although the structure of the atom was understood in the light of Bohr's atomic theory, the distribution of the electrons was not yet known for all the elements, including the rare earth metals. Moseley [53, 54], too, had not yet included all the known elements in his X-ray diagram. As to the rare earth metals, an X-ray spectrum was still not available for terbium, dysprosium, thulium, ytterbium and lutetium, with the result that the sequence of elements 66 and 67, holmium and dysprosium, was established incorrectly and two elements[19] were called thulium, Tm I (69) and Tm II (70), which meant that Yb and Lu were each advanced by one place

[17] Henri Gwyn Jeffreys Moseley (1887–1915).
[18] For a discussion of this subject see 13.5.
[19] Here Moseley used data from a paper published by Auer von Welsbach [55] in 1911, claiming the demonstration of three elements as thulium on the basis of spectroscopic measurements.

References p. 282

and that no vacant space was left at No. 72. Moseley could not place keltium[20], which had been announced by Urbain in 1911; he was evidently not aware that this element and lutetium were identical. Certainly this gifted scientist would have contributed further to the development of the atomic theory, if he had not been killed at the age of 27 during the battle of the Dardanelles in 1915.

Rydberg [56] gave the correct sequence despite the fact that he based his conclusions about the number of places and the atomic weights on his formula $(4n^2)$ for the numbers of elements in a period (12.2), which caused a shift of two places (13.5) for all elements except H and He.

It was therefore clear after 1913 that the rare earth metals as a group must occupy a more or less separate place in the periodic system, as the systems of many investigators show. The relative position of these elements, however, was not assigned on a uniform basis. The small difference between the atomic weights of neodymium and praseo-dymium and the absence of elements 61 and 72 (the latter of which—the future hafnium—did not belong to the series of the rare earth metals) resulted in differences of opinion about the sequence and the number of the rare earth metals[21].

It was not until after 1922, when Bohr put forward his detailed electronic theory, that the long unexplained exceptional position of the rare earth metals could be satis-factorily elucidated. Bohr himself [57, 58, 59] immediately constructed a perfect system (Fig. 55, p. 156), based on his spectrographic measurements (see also 10.10 and 10.11). Those who designed systems could not draw on the work of Van den Broek [60], for instance, because he introduced essential modifications[22] of the system he produced in 1911 (Fig. 86, p. 188), in which the rare earth metals were still interpreted as homologues of other elements. To achieve symmetry in his system Van den Broek was forced to assume three elements as gadolinium (Gd_1, Gd_2 and Gd_3), three ele-ments as dysprosium (Dy_1, Dy_2 and Dy_3), three elements as thulium (Tu_1, Tu_2 and Tu_3) and two elements as terbium: Tb_1 and Tb_2. Evidently he, too, accepted Auer von Welsbach's reported discovery [55] of three forms of thulium.

Of the systems developed between 1913 and 1922, mention of the following[23] will suffice. Although Kohlweiler [62] did not incorporate the atomic numbers into his system numerically (Fig. 87, p. 189), he did class the elements according to increasing

---

[20] Also called celtium.

[21] An example of a peculiar grouping of elements is found in the system devised by Hackh in 1918 which, as we have already stated (6.14), must be seen not as a *periodic* classification but as an arrange-ment (Fig. 81, p. 184) according to atomic number. The series of the rare earth metals ends with the elements Ad (69), Cp (70), Yb (71) and Lu (72). The first two elements are aldebaranium and cassi-opeium, from which Auer von Welsbach thought he had separated ytterbium in 1907. Lutetium is the element isolated from ytterbium by Urbain in the same year. It soon appeared, however, that cassi-opeium and lutetium were one and the same element, and that aldebaranium was identical with ytterbium. That Hackh nevertheless arrived at a total number of 14 rare earth metals is due to the fact that he omitted thulium and gave lutetium the atomic number of 72.

[22] His 1913 system includes so many incorrect and unfilled spaces that the atomic number of ura-nium is given as 118 [61].

[23] For the literature on this subject, the reader is referred to Chapter 6.

atomic number. His result includes 14 rare earth metals. A vacant space is left for the still to be discovered promethium, but the sequences of Pr and Nd and of Dy and Ho are reversed, aldebaranium is inserted in the place of ytterbium[21], two elements occur as thulium[19] (Tm I and Tm II), and no vacant space is left for hafnium[24].

The three-dimensional system of Harkins and Hall (Fig. 67, p. 168) (see also 6.10) suggests that they regarded the rare earth metals as isotopes of other elements. This system also has peculiarities: cerium is taken as a homologue of zirconium; lanthanum, on the contrary, as a rare earth metal, and here again thulium is plit in two.

Vogel [64] represented all the rare earth metals in their proper order on a loop (Fig. 125). He visualized the whole system as helical, with both the rare earth metals and the transition elements on separate internal spirals. The groups of analogous elements are located on straight lines drawn parallel to the axis through points 0 to VIII, i.e., perpendicular to the plane of the diagram.

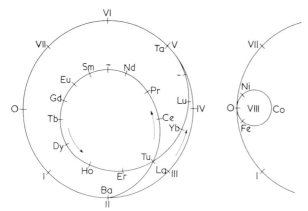

Fig. 125   Arrangement of the rare earth metals by Vogel.

Nodder [65], too, considered the rare earth metals to be a separate section of the period between the columns of elements, but between the fourth and fifth rather than the third and fourth columns (Fig. 69, p. 171). Cerium is placed in the fourth column, and no vacant space is left for hafnium. The total number of elements is, indeed, 14, but the last element, Ke (evidently keltium or celtium discovered by Urbain in 1911), seems to be identical to lutetium (discovered by him four years earlier).

Soddy and Fajans, who did such important work on the investigation into isotopes and in 1913 found the displacement law for the radioactive series (see Chapter 13), are mentioned last because they practically neglected the rare earth metals in their systems. Soddy [66] did not even place these elements in his spiral system (Fig. 72, p. 174) and, as late as 1921, Fajans [67] failed to assign to these elements a place of their own in the

[24] Nernst [63] defended a similar grouping in 1921.

periodic system. He, too, assumed a second thulium and inserted it at No. 72, later required for hafnium. Thulium II existed only on paper, and no atomic weight was ever suggested for it. In a certain sense, the long delay in the solution of the problems of the rare earth metals must be ascribed to certain mistaken "discoveries" announced by Auer von Welsbach [55], as well as to the vacant spaces left in his system for undiscovered elements in the vicinity of the three forms of thulium, for which there was no justification.

10.9. Analogies within the series of the rare earth metals

As was stated earlier (10.6), the system Brauner published (Fig. 122, p. 270) in 1908 already expressed the agreement in main valence he had found between lanthanum and gadolinium. Renz [68], who apparently took his inspiration from Soddy's system (10.8), placed the rare earth metals on two spiral rings (Fig. 126) drawn almost concentrically between the third and fourth groups, thus indicating an analogy between the first and second sets of seven elements:

La	Ce	Pr	Nd	61	Sm	Eu	Gd
	Tb	Dy	Ho	Er	Tu	Yb	
Lu							

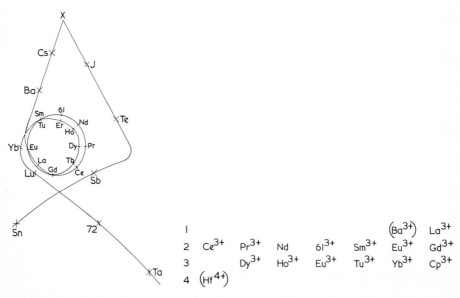

Fig. 126   Arrangement of the rare earth metals in the system of Renz.

Fig. 127   The rare earth metals arranged by Klemm.

This analogy concerned, among other things, the degree of hydrolysis and the solubilities of the salts.

At the end of the 1920's, Klemm [69] did much practical work on the magnetic properties of the rare earth metals and confirmed Brauner's suggestion, provisionally only for the ions (Fig. 127). As the structure of the atoms was by now better understood, Klemm was able to conclude that, while in the case of the lanthanum atom, the energy level was characterized by the quantum numbers $n = 4$ and $l = 3$ (i.e. the $4f$-orbit), for gadolinium the level was half occupied by 7 electrons and that of lutetium fully occupied. This explained the existence of the stable ions Eu^{2+}, Gd^{3+} and Tb^{4+}, a series which is comparable to the series Ba^{2+}, La^{3+} and Ce^{4+}, and Yb^{2+}, Lu^{3+} and Hf^{4+}. None of these ions have electrons in the two $6s$-orbits and the $5d$-orbit.

The same era also saw the first spectrographic investigations into the structure of the $4f$-orbit which, because of the extremely small differences between the wavelengths of the spectral lines, resulted in various concepts concerning the occupation of the levels.

10.10. The exceptional position of the rare earth metals[25]

The exceptional position of the rare earth metals only began to become really clear when it was found that in their atoms the fourth shell, to which no electron had been added from Ce, came to be gradually occupied.

In 1921 Bury [70], lecturer in physical chemistry at University College of Wales, Aberystwyth, elaborating Langmuir's theory[26] on the stability of 8-electron configuration [71], postulated that in the rare earth metals series the 18th (or 19th) through the 32nd electron was located in the $4f$-orbit, with which the fourth shell was completed. Thus, for instance, Ce has the configuration 2, 8, 18, 18, 8, 4, or 2, 8, 18, 19, 8, 3, and Gd 2, 8, 18, 25, 8, 3. Bury did not give a subdivision of the shells; he certainly could not know that the sixth shell has 2 instead of 3 electrons and, consequently, the fifth shell 9 instead of 8 electrons.

One year later, Bohr [57, 58, 59] was able to give some kind of subdivision of the shells into subshells on the basis of his spectroscopic measurements. He split the fourth shell into subshells[27] 4_1, 4_2, 4_3 and 4_4, on the assumption that, after each of the first three orbits had been filled with 6 electrons, subshell 4_4 was filled after lanthanum with 8 electrons. Subshells 4_1, 4_2 and 4_3 would obtain their maximum occupation with two more electrons. Bohr did not describe this process in detail (see also reference 73), but

---

[25] For the characteristics s, p, d and f, see 12.2.

[26] In 1919 Langmuir, the future Nobel prizewinner (1932) and former student of Nernst's, did not yet accept the rare earth metals as a separate group of elements, and classified Tb through Lu as homologues of Cs through Sm.

[27] Only in 1924 was the correct subdivision of the shells made as 2, 2 + 4, 4 + 6, and 6 + 8 electrons by Stoner [72], lecturer in physics at the University of Cambridge and a former student of Rutherford's.

the structural scheme was given as follows:

	4_1	4_2	4_3	4_4
Ce	6	6	6	1
Pr	6	6	6	2
↓	↓	↓	↓	↓
Cp	8	8	8	8

In 1924 the full description of a subdivision of the 4_4-subshell was provided by Main Smith [74]. In this subshell he included 14 electrons, divided into two groups, one of 8 electrons and one of 6:

Ce	Pr	Nd	—	Sa	Eu	Gd	Tb	Ds	Ho	Er	Tm	Yb[28]	Lu
1	2	3		5	(5)	6	7	8	8	8	8	8	8
					(1)	1	1	1	2	3	4	5	6

In 1926 McLennan and his co-workers [76] concluded from the spectra a regular occupation of the 4_4-orbits by 14 electrons. Goldschmidt and his group [77, 78] made a very positive contribution to the demonstration of the exceptional position occupied by the rare earth metals, by showing that the ionic radii of the lanthanides decrease with the atomic number, the so-called contraction of lanthanides.

The quantum-mechanical calculation determining the starting position of this series also dates from this decade. Using the X-ray terms of the elements, Sugiura and Urey [79] calculated cerium to be the first representative. According to calculations dating from 1941 by Mrs. Goeppert-Mayer [80], the future Nobel prizewinner, the 4f-position is first occupied at element 60 or 61. Since this problem concerns extremely small differences in the results of measurements and the subsequent calculations, the arrangements of electrons arrived at by several investigators were divergent, with the result that in some cases the series of lanthanides started with lanthanum. To illustrate this situation, some of the results are shown in Table 5. It was certain that half of the 4f-orbit was prefilled. The element europium, with which this occurs, then has increased stability in the ionic state. Cap [81] calculated the ionization energy of the atoms by the Slater quantum-mechanical method, and concluded that cerium is not a real rare earth metal.

The most probable electron distribution is the one given in the third row of Table 5 [82].

---

[28] Swinne [75], who in 1926 subdivided the 4_4-subshell into 6 + 8 orbits, preferred aldebaranium, reported by Auer von Welsbach in 1907, but later found to be identical with ytterbium, discovered by Marignac in 1878.

TABLE 5

$4f$ ELECTRON STRUCTURE OF THE RARE EARTH METALS PROPOSED BY SEVERAL INVESTIGATORS

Number	Name	Quill [a] 1935	Meggers [b] 1947	Finkelnburg [c] 1964
57	La	O		0
58	Ce	(1)	2	1
59	Pr	(2)	3	3
60	Nd	(3)	4	4
61	Pm	(4)	5	5
62	Sm	6↓	6	6
63	Eu	7	7	7
64	Gd	7	7	7
65	Tb	(8)	9	9
66	Dy	(9)	10	10
67	Ho	(10)	11	11
68	Er	(11)	12	12
69	Tm	13	13	13
70	Yb	14	14	14
71	Lu	14	14	14

( ) uncertain

[a] L. L. QUILL, *Chem. Rev.*, 23, (1938) 87.
[b] W. F. MEGGERS, *Science*, 105 (1947) 514.
[c] W. FINKELNBURG, *Structure of Matter*, New York-London 1964, p. 132.

10.11. Systematization of the rare earth metals after 1922

Bohr's system [57, 58] (Fig. 55, p. 156) of 1922, which we have considered in *10.8*, represented a perfected version of the systems composed by Thomsen (Fig. 53, p. 151) and Werner (Fig. 54, p. 153). In this system, which because of its form has also been called the *temple of the elements* [83], special places are left for both the lanthanides and the actinides, so that the properties of the elements of the two parallel series are not compared with those of other elements of the system. There is one exception to this; Bohr did not include lutetium among the lanthanides but rather, like lanthanum, among the homologues of yttrium. This left 13 rather than 14 intervals for the rare earth metals in Bohr's system, because lanthanum is considered to be a homologue of yttrium. The order required no further alteration, however. The information about isotopes and the radioactive substances, accumulated during the preceding decade, also helped to make it possible to assign each element its proper place in the periodic system.

We shall conclude by discussing a few systems in which spaces are left vacant for the rare earth metals. Many investigators and teachers put forward ingenious ideas for providing each element with the most accurate and best-founded place. For a more extensive treatment of some of these systems, the reader is referred to Chapter 6.

References p. 282

It should be mentioned in advance that, after Bohr, no essential changes were made in the periodic system. The systems now to be treated are discussed only to demonstrate the increasing attention paid to the rare earth metals. We may start with the system of Courtines (Fig. 78, p. 182). Its characteristic form has already been described, the rare earth metals being shown three-dimensionally on an accordion figure. In essence, the 14 elements are incorporated as homologues of yttrium. Later, Stedman [84] expressed these elements in a projection (Fig. 82, p. 185).

The position of the rare earth metals was given much attention by Scheele and Clark. Scheele gave them and also the actinides a proper place in his circular system (Fig. 71, p. 173). He may even have carried matters a little too far by also classifying the series of the lanthanides and actinides, like the elements of the main and secondary series, in eight groups. The valences of these elements show a certain progression but this does not make it justifiable that, for instance, in both series Group IVc contains three elements and Group VIIIc five. Scheele accentuated the analogies between the elements of the a-, b-, and c-groups in his modification of Von Antropoff's system (Fig. 60, p. 160; see also 6.9). Clark constructed an analogous system in the form of a projected "arena" (Fig. 80, p. 183). He, too, gave each element a specific place. Now, the relationship between the elements of the La-series and the Ac-series is such that Group IVc is not subdivided (containing only Ce and Th), so that the properties of the elements of Groups IVc, Vc, VIc, and VIIc are realistically compared with those of the respective a- and b-groups. Therefore, Group VIIIc, which Clark called *subperiod transition*, contains twice seven elements. Objections may be raised to the inclusion of thulium in Group Ic and ytterbium in Group IIc, as Scheele also did.

As we shall see later, when we consider the actinides, several effective systems had already been published twenty years earlier, from 1927 to 1929, by Janet [85, 86, 87, 88] (Figs. 70, 73, 74, pp. 172 and 175, respectively, and 128), in which the two series of the rare earth metals are handled consistently throughout. The system (Fig. 56, p. 157) Romanoff prepared in 1934 to celebrate the occasion of Mendeleev's centenary deserves special attention. This was a simple two-dimensional system in which each element had been given its proper place, and was used by the present author [89] as an example to bring the periodic system up to date (Figure on the cover). Fig. 77 (p. 180) in which Romanoff also wrote out the rare earth metals, is less convenient.

In Zmaczynski's fan-shaped system (Fig. 57, p. 158), constructed on the same principle (see also 6.8), the actinides are lacking but lutetium has been placed as a homologue of scandium because an unusual division into main and secondary series was used. Achimov [90] properly included the actinides as homologues of the lanthanides (Fig. 58, p. 159).

Another means of emphasizing the exceptional position of the rare earth metals was used by Le Roy [91], who placed them (shown in small type) and the transuranium elements then known at the extreme right-hand side of his system (Fig. 129). The elements of the two series therefore have very few points of contact with the other elements.

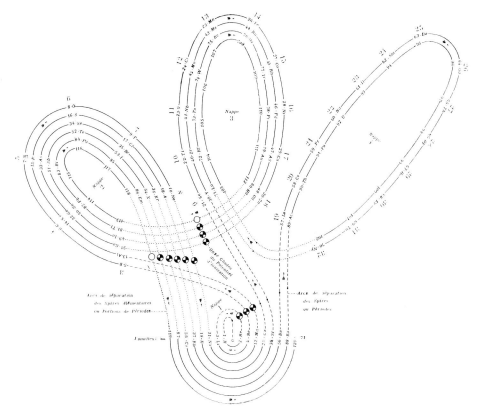

Fig. 128   System of Janet (1928).

After the appearance of the quantum theory, several systems were based on the quantum theory. Although the rare earth metals received their due place, the factor of increasing atomic weight was not applied, and these systems may therefore not be qualified as *periodic* (see, for instance, Figs. 107 and 108, p. 207). The rare earth

```
Li
Na
K     Ca   Sc
Rb    Sr   Y
Cs    Ba   La  Ce  Pr  Nd  Il   Sm  Eu  Gd  Tb  Dy  Ho  Er  Tm  Yb  Lu
?     Ra   Ac  Th  Pa  U
```

Fig. 129   Right side of Le Roy's system.

metals, as well as the other elements with a total quantum number $n = 4$, are included in the fourth period.

An example of systems in whose design the rare earth metals were considered as intermediate elements is that of Shemiakin [92] (Fig. 130), in which they are placed stepwise between periods, in some cases together with other elements. In addition to the sound classifications there were many systems in which the rare earth metals were incorrectly placed. Consideration of the latter and analysis of their error would carry us beyond the scope of this volume.

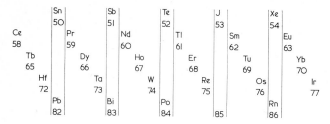

Fig. 130   Arrangement of the rare earths by Shemiakin.

10.12. Conclusion

Even before Bohr's atomic theory and before the atomic number had been substituted for the atomic weight in the construction of their periodic system, some investigators had already given the rare earths their proper place. However, an insufficient number of these elements were known before the turn of the century to permit of a definitive classification. This makes it all the more remarkable that Bassett, Bayley, Thomsen and Werner saw the rare earths as elements that were unequivocally entitled to a place of their own in the system even though they could not be classified on the basis of properties analogous to those of other elements. That the discoverers of the periodic system encountered great difficulties in their attempts at classification is hardly surprising, because in 1871 only 3 of the 14 elements had been identified. For example, didymium was proved to be a mixture as late as 1878, and even then this analytical result was not universally accepted.

After the electronic theory had been developed during the years 1913 to 1926 and all rare earth metals, except for promethium, had been discovered, however, systems of elements were still put forward in which these elements were not given the places required by their electronic structure.

References

1 D. MENDELEJEW, *Bull. Acad. Imp. Sci. (St. Petersburg)* [3], 16 (1870) 46.
2 D. MENDELEJEFF, *Ann.*, Suppl. VIII (1871) 133.
3 C. RAMMELSBERG, *Ber.*, 6 (1873) 84.
4 D. MENDELEJEFF, *Ann.*, 168 (1873) 45; *Ber.*, 6 (1873) 558.
5 L. MEYER, *Die modernen Theorien der Chemie*, 3rd ed., Breslau (Wroclaw) 1876, p. 293.
6 L. MEYER, *op. cit.*, 2nd ed., Breslau (Wroclaw) 1872, pp. 342, 345.
7 W. F. HILLEBRAND, *Ann. Physik (Pogg.)*, 158 (1876) 71.
8 H. WILDE, *Chem. News*, 38 (1878) 66, 96, 107.

9 T. GRAHAM and F. W. R. OTTO, *Lehrbuch der Chemie*, 3rd ed., Brunswick 1885, Vol. I, 2nd section, p. 162.

10 L. MEYER, *Ber.*, 11 (1878) 576.

11 L. F. NILSON and O. PETTERSSON, *Ber.*, 11 (1878) 38.

12 W. CROOKES, *J. Chem. Soc.*, 53 (1888) 487.

13 W. CROOKES, *Die Genesis der Elemente*, A lecture for the Royal Institution at London on 18 February, 1887, 2nd Germ. ed. by W. Preyer, Brunswick 1895.

14 W. CROOKES, *Chem. News*, 54 (1886) 115; 55 (1887) 83, 95.

15 L. MEYER, *Die modernen Theorien der Chemie*, 4th ed., Breslau (Wroclaw) 1880, p. 138.

16 W. PREYER, *Ber. Deut. Pharm. Ges.*, 2 (1892) 144; *Verhand. Phys. Ges. Berlin*, 10 (1891) 85; *Naturw. Wochenschr.*, 6 (1891) 523; 7 (1892) 4, 11.

17 B. BRAUNER, *Monatsh.*, 3 (1882) 1; *J. Chem. Soc.*, 41 (1882) 68; *Ber.*, 15 (1882) 109, 115.

18 L. F. NILSON and O. PETTERSSON, *Ber.*, 13 (1880) 1459.

19 D. I. MENDELEEV, *Zhur. Russ. Fiz. Khim. Obshchestva*, 13 (1881) 517; *Ber.*, 14 (1881) 2821.

20 L. F. NILSON, *Ber.*, 13 (1880) 1430, 1439.

21 L. F. NILSON and O. PETTERSSON, *Ber.*, 13 (1880) 1451.

22 B. BRAUNER, *Chem. Ztg.*, 5 (1881) 791.

23 B. BRAUNER, *Tageblatt Naturf. Ver. Salzburg*, (1881) 49; *Chem. Zentr.* [3], 13 (1882) 84.

24 G. A. BARBIERI, *Atti Accad. Lincei*, 23 I (1914) 805.

25 F. H. LORING, *Chem. Products*, 8 (1945) 54.

26 B. BRAUNER, *Bull. Soc. Chim. France* [2], 38 (1882) 176; *Chem. News*, 46 (1882) 249.

27 T. BAYLEY, *Phil. Mag.* [5], 13 (1882) 26; summary in *J. Chem. Soc.*, 42 (1882) 359.

28 H. BASSETT, *Chem. News*, 65 (1892) 3, 19.

29 J. THOMSEN, *Oversigt Kgl. Danske Videnskab. Selskab*, (1895) 132; *Z. Anorg. Chem.*, 9 (1895) 190.

30 F. W. CLARKE, *J. Am. Chem. Soc.*, 16 (1894) 179.

31 W. PREYER, *Das genetische System der Elemente*, Berlin 1893.

32 W. PREYER, *Ber.*, 29 (1896) 1040; summary in *Chem. News*, 73 (1896) 235.

33 C. BENEDICKS, *Z. Anorg. Chem.*, 22 (1900) 393.

34 T. W. RICHARDS, *Am. Chem. J.*, 20 (1898) 543; *Chem. News*, 78 (1898) 182, 193.

35 F. RANG, *Chem. News*, 72 (1895) 200.

36 J. WALKER, *Introduction to Physical Chemistry*, 2nd ed., London 1901, p. 45.

37 G. J. STONEY, *Phil. Mag.* [6], 4 (1902) 411, 504.

38 B. BRAUNER, *Verhandl. Ges. deut. Naturf. Aerzte*, 71 (1899) 131.

39 B. D. STEELE, *Chem. News*, 84 (1901) 245.

40 B. BRAUNER, *Z. Anorg. Chem.*, 32 (1902) 1.

41 B. BRAUNER, *Zhur. Russ. Fiz. Khim. Obshchestva*, 34 (1902) 142.

42 B. BRAUNER, *Z. Elektrochem.*, 14 (1908) 525.

43 F. H. LORING, *Chem. News*, 100 (1909) 281.

44 N. N. BEKETOV, *Tageblatt Russ. Naturf. Vers.*, (1902) 71.

45 H. BILTZ, *Ber.*, 35 (1902) 562, 4241.

46 R. MARC, *Ber.*, 35 (1902) 2382.

47 W. MUTHMANN and L. WEISS, *Ann.*, 331 (1904) 1.

48 C. BENEDICKS, *Z. Anorg. Chem.*, 39 (1904) 41.

49 G. RUDORF, *Das periodische System, seine Geschichte und Bedeutung für die chemische Systematik*, Hamburg–Leipzig 1904, p. 240.

50 A. WERNER, *Ber.*, 38 (1905) 914, 2022.

51 A. GUTBIER and F. FLURY, *J. prakt. Chem.* [2], 75 (1907) 99.

52 R. J. MEYER, *Naturwissenschaften*, 2 (1914) 781.

53 H. G. J. MOSELEY and C. G. DARWIN, *Phil. Mag.* [6], 26 (1913) 210.

54 H. G. J. MOSELEY, *Phil. Mag.* [6], 27 (1914) 703.

55 G. AUER VON WELSBACH, *Sitzber. Akad. Wiss. Wien Math. Naturw. Klasse*, Abt. IIb, (1911) 120, 193.

56 J. R. RYDBERG, *Lunds Univer. Arsskrift* [2], 9 (1913) No. 18; transl. in *J. Chem. Phys.*, 12 (1914) 585.

57 N. BOHR, *Drei Aufsätze über Spektren und Atomen*, Brunswick 1922, p. 132.

58 N. BOHR, *The Theory of Spectra and Atomic Constitution, three Essays*, Cambridge 1922, p. 70.

59 N. Bohr, *Fysisk Tidsskrift*, 19 (1921) 153; transl. in *Z. Physik*, 9 (1922) 1.

60 A. van den Broek, *Physik. Z.*, 12 (1911) 490.

61 A. van den Broek, *ibid.*, 14 (1913) 32.

62 E. Kohlweiler, *ibid.*, 21 (1920) 203, 311.

63 W. Nernst, *Theoretische Chemie*, 8th–10th ed., Stuttgart 1921.

64 R. Vogel, *Z. Anorg. Allgem. Chem.*, 102 (1913) 177.

65 C. R. Nodder, *Chem. News*, 121 (1920) 269.

66 F. Soddy, *The Chemistry of the Radio-elements*, Vol. II, *The Radio-elements and the Periodic Law*, London 1914.

67 K. Fajans, *Radioactivität und die neueste Entwicklung der Lehre von den chemischen Elementen*, Brunswick 1921, pp. 99, 100.

68 C. Renz, *Z. Anorg. Allgem. Chem.*, 122 (1922) 135.

69 W. Klemm, *Z. Anorg. Allgem. Chem.*, 184 (1929) 345; 187 (1930) 29; 209 (1932) 321.

70 C. R. Bury, *J. Am. Chem. Soc.*, 43 (1921) 1602.

71 I. Langmuir, *J. Am. Chem. Soc.*, 41 (1919) 868.

72 E. C. Stoner, *Phil. Mag.* [6], 48 (1924) 719.

73 H. Thirring, *Oesterr. Chem. Ztg.*, 27 (1924) 169.

74 J. D. Main Smith, *Chem. & Ind.*, 1 (1924) 323.

75 R. Swinne, *Z. Techn. Physik*, 7 (1926) 166, 203.

76 J. C. McLennan, A. B. McLay and H. Gray Smith, *Proc. Roy. Soc. (London)*, A 112 (1926) 76.

77 V. M. Goldschmidt, F. Ulrich and T. Barth, *Skrifter Norske Videnskaps-Akad. Oslo, I Mat.-Nat. Klasse*, (1925) No. 5.

78 V. M. Goldschmidt, T. Barth and G. Lunde, *ibid.*, (1925) No. 7, p. 13.

79 Y. Sugiura and H. C. Urey, *Kgl. Danske Videnskab. Selskab*, 7 (1926) No. 13.

80 M. Goeppert-Mayer, *Phys. Rev.*, 60 (1941) 184.

81 F. Cap, *Experientia*, 6 (1950) 291.

82 W. Finkelnburg, *Structure of Matter*, New York–London 1964.

83 J. Gilles, *Natuurw. Tijdschrift*, 10 (1928) 47.

84 D. F. Stedman, *Can. J. Research*, 25B (1947) 199.

85 C. Janet, *Chem. News*, 138 (1929) 372, 388.

86 C. Janet, *La Structure du Noyau de l'Atome considérée dans la Classification périodique*, Beauvais 1927.

87 C. Janet, *La Classification hélicoïdale des Eléments chimiques*, Beauvais 1928.

88 C. Janet, *Essai de Classification hélicoïdale des Eléments chimiques*, Beauvais 1928.

89 J. W. van Spronsen, *Chem. Weekblad*, 47 (1951) 55.

90 E. I. Achimov, *Zhur. Obshcheĭ Khim.*, 17 (1947) 1241.

91 R. H. Le Roy, *J. Chem. Educ.*, 8 (1931) 2052.

92 F. M. Shemiakin, *Zhur. Obshcheĭ Khim.*, 2 (1932) 62.

Chapter 11

The transition metals

11.1. Introduction

The difficulties connected with the incorporation of the elements of the eighth group, that of the transition metals, are easy to understand. These elements have many chemical and physical properties in common, but their atomic weights show relationships differing from those between the atomic weights of other elements with analogous properties. They show few differences with respect to triads, but when the regularities in these elemental attributes were first investigated, most of the atomic weights had been incorrectly determined and the discernment of a law was still far off.

Not until the periodic system had been set up and its significance with regard to the prediction of such properties as atomic weight was understood, could the transition metals be assigned a place in the system. Even so, difficulties continued to arise just because of the small differences between the atomic weights, especially as cobalt and nickel proved not to occur in the proper place according to increasing atomic weight. Some investigators regarded the eighth group not only as a transition group but also as a series of elements with a collegative (connecting) function by which the metals with a valence of 7 are bound to the univalent (half) noble metals [1, 2].

11.2. The transition elements before the discovery of the periodic system

In 1829 Döbereiner had already placed some of these elements in the triads he had set up, but not in separate triads. They are found together with other heavy metals such as manganese, chromium, copper, etc. (see *4.2*). In 1853, however, Gladstone [3] regarded the transition metals as elements belonging together. According to him, chromium and manganese also belong to one of the three groups of the category whose elements possess almost the same atomic weight (see *4.8*):

Cr	Mn	Fe	Co	Ni,
26.7	27.6	28	29.5	29.6

Pd	Rh	Ru,	and
53.3	52.2	52.2	

Pt	Ir	Os.
98.7	99	99.6

Gladstone evidently wondered whether, because of the close resemblance in properties, the atomic weights of these elements would be found to have the same value, but he was unable to obtain evidence in support of his speculations. When Lenssen [4] set up his triads in 1857, some modified atomic weights were in use, making the elements in most triads, including transition metals, occur in a different order; for example:

$$\frac{Pd + Rh}{2} = Ru \quad and \quad \frac{Os + Ir}{2} = Pt \text{ (see also Fig. 7, p. 81).}$$

11.3. The transition elements considered by the discoverers of the periodic system[1]

In the graphical representation used by Béguyer de Chancourtois (Fig. 16, p. 99) the sequence of the metals of the transition group is generally incorrect for two reasons. Firstly, the principle of classification admitted of only one distinction, viz., the recurrence of the properties of the elements after every 16 atomic weight units; secondly, the value of some atomic weights was doubtful. For instance, the assumed atomic weight for osmium had twice the actual value.

It was not until 1865 that Newlands began to place the transition metals (Fig. 24, p. 108), some of them in pairs: Co and Ni, Ro (= Rh) and Ru, and Pt and Ir. This procedure shows his uncertainty about how to deal with these elements. He classified them, like the other elements of the transition period, as homologues of the remaining elements, among them the halogens, to which they of course bear no relationship.

In 1864 Odling closely approached an ideal classification. Although the second and third triads were placed as each other's homologues, the elements Fe, Co and Ni form, together with Cu, a "tail" on the other side of his system (Fig. 27, p. 114).

Hinrichs did not consider that the elements concerned belonged to a separate group. His 1867 system (Figs. 30 and 30a, p. 120) therefore includes Fe, Rh and Ir, and Pd and Pt as homologues of other elements, i.e., Al, and Si and Ti, respectively. In 1869 Hinrichs was still inclined to regard the elements Fe, Ni and Co, together with Cr, Mn and U, as metals for which there was no valid place in the system (Fig. 34, p. 124); at that time he regarded Rh and Ir, together with Mn, as homologues of Al (Fig. 31, p. 122).

The elements that Meyer arranged in 1864 in groups outside his system according to their valence include the transition metals as quadrivalent elements. Incorrect values for most of the atomic weights, however, made it impossible for him, too, to arrive at a correct arrangement; only Ru, Rh and Pd are in their proper order. The atomic weights of these elements had been fairly accurately determined before 1862. It is therefore even more regrettable that Meyer, although he gave the correct order Fe, Co, Ni, four years later placed Ni elsewhere separately and regarded Mn as a homologue of Ru and Pt (Fig. 37, p. 128).

Almost the same arrangement was made by Mendeleev in 1869 (Fig. 39, p. 129), but he reversed the order of Rh and Ru and assigned one place to Ni and Co. He cannot

[1] For literature references, see the corresponding sections of Chapter 5.

have done so because both elements have the same atomic weight, for then Rh and Ru should also share one place; the reason is to be sought in the missing homologues of manganese, the still undiscovered elements technetium and rhenium. Consequently, like other investigators after him, Mendeleev was compelled, if he was to avoid introducing two more unknown elements, to place manganese in the transition group, thus forcing nickel and cobalt to share a single place.

In his second publication of 1869, however, Mendeleev [5] incorporated the eighth group (Fig. 131). Two years later he was in doubt about whether to include the ele-

Li	Be	B	C	N	O	F			
Na	Mg	Al	Si	P	S	Cl			
K	Ca	–	Ti	V	Cr	Mn	Fe	Co	Ni
Cu	Zn	–	–	As	Se	Br			
Rb	Sr	–	Zr	Nb	Mo	–	Rh	Ru	Pd *)
Ag	Cd	–	Sn	Sb	Te	J			
Cs	Ba	–	–	Ta	W	–	Pt	Ir	Os

*) In the original system: Pl

Fig. 131   Another system of Mendeleev (1869).

ments of the copper group in the eighth group or in that of the alkali metals. He therefore provisionally reserved two places for them (Fig. 45, p. 137). Even in 1891 Mendeleev [6] had not come to a decision. As a result of assuming an almost empty period for elements with atomic weights lying between about 140 and 180, he predicted four unknown elements in the transition group, i.e., elements belonging between the ruthenium and the osmium groups. The elements of the transition group now had their correct order, due among other things to the fact that Mendeleev then assumed a smaller atomic weight for osmium. Mendeleev remained uncertain about whether this assumption was justified. On reversing the order of the elements Os and Ir he had already [7] in 1870 actually done the same thing as Meyer, who in the same year placed Os before Ir and Pt (Fig. 38, p. 129), even though a greater atomic weight had been determined for this element. The octavalence[2] of osmium was not a factor in classification at that time, although it followed from vapour-density measurements of osmium tetroxide made in 1859 by Sainte-Claire Deville and Debray [8]. It was Mendeleev who first made the iron group a true transition group. In 1869, Mendeleev had already classified Co and Ni—which Meyer put in one place in 1870—in reverse order with respect to their atomic weight (Fig. 131). In 1872 Meyer did likewise (Fig. 41, p. 132).

These inversions by Meyer and Mendeleev were made for reasons of analogy. Later, it would be found that the atomic weights of Ir, Pt and Os had not been correctly determined. The Co–Ni inversion was for a considerable time a problem analogous to that of the Te–I and Ar–K reversal (see 8.3).

(Even before Mendeleev and Meyer, Kremers had assigned a separate place in his

---

[2] The octavalence of osmium was not definitely proved until the beginning of this century [9].

system to cobalt and nickel (Fig. 49, p. 140). However, he did not apply the principle of analogy.)

We thus see that the transition elements could not be correctly incorporated in the periodic system until its form had been established and some experience had been gained with it. This again shows how useful the system is: the importance of the numerical value of the atomic weights derived by experimentation was outweighed by the demands of a logical arrangement.

The atomic weights of iridium and platinum were given correct values ten years after the discovery of the periodic system. Seubert, who did outstanding work to prove the validity of the system, came to the correct atomic weights of these two elements experimentally, his firm belief in the validity of the system having led him to distrust the old atomic weights. In 1878 he found 193 for iridium [10] and in 1881 195 for platinum [11]. It was he, too, who in 1888 assigned the correct atomic weight to osmium [12].

11.4. The transition group and the further development of the periodic system

After the shape of the periodic system had been established and the elements of the transition group classified, most scientists accepted it. We may mention as one instance the system constructed by Bayley (Fig. 51, p. 149) in 1882, in which the intermediate character of the transition group is particularly well expressed. In the system evolved by Thomsen (Fig. 53, p. 151) in 1895, which has already been cited, the transition group shows to less advantage, and the same may be said of Richards' system[3] (Fig. 119, p. 269), although the order of the elements of this group is correct. Nevertheless, there were reasons for re-examining the problem of the transition metals. One of these reasons was the tellurium–iodine inversion, and another was the discovery of the noble gases. When Jan Willem Retgers (1856–1896), who made an intensive study of crystal structures in matter, attempted to solve the former problem in 1891, he saw no other solution but to include tellurium in the transition group [14] (see 8.2), and several years later [15] collected the 6th, 7th and 8th groups of the periodic system in a single group to which he added thorium. As a result, the transition metals no longer had a position separate from the other elements and were even no longer considered homologues of main or sub-group elements, which distorted the periodic system (Fig. 132).

Crookes [16] also confused the principle of analogy. Classifying according to magnetic properties, he took Ir–Pt–Os for the sequence of the elements of the third

---

[3] In 1911 Arthur John Hopkins [13], professor of chemistry at Amherst College, Massachusetts (U.S.A.), referred to this system. He did not understand, however, that he introduced a distortion with respect to the elements of both the transition group and the rare earth series by regarding the former and some three of the latter as homologues of silicon, which Richards did not.

transition triad. He also included both fluorine and manganese in the transition group (Fig. 75, p. 178)[4].

Two years after the discovery of krypton, but before its incorporation in the system, Howe [17], who in 1897 had published a detailed study of the transition metals, included this element in the transition group as the middle member of a triad whose extreme elements had atomic weights of 82 and 84. Elements with atomic weights of 130[5] and 150 were also postulated for this transition series. The latter triad would then fall into the rare earth series (see 9.4). By his arrangement of the transition metals and the zerovalent elements, Howe, of course, violated one of the laws essential to the periodic system of elements, namely the principle of analogy.

$\frac{O}{S}$	
Cr,	Mn,
Fe, Co,	Ni
Se	
Mo,	Ru,
Rh,	Pd
Te	
W,	Os,
Ir,	Pt
Th,	U

Fig. 132   Contracted columns of Retgers.

Around the turn of the century, many investigators regarded the incorporation of the transition metals as an unsolved problem. Hermann Christian Staigmüller [18] (1857–1908), who taught mathematics at the Stuttgart Grammar School, classed manganese in the eighth group, which isolated cobalt, without homologues, from the series of the transition metals. Biltz [19] countered this by placing manganese and iron, which certainly share properties with this element, in one place. Staigmüller's treatment caused an unusual effect in analogy: gold became a homologue of cobalt and mercury a homologue of copper and silver:

Cr	Mn	Fe	Ni	Co	Cu
Mo	Ru	Rh	Pd	—	Ag
	Os	Ir	Pt	Au	Hg[6]

It is therefore incomprehensible that Zenghelis [21], who regarded the various

[4] The two brackets below Ru and a line below Pt in the figure may imply that Crookes still expected elements in those places.

[5] Howe could here have taken xenon, which was in fact found to have this atomic weight.

[6] Sir Prefulla Chandra Rây (1861–1944) [20], professor of chemistry at Calcutta, defended this location of mercury on the grounds that in its univalent state this element shows many analogies with copper and silver, which had the same consequence of bringing gold into the transition group.

systems with great scepticism ("für den praktischen Zweck des Unterrichts zwecklos, wenn nicht geradezu schädlich"), should consider precisely this system of Staigmüller, which closely resembles Otto's *[22]* (Fig. 116, p. 265), to be the most successful attempt of all.

Biltz had another reason for making manganese and iron share the same place, namely that no homologues of manganese were known. But there was no justification for incorporating manganese in the eighth group:

Mn, Fe	Co	Ni
Ru	Rh	Pd
Os	Ir	Pt

In the heptavalent state this element could have headed a group of still undiscovered elements. Having two elements in one place was never an ideal solution, however. But this was not Biltz' only difficulty and he readily subscribed to the statement made by Ramsay and Travers *[23]* not long after they had discovered the noble gases: "......a study of this arrangement (the periodic) is a somewhat tantalising pleasure".

Reynolds' conclusion *[24]* that the manganese group[7] lacked homologues and therefore manganese had to be placed in the transition group, was also very premature, the more so as the recent discovery of the noble gases and several radioactive elements had shown that not all the elements were known. Reynolds' solution only created another problem, since it resulted in two vacant spaces below cobalt.

The discovery of isotopes also influenced the nomenclature of the transition metals. In 1913, the year in which isotopy was discovered, Büchner *[26]* (1880–1967), who later held a lectureship in inorganic chemistry at the Municipal University of Amsterdam, had represented the isotopes of niton, radium and uranium in his system as Σ Nt, Σ Ra and Σ U, respectively, and arranged the transition metals in the same manner, i.e. like the sigmads Σ Fe, Σ Ru and Σ Os, in the seventh group[8]. This artificial solution would probably have been avoided if Büchner had not reached his conclusions before the problem of isotopy was conclusively solved by Soddy in the same year. Nevertheless, Büchner did note a difference between the sigmads of the transition elements and those of the rare earth metals with the radioactive pleiads, but did not regard it as essentially significant. The difference in properties shown by the members of the iron pleiad but not by those of the uranium pleiad, was ascribed by Büchner to the relatively large value of the atomic weights of the latter elements.

A very satisfactory approach was contributed by Rita Brunetti *[27]* (1890–1942),

---

[7] Other investigators also connected the incorporation of the transition metals with the missing homologues of manganese, among them Bichowsky *[25]* (b. 1889), later professor of physics at Johns Hopkins University in Baltimore, who did so as late as 1918, seven years before rhenium was discovered.

[8] Büchner treated the rare earth metals similarly. They were gathered together in groups ΣZ^{III} and ΣZ^{IV}, as the ter- and quadri-valent homologues of yttrium and zirconium, respectively.

lecturer and future professor of experimental physics in Ferrara and Pavia. She incorporated each of the transition metals in the place to which it belonged without involving either elements of the series of the noble gases or any other elements. In this system the noble gases and the transition metals are on the extreme left:

He				Li
Ne				Na
Ar				K
	Fe	Co	Ni	Cu
Kr				Rb
	Ru	Rh	Pd	Ag
X				Cs
	Os	Ir	Pt	Au

11.5. Conclusion

Although the highest valence was one of the preferred criteria for classification, the close resemblance between these elements created insurmountable difficulties and, in addition, they formed three triads whose members differed very little in atomic weight. From the electronic theory it follows that the number of electrons outside the noble gas configuration is too large to cause them to be valence electrons in all cases. Only for osmium has a valence of 8 been demonstrated. The values of the transition metals are 2, 3 and 4 rather than 8, 9 and 10. They consequently occupy a position in the periodic system between the elements forming positive ions with the noble gas configuration and those forming such ions with an outer 18-electron structure. They thus form a natural transition group. Within this group regularities do indeed occur, i.e., with respect to the electronic structure of the metals and, in the formation of complexes, in the filling-up of empty electronic orbits.

References

1 B. BRAUNER, *Das periodische System der Elemente von Mendelejew, Erklärende Einleitung für den Gebrauch der Wandtafel*, Leipzig 1908, pp. 5, 10.
2 A. PICCINI, *Z. Anorg. Chem.*, 19 (1899) 295.
3 J. H. GLADSTONE, *Phil. Mag.* [4], 5 (1853) 313.
4 E. LENSSEN, *Ann.*, 104 (1857) 177.
5 D. MENDELEEV, *Proceedings 2nd Meeting of Scientists* (Russ.), 23 Aug. 1869, p. 62; see also B. N. MENSCHUTKIN, *Nature*, 133 (1934) 946.
6 D. MENDELEJEFF, *Principles of Chemistry*, transl. from the Russian, 5th ed., London 1891, Vol. II p. 19.
7 D. MENDELEJEW, *Bull. Acad. Imperiale Sci. (St. Petersburg)*, [3], 16 (1870) 46.
8 H. SAINTE-CLAIRE DEVILLE and H. DEBRAY, *Ann. Chim. Phys.* [3], 56 (1859) 403.
9 See e.g. F. KRAUS and D. WILKEN, *Z. Anorg. Chem.*, 137 (1924) 349.
10 K. SEUBERT, *Ber.*, 11 (1878) 1767.
11 K. SEUBERT, *Ann.*, 207 (1881) 1.

12 K. SEUBERT, *Ber.*, 21 (1888) 1839.
13 A. J. HOPKINS, *J. Am. Chem. Soc.*, 33 (1911) 1005.
14 J. W. RETGERS, *Z. Physik. Chem.*, 8 (1891) 6.
15 J. W. RETGERS, *Z. Physik. Chem.*, 16 (1895) 577.
16 W. CROOKES, *Proc. Roy. Soc. (London)*, 63 (1898) 408; *Z. Anorg. Chem.*, 18 (1898) 72; *Chem. News*, 78 (1898) 25.
17 J. L. HOWE, *Chem. News*, 82 (1900) 15, 30, 37, 52.
18 H. STAIGMÜLLER, *Z. Physik. Chem.*, 39 (1901) 245.
19 H. BILTZ, *Ber.*, 35 (1902) 562.
20 P. C. RÂY, *Chem. News*, 109 (1914) 85.
21 C. ZENGHELIS, *Chem. Ztg.*, 30 (1906) 294, 316.
22 T. GRAHAM and F. W. R. OTTO, *Lehrbuch der Chemie*, 3rd ed., Brunswick 1885, Vol. I, Part II p. 162.
23 W. RAMSAY and M. W. TRAVERS, *Phil. Trans. Roy. Soc.*, A 197 (1901) 47.
24 H. REYNOLDS, *Chem. News*, 96 (1907) 260.
25 F. R. V. BICHOWSKY, *J. Am. Chem. Soc.*, 40 (1918) 1040.
26 D. H. BÜCHNER, *Chem. Weekblad*, 12 (1915) 336.
27 R. BRUNETTI, *Nuovo Cimento* [6], 22 (1921) 1.

Chapter 12

Shape of the periodic system; main and sub-group elements

12.1. Introduction

The periodic system (Figure on inside cover) is composed of groups of analogous elements, each of which belongs to one of a group of periods. Actually, it was the latter type of arrangement of elements—perpendicular to that of the groups of analogous elements—that was responsible for most of the difficulties encountered during and after the years of discovery of the periodic system.

Only after 1913 was it appreciated that elements belonging to the same period had a common feature, viz. the inner part of their electronic structure. Admittedly, many of these periods had already been composed on the basis of chemical and physical properties before that time. The length of the periods could not be calculated, however, until the electronic structure of the atoms of all the elements was understood several years later. Until then, it remained difficult not only to distinguish the various kinds of periods but to determine also the number of elements within each period. As late as 1894 scientists were surprised by the discovery of argon, which, as the first[1] noble gas to be discovered, led to a new group of elements, thus extending the length of the short periods to eight elements.

In 1906 Rydberg [1], an authority on the atomic structure of elements, stated that the length of each period could be described by the formula $2n^2$. Seven years later [2], knowing that most periods occur doubled, he proposed the formula $4n^2$. This had grave consequences for the first period, which would then have to contain four elements! Two of these four places were filled by the electron and the hydrogen atom, leaving two empty spaces, since helium was placed in the second period (see also *13.5*). Needless to say, Rydberg's system contained 95 rather than 92 spaces for elements. Quite apart from his creation of a superfluous period, Rydberg violated the principle of periodicity when he regarded the two periods of eight elements (He through F and Ne through Cl) as one entity of $16 (= 4 \times 2^2)$, which also tore apart pairs of closely analogous elements. And these two periods of eight elements[2] were precisely those with such a long and well-established history.

As we have seen, certain elements proved to be extremely difficult to classify. In the foregoing we have discussed the noble gases, rare earths and transition elements; the

---

[1] At least of those on earth; helium had already been characterized as a solar element.
[2] Before 1894: seven elements.

actinides will be discussed in Chapter 14. Curiously enough, the position of two elements with low atomic weights, viz., hydrogen and beryllium, also posed problems that were hard to solve. Interest in these problems seems to have flared up from time to time, for no apparent reason, and then faded out again, even in the work of the same investigators.

12.2. Deduction of the lengths of the periods and the shape of the system

As shown in Chapter 6, long before the creation of Bohr's atomic theory an almost ideal system—certainly for that time—had been set up by several investigators. In their systems four types of periods can be distinguished, viz., one of 1 or 2, two of 7 or 8, two of 17 or 18 and one of about 32 elements, although these numbers had little theoretical significance. The first period to be observed was that of seven elements (law of octaves), and there was a tendency to maintain this periodicity, especially since there were relatively few exceptions to this rule. As early as 1904, Abegg [3] drew on the electronic theory to explain the fact that the highest positive and negative valence of each element was 8. After the atomic theory had been developed it was found that, in spite of the theoretical justification of the other lengths, a period of eight elements was preferable. The noble gas configuration was responsible for this conclusion. Many investigators preferred, when possible, to divide longer periods into periods of eight (or seven) elements, a procedure which also had theoretical support. The following brief theoretical exposition is essential to a clear understanding of this point.

The true periods of eight elements contain atoms whose outermost shell possesses only either s-electrons or s- and p-electrons[3]. The last element with $2s$- and $6p$-electrons in its outermost shell, i.e., with a complete s-level and a complete p-level, is always a noble gas. The periods of 18 elements contain, in addition to these eight elements, another ten elements having d-electrons in the penultimate shell. The periods of 32 elements furthermore contain 14 elements possessing one or more f-electrons in the atomic shell lying two levels inwards.

The very first period contains elements with only s-electrons, and there are two of these elements. The periods can therefore contain 2, 8, 8, 18, 18, 32 .... elements. Four kinds of periods can be distinguished in the present periodic system with its 104 elements. The periods containing elements with only s-, or s- and p-electrons in their outermost atomic shell and a still uncompleted orbit with d-electrons, belong to the main groups. The sub-group elements possess atoms whose d-levels are in the process

---

[3] s, p, d, f, etc., refer to the energy state of the electrons in the electron shells. Not more than $2s$, $6p$, $10d$, $14f$, etc. electrons can occur per shell. The K, L, M, N, etc. shells cannot contain more than $2s$, $2s + 6p$, $2s + 6p + 10d$, $2s + 6p + 10d + 14f$, etc. electrons, respectively. A detailed treatment of this subject can be found in Hund [4], for example, or in one of the many modern works on atomic structure.

of being filled. The elements whose 4*f*- or 5*f*-levels are filled are called lanthanides and actinides, respectively.

To elucidate the four periodicities, i.e., the recurrence of the properties of the elements after every 2, 8, 18 and 32 elements, we must consider the atomic theory in a little more detail. The successive shells K, L, M, N, O, P, Q can contain 2, 8, 18, 32, 50, 72, 98 electrons, respectively. The general formula that these numbers of electrons obey is $2n^2$ ($n = 1, 2, ...$). Table 6 shows both the ideal complements of the shells that would result from a regular filling of the orbits and the real lengths of the periods.

If the shells were regularly filled, more than four kinds of periods could be distinguished and no period would occur as "double" in the periodic system. This point requires closer consideration. The components of the various shells, i.e., the *s*-, *p*-, ... orbits, are normally filled in the K- and L-shells. The irregularity begins in the M-shell, where the filling of the *d*-orbits is delayed until the 4*s*-orbits, i.e., the *s*-orbits of the N-shell, have already been occupied. The remaining irregularities can be seen from the

TABLE 6

ELECTRON DISTRIBUTION OVER ATOMIC SHELLS AND ORBITS

Filling of the periodic system		Filling of the shells		
Periods	Number of elements	Letter of the shell	Total number of possible electrons; formula 2 n²	Subdivision by orbits
1	2	K	2; s	2
2	8	L	8; s+p	2+6
3	8	M	18; s+p+d	2+6+10
4	18	N	32; s+p+d+f	2+6+10+14
5	18	O	50; s+p+d+f+g	2+6+10+14+18
6	32	P	72; s+p+d+f+g+h	2+6+10+14+18+22
7	32(?)	Q	98; s+p+d+f+g+h+i	2+6+10+14+18+22+26

periodic system on the cover. The periodicity of the system has this simplicity precisely because these irregularities in the building-up of the electron orbits show a striking regularity! It may be more or less a matter of chance that the resulting periods satisfy the same rule, $2n^2$, by which all but the first[4] period occur as doublets. The periods contain 2, 8, 8, 18, 18, 32, etc. elements, in that order. This is due to the fact that, as the shells become occupied, only a given number of subshells are filled after the

---

[4] As has been shown (*12.1*), Rydberg also took the first period double. Janet [5] also made two periods occur with 2 elements. He could do so by making a shift after these two periods (H–He and Li–Be) of two places in the remaining periods with respect to the generally accepted periods.

utilization of a subsequent shell, so that each time there are two identical combinations of $s, p, d, \ldots$ electrons. At the end of the 4th period, both the s- and p-orbits of the N-shell and the d-orbits of the M-shell are complete; with the 5th period both the s- and p-orbits of the O-shell and the d-level of the N-shell are filled; these two periods thus have the same combination of s, p, d with $2 + 6 + 10 = 18$ electrons and the corresponding 18 elements. This phenomenon is explained by the energy-levels of the different s-, p-, d-, $\ldots$ states. A more detailed exposition would take us beyond the scope of our subject.

In its present form the periodic system contains seven periods. The first period consists of 2 elements, the second and third each consist of 8 elements, the fourth and fifth each contain 18 elements, and the sixth period has 32 elements; the seventh period has not yet been filled but should also contain 32 elements. According to the continuity of the structural irregularities of the atomic shells, element 104 will have a $6d^2$-configuration. In this element the d-orbit of the 6th shell will have been completed; in element 113 a start will be made in the occupation of the $7p$-orbit, which will be complete at element 118, a noble gas. This prediction therefore takes into account a retardation in the completion of the $5g$-orbit. The filling of the 5th shell, which can contain 50 electrons, is interrupted at the element lawrencium. Probably only at about element 123 will a commencement of the $5g$-orbit be made, when one $7d$- as well as one $6f$-orbit becomes occupied[5]. The eighth period could then contain 50 elements.

12.3. Influence of the shape of the periodic system on its discovery

In the previous section it was noted that the filling of the atomic shells with electrons is erratic. This must be considered as the decisive factor in the discovery of the periodic system of elements at a time when not all its secrets were yet known. If the filling were regular in its course, there would be elements belonging to one of at least 5 different periods. It would also have been difficult to distinguish these 5 periods, especially when more than one element of some of these periods was unknown, because with a regular occupation each successive electron would be brought into the outermost atomic orbit and the specific properties of the successive elements would have been very different. There would have been no rare earth metals, for instance. It was the very occurrence of these elements which enabled an early periodic system to be set up; for the small differences in their chemical properties would only have confused the picture. Those already known in the 1860s and 1870s were put aside for the time being. In any case it was at first unnecessary to reserve a special period for the rare earths, at least as long as no properties analogous to those of the remaining elements

---

[5] Seaborg [6] assumed that the $5g$-level develops its filling-up immediately after the $8s$-orbits, i.e., with element 121.

were ascribed to them[6]. But the reservation of a special subdivision of a period by Bayley and Bassett, among others, testifies to a consistent application of the principle of analogy. Had these elements had no properties in common, the chemists of the last century could not have set up a periodic system because of the complexity of analogies between the elements; nor would the other elements have shown such simple relationships. The discovery of the periodic system might then have been delayed until 1913, when the theoretical background became known. This is not meant to imply that no difficulties were encountered in incorporating the rare earths (see further Chapter 10). But these were specific problems that did not essentially effect the form of the periodic system. An essential modification would have been required if these elements had not been rare in both their occurrence in nature and their properties.

12.4. Classification of elements before and during the period of discovery

Although no distinction could be made between main and sub-group elements in the years preceding the discovery of the periodic system, virtually the same problem was encountered: the classification of elements according to valence and analogous chemical properties in more than one group. When, in spite of a resemblance in valence, the properties of the elements distinctly belonged to two or more groups, this usually presented no problem. The element magnesium, for one, caused some difficulties. According to its properties, this element could be classified as a member of both the calcium and the zinc groups. Cooke, Lenssen and Odling believed that the resemblance of magnesium and zinc determined the classification; Gmelin thought of that of magnesium and calcium. In the decade of discovery, the classification of magnesium became very important and modified the form of the periodic system. Five of the six discoverers worked on this problem. Béguyer de Chancourtois was prevented by the rigidity of his *Vis tellurique* from changing the place of magnesium determined by this graphical representation, in which it occurred with Be, Ca and Sr on one vertical line; barium is missing from the system.

Newlands quite consistently favoured two possibilities, viz., to consider magnesium as the second[7] member of either the calcium or the zinc group. In his 1864 classification (Fig. 22, p. 107), which cannot yet be called a periodic system, Newlands already created a division, albeit incomplete, into specific groups of elements. One year later, after discovering his law of octaves, Newlands no longer made a distinction between different kinds of groups. He tried to distribute almost all the elements over 7 columns (Fig. 24, p. 108).

---

[6] Mendeleev classified only a few rare earth metals. He did, however, reserve places in the 8th and 10th periods of his system (Fig. 45, p. 137) and he also created an empty period (the 9th). This procedure gave his system long periods, some of which had many vacant places (see also *10.3*).

[7] Newlands regarded lithium as the first member (see *5.3*).

Odling carried through a stricter division as early as 1864 (Fig. 27, p. 114). After calcium, his system splits up into three parts. The broad central part chiefly contains elements of the main groups; below them, the periods contain the sub-group elements and, above, the transition metals. Four years later, in 1868, Odling unfortunately partly abandoned this very well thought-out arrangement. In this latter system (Fig. 28, p. 115), for example, the elements Zn, Cd, Hg and Ag are included in the main series and the elements of the K- and Ca-groups excluded from it. Magnesium, formerly classified as a member of the calcium group, was now placed together with beryllium as a homologue of zinc.

Although Hinrichs evidently believed that magnesium has more properties in common with zinc than with calcium (Figs. 31, 33, and 34, pp. 122-124), in his 1867 system (Figs. 30 and 30a, p. 120) he placed magnesium at the head of both the calcium and the zinc groups. At magnesium Hinrichs split the radius bearing this element into two parts. This ambivalence is probably to be ascribed to the simple numerical relation he found in 1866 between the atomic weights of magnesium, calcium, strontium and barium (Fig. 29, p. 118).

Meyer always classified magnesium as a member of the main group. In 1868 he carried out a consistent division into short and long periods (Fig. 37, p. 128). Although his system shows some imperfections, he followed a logical method. It is therefore the more to be regretted that Meyer in 1870 nullified the division by combining the groups (Fig. 38, p. 129).

In Mendeleev's first system[8] (Fig. 39, p. 129) the calcium group was excluded from the main series. In his unsplit system (Fig. 131, p. 287) of the same year, however, this was eliminated by the combination of the groups. In this subsequent system, in which he made a distinction between odd and even elements (see e.g. Fig. 47, p. 138), Mendeleev consistently took magnesium as a homologue of zinc.

It will be useful here to go into the question of whether it makes a fundamental difference to consider magnesium as a homologue of calcium or as a homologue of zinc. As we have already seen, magnesium has properties in common with both elements, so that, as far as analogy is concerned, there is no valid reason for preferring either of the two. A closer examination of several systems shows that the requirement of increasing atomic weight is also satisfied by both arrangements. Furthermore, closed blocks of groups and of periods of elements can be obtained in either case. Only when the characteristic difference of 16 atomic weight units between the elements of some periods is considered to be fundamental, must magnesium and calcium be placed as each other's homologues. Many investigators tried to solve this last problem. The incorporation of neighbouring elements of magnesium, such as sodium, aluminium, etc., should have caused fewer difficulties because of the more pronounced analogy between properties of these and certain other main group elements, but it was not found to be a simple problem by all those who worked on it.

[8] See also 5.8.

12.5. The subdivision of the periodic system after 1871

It will by now be clear that the division of the periodic system into groups and periods continued to be a stumbling block after 1871. Some of the most important aspects of this phase should be briefly reviewed. The first system in which the division into main and sub-group elements is satisfactorily expressed and in which the transition metals and rare earths occupy a sufficiently well-founded place, was designed by Bayley (Fig. 51, p. 149). We have already discussed this important system exhaustively (*10.5*) and therefore need only mention here that the elements of the long periods, not including the transition metals, were treated in pairs as homologues of the elements of the short periods. Bayley indicated this by means of connecting lines. This system may be seen as a forerunner of those produced by Bassett, Thomsen, Werner and Bohr.

If all elements with analogous properties are to be placed in one group, which does not permit a system in which the relationship of the elements of the short periods is expressed by the elements of both the main groups and the sub-groups, a division must be made in the short periods. The location of this division indicates which elements of the long periods belong to the sub-group and which to the main group.

More than one preferred solution came into vogue. Otto, in his system of 1885 (Fig. 116, p. 265), split the two short periods, one after C and the other after Si, as did Rang seven years later (Fig. 98, p. 197). This brought the elements from vanadium through germanium and their homologues into the sub-group. This division was still favoured many years later, when it was known to be in conflict with the division according to electron configuration [7,8]. In 1892 Bassett considered the Cu-group and the Zn-group to belong to the main group[9] (Fig. 52, p. 150). He apparently attached greater value to a system with closed groups than to the principle of analogy. As a result, both the group of the alkali metals and that of the alkaline earth metals were entirely disrupted, to the disadvantage of this otherwise very convenient system.

In his pretzel-shaped three-dimensional system (Fig. 75, p. 178) Crookes regarded the first two periods as one large period; the elements of the first period therefore have quite different homologues from those of the second period. Beryllium came to occupy a place as a homologue of the elements of the calcium group, but magnesium became a member of the zinc group. The elements of the pairs boron–aluminium, carbon–silicon, etc., also have different homologues. In 1901 Walker (Fig. 120, p. 270) split up his system after beryllium and magnesium to achieve a better-founded division into main and sub-groups. This procedure later proved to be theoretically sound. More and more investigators, including Werner (Fig. 54, p. 153), Brauner [9], Pfeiffer [10, 11] and some others [12], came to regard this division as the final solution.

A very fine system (Fig. 56, p. 157) is the one constructed by Romanoff, but it too dates from 1934. The splitting of the short periods is made after Be and Mg, respec-

[9] As Zmaczynski still did in 1937 (Fig. 57, p. 158).

tively; the one in the long periods after Sc and Y respectively (6.7).

Of more recent date (1952) is the proposal put forward by Szabó and Lakatos [13] to consider beryllium and magnesium as belonging to the zinc group. Zinc, cadmium and mercury do indeed have 2s-electrons, but their ionic configurations are different; this, however, is also true for the remaining elements of the first and second periods. Beryllium and magnesium remain transition elements as far as their properties are concerned.

Emerson believed he had found a solution to the dilemma by placing beryllium and magnesium as homologues of all the elements between calcium and zinc. To this end, he made use of the period vacancies resulting from a circular design (Fig. 66, p. 167). This procedure has no chemical significance, however.

During the past few decades, the scandium group has occasionally been regarded as belonging to the main group, and sometimes carbon and silicon as belonging to both the main and sub-groups [14–17]. The inclusion of beryllium in the main group and magnesium in the sub-group is exceptional [18], as is the reversed idea [19, 20]. Holmes [21] believed to have found another solution. In his system, in which both the main and sub-group elements appear in the same column (after the short period alternately on the left and right sides of the column), he placed the elements of the first and second periods further to the right in the column as they occur further to the right in the period. His intention was to treat the elements Li and Na, Be and Mg, and B and Al rather as homologues of the K-, Ca- and Sc-groups, respectively; N and P as those of As; O and S as those of Se; and F and Cl as those of Br. C and Si then occupy an intermediate position. In other words, a different way of expressing an idea of, for example, Von Antropoff (Fig. 59, p. 160). Another variant of this arrangement was given by Sears [22], who considered Be, B, Al, C and Si to belong to both the main and sub-groups, but Mg only to the main group.

Lastly, we may mention an example of a system in which the relationships between elements are expressed as completely as possible, notably Clark's system, which is shown in Fig. 79 on p. 182.

12.6. Beryllium

The difficulties encountered in the classification of beryllium can be reduced to the differences in the valences assigned to it, depending upon whether this element (in the elementary state or as a component of compounds) is compared with magnesium or aluminium. If 9 is taken for its atomic weight, beryllium is easy to classify if bivalence is accepted. But there was no immediate agreement on this point. Beryllium oxide has properties in common with both the oxide of bivalent zinc and that of tervalent aluminium, and many investigators considered the amphoteric character of both beryllium oxide and aluminium oxide to be decisive. Consequently, from the very start there was uncertainty as to the valence of beryllium (see also [23]). This uncertainty was expressed in the classification of homologous elements in separate groups as well as in the periodic system.

De Chancourtois, Meyer, Odling (in 1868) and Mendeleev incorporated beryllium in the magnesium group, but Hinrichs [24] could not decide between a position as first member of the cadmium group or as first member of the iron group.

Mendeleev [25] studied the problem, including the aspect of tervalence, and came to the conclusion that the ideal place was above Mg. In this connection it should be noted that Odling and Mendeleev placed beryllium and magnesium not in the series of alkaline earths but in the zinc group, as the result of a view of analogy long adhered to by many scientists[10]. Assuming beryllium to be tervalent, Kremers and Baumhauer gave it a place above aluminium.

The main discussion about the incorporation of beryllium began in 1879. Some investigators agreed with Lars Frederik Nilson [26] (1840–1899), discoverer of scandium and professor of analytical chemistry at Uppsala, who with Pettersson investigated beryllium and the rare earths and thought he could deduce the former's tervalence from its specific heat value. To satisfy the law of Dulong and Petit, the atomic weight of 9.2 had to be changed to 13.8, the latter value implying the tervalence of beryllium. The specific heat being 0.4079, the value for the atomic heat $0.4079 \times 13.8 = 5.63$ was reasonably satisfactory. Meyer [27] immediately realized the ensuing difficulty. On the one hand, an element with an atomic weight of 13.8 would come to occupy a place in the system between C and N, and, on the other, beryllium as tervalent metal or semi-metal would become the first member of the Al group, i.e., in the place occupied by boron. It is surprising that Meyer finally tried to classify beryllium with an atomic weight of 11 or 11.5 between B and C, thus creating an entirely new group in which he also included a few rare earths that had been difficult to classify (see 10.4).

Brauner [28] considered the classification of beryllium to be a vital question. His solution approached the actual situation more closely. With the atomic weight[11] 9.2, beryllium was a bivalent element exactly fitting into the system above Mg.

Nilson and Pettersson [30] in 1880 could agree with the proposed classification of beryllium as a bivalent metal, but believed that the atomic weight should be 8 because of the difference between its atomic weight and that of its congener magnesium (24). This difference also exists between the neighbouring pairs Li (7)–Na (23) and B (11)–Al (27). Nevertheless, the incorporation of beryllium continued to be regarded as a problem[12], as did the Te–I inversion. These investigators apparently did not have much confidence in the periodic system. They applied the old theory of triads in

[10] See also 12.4 and 12.5.

[11] A few years later [29] he determined this atomic weight as 9.1.

[12] Armstrong's classification [31], based on his hypothesis that the numbers from 1 to 240 all represent atomic weights, is hardly worthy of consideration. Armstrong thought that the element with an atomic weight of 8 was an admixture of beryllium's homologue magnesium, and that beryllium itself should head a new group. We have already discussed Armstrong's views (see p. 253).

determining atomic weights, and their confidence was weakened even further by the difficult position of the rare earths *[32]* (see *10.4*).

In 1904 Wetherell *[33]* took advantage of the exceptional position of beryllium to apply his theory of satellite elements as he had done before with tellurium and argon (*8.2* and *8.4*). The satellite element as an admixture of beryllium should lead to an excessively high atomic weight for the pure element. Its deviation from both the most likely atomic weight value of 8 and from the properties of magnesium are, according to Wetherell, ascribable to this *planetary satellite character* of beryllium. Loring *[34]* also accepted that "beryllium" consisted of more than one element. These hypotheses had no scientific value, however.

For a long time several scientists could not decide to regard beryllium as a homologue of magnesium, in spite of the resulting distortion of the periodic system by not doing so and the many properties shared by these elements, and continued to hesitate even after the electron structure of beryllium was known. For example, in 1944 Emerson *[35]* believed beryllium to belong to the copper and zinc group (Fig. 66, p. 167; see also *12.5*).

The fact that the properties of beryllium diverge slightly from those of its homologues does not mean that the classification of this element should involve any doubts about the validity of the periodic system, even though beryllium does not satisfy the law of Dulong and Petit. Indeed, its exact atomic weight (9.013) had been known for a long time *[36]*.

12.7. Hydrogen

The main and persistent difficulty offered by hydrogen was whether it should be placed as a homologue of lithium or of fluorine. Both possibilities had their supporters. Although its electron configuration ($1s^1$) links it to the alkali metals, its analogy with the halogens is strong. As hydrogen forms halides, as do the alkali metals, and since there is sometimes great disparity between alkali hydrides and halides, it seemed proper to place it in the 1st column of the periodic system rather than in the 7th column. When hydrogen was liquefied for the first time, many scientists expected the liquid to have metallic properties.

In the years prior to the construction of the periodic system, little attention was paid to hydrogen. Cooke was the first to include it among the alkali metals (Fig. 6, p. 79). During the years of discovery of the system, Newlands and Meyer were not concerned about this element. Odling placed hydrogen in his system in 1864 (see Fig. 27, p. 114), thereby implying that he regarded it as a homologue of the univalent metals and also assumed two still unknown homologues. However, he did not make any explicit statement about this.

In Hinrichs' 1867 system (Figs. 30 and 30a, p. 120) hydrogen forms, as it were, the centre, in agreement with Hinrichs' atom concepts (see *5.6*). In 1869 Hinrichs classified this element without homologues (Fig. 31, p. 122).

In 1869 Mendeleev hesitated about where to place hydrogen, because of its low atomic weight. He favoured the copper group because of the analogy in peroxide formation (Fig. 39, p. 129). Mendeleev might also have come to this conclusion on the basis of the univalence of hydrogen or by realizing that this element combines with the alkali metals as well as with the halogens. In 1871 Mendeleev returned hydrogen to the group of the alkali metals (Figs. 43 and 48, pp. 134 and 139), which again implied 6 undiscovered elements (see *5.11* and *7.6*).

In 1872 Meyer placed hydrogen without considering analogy (Fig. 41, p. 132). His uncertainty with respect to this problem is shown by his remark that hydrogen occupies an exceptional position *[37]*. Newlands *[38]* became more positive, at any rate a few years after his discovery of the law of octaves. In 1872 he regarded the element as a halogen.

In the systems set up later on, hydrogen was either omitted or placed above lithium. In exceptional cases it was regarded as a halogen and occasionally as a central element[13] (see e.g. Figs. 51, 53 and 119, pp. 149, 151 and 269). Discussion about the singular properties of this element then subsided until 1896, when Masson *[40]* defended its location above fluorine on these grounds:

(1) The difference between the atomic weights of hydrogen and fluorine, viz. 18, is about equal to that between the other homologous elements, i.e., between 15 and 20;

(2) There is now no need for the 6 new elements required by placing hydrogen above Li; and

(3) Hydrogen has no metallic character apart from being replaceable by metals.

In answer to Masson's publication, Newlands *[41]* remarked that he had realized these points already in 1872, but that the difficulties with hydrogen still remained. Hydrogen does not bear such a close resemblance to the halogens, as noted by Martin *[42]*. Following Masson, Crookes *[43]* and Wilhelm Ostwald *[44]* wanted to place hydrogen as a homologue of fluorine, but to do so Crookes shifted fluorine to the transition group to preserve the shape of his system, which demands equal periods (see *6.13*). Brauner *[45]* maintained hydrogen in the place given it by Mendeleev, even though he noticed a resemblance to chlorine: H_2O and Cl_2O. Brauner thought the fact that hydrogen cannot attain the highest valence of chlorine, viz. 7, argued against its consideration as a homologue of chlorine. Rudorf *[46]* was of the opinion that hydrogen belonged to neither of these two groups but rather to that of the noble gases, but his reasoning is not clear.

The differences of opinion about analogies between the properties of hydrogen and other elements were based on a fundamental disagreement. The approach to the problem can be made from the properties of hydrogen as a simple substance (thus making the physical properties principal) or as components of compounds (thus as a

---

[13] Rang *[39]* thought it could be considered as a homologue of gallium (Fig. 98, p. 197). Unfortunately, he did not explain this singular choice.

real element in the philosophical sense; see *3.7*). But even after a choice is made the problem is not solved: taking hydrogen as an element in a compound can also lead to different conclusions. This is illustrated by a comparison of the series HF, LiF, NaF, with the series CH_4, CF_4, CCl_4, etc.

The reason for all these difficulties lies in the fact that the hydrogen atom can either lose one electron (H^+) or take up one (H^-) (see also *13.7*); in the former case the resulting ion is not so different from the alkaline metals, whereas the H^- ion can be compared with the F^- ion. Hydrogen was also a problem for Bohr, who wanted to base his system on electron structure.

12.8. Conclusion

Once it is accepted that the properties of elements are mainly determined by the electrons in the two outer atomic shells, the problem arises as to how to arrange elements identical as regards outer electrons but different with respect to the inner situation. The latter may correspond either to a noble gas configuration or to an 18-electron system (8- and 18-electrons, *s, p* and *s, p, d* respectively). A number of examples may clarify the problem (see also *[47, 48]*).

Bivalent metals: Ca–Zn, Sr–Cd, Ba–Hg, in which the first of each pair upon losing two electrons acquires a noble gas configuration (corresponding to Ar, Kr and Xe, respectively); the second produces an 18-electron configuration. Monovalent metals: K–Cu, Rb–Ag, Cs–Au, where we have an analogous situation.

It seems logical to make a subdivision within each of the two metal groups mentioned in order to account for the differences and analogies indicated. Thus, the elements, whose electron configuration is related to that of the noble gases, are placed in so-called main groups, those related to closed 18-electron systems in sub-groups. However, this system can be used only for univalent and bivalent metals. In the case of tervalent elements or elements of even higher valence, a different division has to be used so as to arrive at a consistent description tallying with their properties. Here, elements, whose positive ions have an 18-electron configuration, are placed in the main group, those with an 8-electron configuration in a sub-group.

To give an effective and simple definition of main and sub-groups nevertheless remains difficult. It is justified to have the elements run parallel to the 8- or 18-electron configurations. The halogens and chalcogens, on the other hand, are inclined to take up electrons, forming a noble gas structure. They are therefore classified as main groups, although their positive ions—starting with Br and Se, respectively—have an 18-electron structure. The elements of the Li- and Na-period of the periodic system are all main group elements. The division into main and sub-group elements starts with the K-period. The most convenient definition for elements of these long periods must be: A main group element is an element the valence electrons of which are *s*-electrons, its ion having a noble gas configuration or *s*- and *p*-electrons, its ion having an 18-electron structure. A sub-group element is an element whose valence electrons are *s*-electrons, its ion having an 18-electron configuration, or *s*- and *d*-electrons, its ion having a noble gas configuration.

The electron structures furnish a good explanation of the resemblances and differences of properties between beryllium and magnesium and the main and sub-group elements of the second group, while the positive and the negative charge of a hydrogen ion can also be explained.

Lastly, we should like to point out that the expression "transition elements" causes confusion. Some scientists, including the author, call the elements Fe, Co, Ni and their homologues transition elements. These elements of the eighth group therefore form a part of the (minor) sub-group. Other scientists, however, equate the whole (minor) sub-group with the transition group. The rare earths are then sometimes called inner transition elements [49]. Instead of main and minor sub-group, some authors use the terms "normal" and "transition(al) sub-group" or "even and odd series" for respectively the first part of the periods through the transition elements, and the second part, i.e. beginning with the elements of the copper group.

References

1 J. R. RYDBERG, *Elektron, der erste Grundstoff*, Lund 1906, p. 11.
2 J. R. RYDBERG, *Lunds Univ. Arsskr.* [2], 9 (1913) No. 18; transl. in *J. Chim. Phys.*, 12 (1914) 585.
3 R. ABEGG, *Z. Anorg. Chem.*, 39 (1904) 330.
4 F. HUND, *Linienspektren und periodisches System der Elemente*, Berlin 1927.
5 C. JANET, *La Classification helicoïdale des Eléments chimiques*, Beauvais 1928.
6 G. T. SEABORG, *Man-made Transuranium Elements*, New Jersey 1963.
7 S. R. BRINKLEY, *Principles of General Chemistry*, New York 1933.
8 B. S. HOPKINS, *General Chemistry for Colleges*, New York 1937.
9 B. BRAUNER, *Das periodische System der Elemente von Mendelejew, Erklärende Einleitung für den Gebrauch der Wandtafel*, Leipzig 1908.
10 P. PFEIFFER, *Z. Angew. Chem.*, 37 (1924) 41.
11 P. PFEIFFER, F. FLEITMANN and R. HANSEN, *J. prakt. Chem.* [2], 128 (1930) 47.
12 A. SMITH, *General Chemisty for Colleges*, New York 1916.
13 Z. G. SZABÓ and B. LAKATOS, *Research*, 5 (1952) 590.
14 F. EPHRAIM, *Inorganic Chemistry*, London 1946.
15 W. NERNST, *Theoretische Chemie*, 8th–10th ed., Stuttgart 1921.
16 E. H. RIESENFELD, *Lehrbuch der anorganischen Chemie*, Zürich 1946.
17 L. C. NEWELL, *College Chemistry*, New York 1925.
18 L. KAHLENBERG, *Outlines of Chemistry*, New York 1915.
19 A. IANDELLI, *Gazz. Chim. Ital.*, 77 (1927) 24.
20 C. K. PALMER, *Proc. Colorado Sci. Soc.*, 3 (1890) 287.
21 H. N. HOLMES, *General Chemistry*, New York 1921.
22 G. W. SEARS, *J. Chem. Educ.*, 1 (1924) 173.
23 H. ROSE, *Ann. Physik* [2], 96 (1855) 436.
24 G. HINRICHS, *Proc. Am. Ass. for Advancement of Science*, 18 (1869) 112.
25 See *Das natürliche System der chemischen Elemente, Abhandlungen von Lothar Meyer und D. Mendelejeff*, ed. by K. SEUBERT, Ostwald's Klassiker No. 68, Leipzig 1895, p. 68.
26 L. F. NILSON and O. PETTERSSON, *Ber.*, 11 (1878) 381.
27 L. MEYER, *ibid.*, 11 (1878) 576.
28 B. BRAUNER, *ibid.*, 11 (1878) 872.

29 B. BRAUNER, *ibid.*, 14 (1881) 53.
30 L. F. NILSON and O. PETTERSSON, *ibid.*, 13 (1880) 1451.
31 H. E. ARMSTRONG, *Proc. Roy. Soc. (London)*, 70 (1902) 86; *Chem. News*, 86 (1902) 86, 102.
32 L. F. NILSON and O. PETTERSSON, *Ber.*, 13 (1880) 1459.
33 E. W. WETHERELL, *Chem. News*, 90 (1904) 260.
34 F. H. LORING, *Chem. News*, 100 (1909) 281.
35 E. I. EMERSON, *J. Chem. Educ.*, 21 (1944) 111.
36 T. JOHANNSEN and O. HÖNIGSCHMID, *Z. Anorg. Chem.*, 253 (1957) 228.
37 L. MEYER, *Ann.*, Suppl. VII (1870) 354.
38 J. A. R. NEWLANDS, *Chem. News*, 26 (1872) 19.
39 P. J. F. RANG, *ibid.*, 67 (1893) 178.
40 O. MASSON, *ibid.*, 73 (1896) 283.
41 J. A. R. NEWLANDS, *ibid.*, 73 (1896) 305.
42 G. MARTIN, *ibid.*, 84 (1901) 154.
43 W. CROOKES, *Proc. Roy. Soc. (London)*, 63 (1898) 408; *Z. Anorg. Chem.*, 18 (1898) 92; *Chem. News*, 78 (1898) 25.
44 W. OSTWALD, *Grundlinien der anorganischen Chemie*, Leipzig 1900, p. 48.
45 B. BRAUNER, *Chem. News*, 84 (1901) 233.
46 G. RUDORF, *Das periodische System, seine Geschichte und Bedeutung für chemische Systematik*, Hamburg–Leipzig 1904, pp. 87, 269.
47 T. CÁCARES, *Anales Soc. Españ. Fis. Quim.*, 9 (1911) 121.
48 J. R. PARTINGTON, *General and Inorganic Chemistry*, London 1946.
49 T. MOELLER, *Inorganic Chemistry*, New York–London 1952, p. 124.

Chapter 13

Radioactivity and modern atomic theory

13.1. Discovery of new elements

The discovery of an appreciable number of new elements, largely as the result of intensive study of the phenomenon of radioactivity, constituted a surprising development.

After Becquerel discovered radioactivity in 1896, the Curies succeeded in isolating polonium and radium only two years later. Even before the atomic weight of the first of this new series of elements had been determined, Willy Marckwald [1] (b. 1864), professor of chemistry in the University of Berlin, who did much research in the fields of organic and radio-chemistry, incorporated it into the periodic system in 1902 as a homologue of tellurium. Gratifyingly, its subsequently determined atomic weight was found to bear out this argument; but it was not until 1913 that this location of polonium was universally accepted. The arrangement of the radioactive elements in the periodic system could only be understood after the concept of isotopy had been introduced by Soddy[1].

Radium was, after some initial disagreement, fairly generally considered to be a homologue of barium. In 1902 Marie Curie [3] determined its atomic weight at 225. With this atomic weight, which differs little from the actual value, radium could conveniently occupy its place directly below barium. However, Carl David Tolmé Runge (1871–1927), professor of physics and photography at the Technical College of Hanover, and Julius Precht [4] (1856–1937), his colleague who taught applied mathematics and investigated spectra, thought in 1903 that their data indicated a value of 258 for the atomic weight of radium. An analogy with barium could not be denied. Therefore, Harry Davy Jones [5] (1865–1911), professor of physics in Johns Hopkins University at Baltimore, who did research on spectra and atomic weights, two years later reduced the value of 258 by 2 to 4 units, accepting Runge and Precht's determinations rather than the value found by Marie Curie. The place of radium as a homologue of barium was not, therefore, altered, but the element was placed in the next period.

Only a year after the discovery of the first two radioactive elements, a third element

---

[1] Soddy first used the term isotope in 1913. He asked himself, however, already in 1911 whether polonium should be regarded as an isotope of tellurium [2].

was demonstrated when Debierne separated actinium from pitchblende in 1899. This new element was not definitely placed as a homologue of lanthanum[2] until 1913, when Alexander Smith Russell [7] (b. 1888), lecturer in physical chemistry at Westminster Trinity College and investigator of radioactive compounds, established the valence of actinium. Nothwithstanding this, Kasimir Fajans [8, 9] (b. 1887), the future professor at Munich and Michigan University at Ann Arbor, who in the same year discovered the displacement law independently of Soddy, could not yet decide between 2 and 3 as the valence. Werner [10] in 1905 had without explanation placed actinium below thulium, which belonged to the series of rare earths; in the space below lanthanum he classified Laα, by which symbol he evidently meant an undiscovered element. But this was before the definite determination of the atomic weight of actinium. In 1899 and 1900 a fourth element was demonstrated independently by several investigators: the radioactive emanation (radon). Even before its precise atomic weight had been determined by Ramsay and Gray [11] in 1910, radon, with an incorrect atomic weight, was regarded as the last of the noble gases (see e.g. Fig. 89, p. 191 and *9.2* and *9.5*).

13.2. Position of radioactive disintegration products; the isotopy concept

Besides these four elements—polonium, radium, actinium and radon—which were really new elements, an additional thirty-five or more "elements" were demonstrated during the first decade of the twentieth century. All these "elements" were later identified as isotopes of well-known elements. Attempts were made to insert these radioactive disintegration products into the periodic system, as exemplified by the system of Van den Broek (Fig. 86, p. 188; see also *6.17*). This author tried to include all the elements, including the disintegration products, in the form of triads. His failure to do so is not surprising. In 1907 Van den Broek [12] placed the α-particle in his system above fluorine. Assuming an atomic weight of two units he regarded the α-particles (half helium atoms, in his view) as the basic component of all elements. The *alphade* system set up by him shows us that he considered all atomic weights to be multiples of 2, which is a direct consequence of the above assumption.

But there were also investigators who soon understood that, no ordinary place being available in the periodic system for all these new elements, several of these substances must therefore be closely related. In 1909 Daniel Strömholm (1871–1950), professor of chemistry in the University of Uppsala, and Theodor Svedberg [13] (b. 1884), lecturer and demonstrator of physical chemistry in the same university, future professor and Nobel Prize winner for chemistry, clearly realized that radium-, actinium- and thorium-emanation were closely related as substances. Radium, actinium-X and

---

[2] A few years before, B. K. Emerson [6] had filled the space under lanthanum in his system with ionium. This disintegration product was found to be an isotope of thorium, however. It was isolated by Hahn and his co-workers in 1907.

thorium-X likewise belonged together. In their system (Fig. 133) these investigators placed the three former elements as homologues of xenon and the latter three as homologues of barium. They were quite certain that these radioactive elements occurred in nature as groups of three, with the same properties, even though they had different atomic weights and could not be separated in the laboratory. It is evident that these investigators had perceived isotopy, but four years would elapse before this phenomenon was defined by Soddy [14]. This problem had occupied his mind for many years. As early as 1910 he had made a distinction [2, 15], though not in those terms, between isotopes and isobars. At the end of 1913, after Van den Broek's conclusion that the atomic number of an element must determine its place in the periodic system, Soddy described all elements possessing the same properties and differing only in atomic weight—so belonging to the same place in the periodic system—as isotopes or isotopic elements [14].

	O row	I row	2 row	3-4 row	
5 Per.	Xe	Cs	Ba	La-Yb	
	Ra – Em	–	Ra	Ionium-(UX-Rad U)	
6 Per.	Akt.– Em	–	Akt X	Radio Act.-Act.	⟩–U
	Th – Em	–	Th X	Rad.Th-MesTh-Th	

Fig. 133 Arrangement of some radio-active decomposition products in the periodic system by Strömholm and Svedberg in 1909.

A convincing indication of the existence of non-radioactive isotopes of "ordinary" elements was obtained in January 1913 from a study of the noble gas neon. In a lecture before the Royal Institution in London, J. J. Thomson reported that he had found in the spectrum of this element a line corresponding to the atomic weight of 22. Since such an element did not fit into the system, he suggested that it was the compound NeH_2 [16]. According to Thomson, a spectral line corresponding to atomic weight 6 would belong to HeH_3. Thomson had a second interpretation of the neon spectrum to fall back upon, namely the existence of two elements, neon and meta-neon, with analogous properties and comparable to the elements of the transition group, for instance Fe and Co, whose properties likewise do not differ greatly [17].

13.3. Positional assignment of the radioactive isotopes

After the first attempts made by Strömholm and Svedberg, many prominent investigators took up the problem of the incorporation of the radioactive isotopes into the periodic system. Most of the isotopes had been studied sufficiently to guarantee their correct locations. The most important of these investigators were Fajans [8, 9, 18, 19], Soddy [20, 21, 22] and Van den Broek [23] who, along with several others [7, 24, 25], were thoroughly occupied with this problem for many years.

Fajans *[8]* came to the conclusion that many of the radioactive disintegration products were alike, except for their atomic weights, and consequently belonged in the same place in the periodic system, for instance RaB, ThB, AktB, RaD, and ThD_2 in the place of Pb and the emanations AktEm, ThEm, and RaEm, which had been demonstrated in three different ways as the last noble gas. Fajans also tried to place the rare earths (see *10.8*) in this way.

He immediately pointed out that an important result of the discovery of isotopy was the solution of the thorny problem of such inversions of elements as that of tellurium–iodine (see *8.2*).

In spite of the conviction with which the solution of the problem of classifying the radioactive disintegration products was put forward, differences of opinion persisted (see e.g. *[21, 24* and *26]*).

13.4. Discovery of protactinium

The discovery of protactinium undoubtedly resulted from the investigation of radioactive substances. This element, the ekatantalum predicted by Mendeleev as early as 1871 (see *7.6*), was found in 1913 as a short-lived isotope. This meant an element with the atomic number 91, which was not easy to identify chemically. Speculation therefore arose about whether there were other such substances, perhaps new long-lived isotopes besides "brevium". Brevium was immediately included in the periodic system (Figs. 66 and 72, pp. 167 and 174, respectively). Five years later, after pitchblende had been examined very thoroughly, a larger quantity of this more persistent element was isolated from it. The most widely known of the investigators who, in 1918, arrived simultaneously at this discovery were Soddy, Hahn and Lise Meitner. The place to be assigned to this element, which was named protactinium, did not present any further difficulties. Even though its atomic weight was smaller than that of thorium, this latter element precedes protactinium in the numerical sequence (order number) *[27]* (see also Chapter 8).

13.5. The atomic number (numerical order) as the basis of the periodic system

The year 1913 was one of the most important in the history of the periodic system. The foregoing material has brought out the fact that early in that year Fajans, while working on systematization, took the step of setting the atomic weight aside. He contended that more than one atomic weight could belong to a single place in the system. The time was thus ripe for further conclusions, which came in rapid succession. It is not our intention to consider which of the following investigators should be credited with priority; the intervals between publications are, furthermore, too brief to permit a decision on this point.

Johannes Robert Rydberg [28, 29, 30] (1854–1919), from 1901 professor of physics at Lund, who had deduced in 1890 the constant which bears his name and had made a study of the relationships between atomic weights as early as 1896 [31] (he distinguished two series of atomic weights[3]: $4n$ and $4n - 1$, which were adopted by Harkins and Hall in 1916), now came to a definite conclusion, although he could not elucidate the principle involved. Rydberg put it as follows: "Als unabhängige Veränderliche nehmen wir also die positiven ganzen Zahlen an und nennen dieselbe, deren eigentliche Bedeutung uns noch volkommen unbekannt ist, bisauf weiteres die Ordnungszahlen der Atome oder der Grundstoffe"[4]. He assumed the existence of two more elements between hydrogen and helium, however, because he thought that each period was doubled and that the number of elements in such a double period could be represented by the formula $4n^2$. Therefore, there were 4 elements for $n = 1$. Consequently, all atomic numbers except that of hydrogen were increased by two in numerical value (see also Chapter 12).

The significance of Rydberg's conclusion became immediately clear when Moseley [33, 34] in the same year (1913) arrived at the concept of atomic number by means of X-ray spectrum analysis. The atomic number had already been introduced, strictly speaking by Van den Broek [23] at the end of 1912[5], but the latter took the nuclear charge as being equal to half the atomic weight, because he believed that the nucleus consisted solely of α-particles. The numerical order of the last element in the periodic system, uranium, therefore became 118, which is also due to the fact that Van den Broek gave most radioactive disintegration products a special place, e.g., in the zero, third and eighth groups (see also 13.2). This simultaneously created new vacant spaces. A few months later [35], Van den Broek realized that this theory had to be modified if it was to be consistent with the experimental results.

13.6. Consequences of the introduction of the atomic number

Once it had become definitely established that rather than the atomic weight, the atomic number or order number of the elements should serve as the basis of the periodic system, mass reversals of elements no longer constituted an infringement of the periodic system (see Chapter 8).

Another consequence was the revival of Prout's hypothesis, which is easy to understand because the atomic weights of the isotopes proved to be the long desired whole numbers [19]. Loring [36, 37] even went a step further and considered the isotopic

---

[3] In which n represents an integer.

[4] This conclusion of Rydberg caused Urbain, the discoverer of lutetium, to say: "Sa loi substituait à la classification un peu romantique de Mendeléeff une précision toute scientifique." [32]

[5] Published in 1913.

mixtures to be composed of a few varieties of isotopes in simple ratios. The structure of chlorine he saw as:

$$Cl = 10 \times 35 = 350$$
$$1 \times 40 = 40$$
$$\overline{}$$
$$390 : 11 = 35.454$$

As could be expected, many investigators objected to this unfounded proposition. The haphazard alteration of atomic weights to obtain whole numbers was condemned. The arrangement of the isotopes in the periodic system could then proceed unimpeded. Fajans [19] placed them in a third direction, perpendicular to the plane of the usual two-dimensional system. King and Fall [38] afterwards made use of a cylinder for this purpose.

13.7. Influence of Bohr's atomic theory[6]

Although this section will be devoted to Bohr's atomic theory and its development, it is not our intention to give a summary of his view or the experiments by which it was confirmed. We shall consider only its influence on the perfection of the periodic system, and indicate the chapters in which it is mentioned. The preceding section of this chapter has already conveyed something of the significance and consequences of the introduction of the atomic number and the discovery of the isotopes. The atomic theory prevented the periodic system from becoming a mere collection of all elementary chemical substances when the discovery of more and more radioactive disintegration products threatened to obscure its unique function.

The theory of the distribution of the electrons over the atomic shells has been discussed in the preceding chapter; the influence of Moseley's rule and the development of Bohr's atomic theory have already been considered in connection with the rare earth metals (see 10.8, 10.9 and 10.10) as well as the determination of the position of the noble gases on the basis of the structure of the electronic orbits (9.10). Some systems based on the atomic theory have also been discussed (6.27).

We therefore need only consider some contributions not discussed elsewhere. Soon after Bohr postulated the model of the planetary system of the atoms in 1913 and Moseley was able to determine, in principle, the definite place of each element by means of their X-ray spectra, Kossel[7] in 1916, after Moseley's death, postulated the stability of the noble gas configurations that comprise either 2 or 8 electrons. The

---

[6] See also Chapter 12.

[7] Walther Ludwig Julius Paschen Heinrich Kossel (1888–1956), son of the Nobel Prize winner Albert Kossel, was at that time an assistant in the Physics Department of the University at Munich. In 1921 he became professor of theoretical physics at Kiel.

electron structure of the element with the greatest atomic weight whose structure was described by Kossel, is Mn, with 2, 8, 8, 7 electrons. Such men as Ladenburg [39], Thomson, Lewis and Langmuir [40] elaborated this view, and a theory about the chemical bond was derived from these structures.

Lewis propounded the octet theory (developed further by Langmuir), assuming a cubical distribution of the eight electrons. Determination of the exact structure of the electron shells and orbits required further investigation of the spectra as well as their interpretation. During this work, which was done in the second decade of the present century, several investigators constructed periodic systems based on the latest data [39–48]. The length of the periods also required explanation. The subdivision of the electron shells and the related arrangement of, for instance, the rare earth metals could not be distinguished from each other at the time of the introduction of the quantum theory, when research was being done by Bohr, Sommerfeld [49], Landé [48], Pauli, Stoner and Main Smith.

In 1922 Bohr was not yet able to give the precise electronic distribution. Not until 1924 could Stoner [50] subdivide the shells into orbits of 2, 2 + 4, 4 + 6, etc. electrons (see also 10.10).

Bedreag was one of those who followed the developments in the atomic field closely; he repeatedly attempted to construct a better system by employing the latest data [49–54]. His basic system was called the *physical system* of elements.

In the middle of the 1920s, when the occupation of the electronic orbits s, p, d and f had been established, hydrogen was rarely classed with the halogens, as it had been previously. As a result of the s^1-electron configuration of hydrogen and the s^2p^5-electron configuration of the halogens, hydrogen forms only a few stable H^- ions in binary compounds, but the halogens very easily form the X^- ion which is extremely stable, particularly in the cases of fluorine and chlorine (see also 12.7).

References

1 W. MARCKWALD, *Phys. Z.*, 4 (1902) 51.
2 F. SODDY, *J. Chem. Soc.*, 99 (1911) 72.
3 M. CURIE, *Compt. Rend.*, 135 (1902) 161.
4 C. RUNGE and C. D. F. PRECHT, *Phil. Mag.* [6], 5 (1903) 476.
5 H. D. JONES, *Am. Chem. J.*, 34 (1905) 467.
6 B. K. EMERSON, *Am. Chem. J.*, 45 (1911) 160.
7 A. S. RUSSELL, *Chem. News*, 107 (1913) 49.
8 K. FAJANS, *Ber.*, 46 (1913) 422; *Le Radium*, 10 (1913) 171.
9 K. FAJANS, *Physik. Z.*, 14 (1913) 136.
10 A. WERNER, *Ber.*, 38 (1905) 2022.
11 W. RAMSAY and R. W. GRAY, *Compt. Rend.*, 151 (1910) 126.
12 A. VAN DEN BROEK, *Ann. Physik* [4], 23 (1907) 199.
13 D. STRÖMHOLM and T. SVEDBERG, *Z. Anorg. Chem.*, 63 (1909) 197.
14 F. SODDY, *Nature*, 92 (1913) 399.
15 F. SODDY, *Ann. Repts. on Progress Chem.*, 7 (1910) 256.
16 See S. DUSHMAN, *Gen. Elec. Rev.*, 18 (1915) 614.

17 F. W. Aston, *Isotopes*, London 1922, pp. 33 ff.
18 K. Fajans, *Naturwissenschaften*, 2 (1914) 429, 453, 543.
19 K. Fajans, *Physik. Z.*, 16 (1915) 456.
20 F. Soddy, *Chem. News*, 108 (1913) 168.
21 F. Soddy, *Nature*, 91 (1913) 57.
22 F. Soddy, *Chem. News*, 107 (1913) 97.
23 A. van den Broek, *Physik. Z.*, 14 (1913) 32.
24 A. Schuster, *Nature*, 91 (1913) 30.
25 N. R. Campbell, *Nature*, 91 (1913) 85.
26 A. Fleck. *Chem. News*, 107 (1913) 95.
27 O. Hahn and L. Meitner, *Ber.*, 52 (1919) 1812.
28 J. R. Rydberg, *J. Chim. phys.*, 12 (1914) 585.
29 J. R. Rydberg, *Acta Regia Soc. Physicograph. Lundensis, Kgl. Fysiograf. Sällskap, Lund, Handl.* 24 (1913) No. 18.
30 J. R. Rydberg, *Lunds Univ. Arsskr.* [2], 9 (1913) No. 18.
31 J. R. Rydberg, *Z. Anorg. Chem.*, 14 (1897) 68.
32 See E. Rutherford, *Proc. Roy. Soc. (London)*, A93 (1917) XXII.
33 H. G. J. Moseley, *Phil. Mag.* [6], 26 (1913) 1024.
34 H. G. J. Moseley, *Phil. Mag.* [6], 27 (1914) 703.
35 A. van den Broek, *Nature*, 92 (1913) 372.
36 F. H. Loring, *Chem. News*, 109 (1914) 169.
37 F. H. Loring, *Chem. News*, 111 (1915) 157, 181.
38 J. F. King and F. H. Fall, *J. Chem. Educ.*, 17 (1940) 481.
39 R. Ladenburg, *Z. Elektrochem.*, 26 (1920) 262.
40 I. Langmuir, *J. Am. Chem. Soc.*, 41 (1919) 868.
41 H. Teudt, *Z. Anorg. Allgem. Chem.*, 106 (1919) 189.
42 O. Hinsberg, *J. Prakt. Chem.* [2], 101 (1920) 97.
43 H. N. Allen, *Nature*, 110 (1922) 415.
44 W. R. Cooper, *World Power*, 1 (1924) 18.
45 Z. Wojnicz-Sianozeçhi, *Roczniki Chem.*, 3 (1923) 261.
46 K. Mahler, *Atombau und periodisches System der Elemente*, Berlin 1927.
47 S. A. Schukarev, *Zhur. Russ. Fiz. Khim. Obshchestva*, 55 (1924) 447.
48 A. Landé, *Naturwissenschaften*, 13 (1925) 604.
49 A. Sommerfeld, *Mem. Proc. Manchester Lit. & Phil. Soc.*, 70 (1925/26) 141.
50 E. C. Stoner, *Phil. Mag.* [6], 48 (1924) 719.
51 C. G. Bedreag, *Bull. Sect. Sci. Acad. Roumaine*, 9 (1925) 158.
52 C. G. Bedreag, *Ann. Sci. Univ. Jassy*, 13 (1924) 64, 315, 346; 14 (1925) 47.
53 C. G. Bedreag, *Compt. Rend.*, 180 (1925) 653.
54 C. G. Bedreag, *Bul. Facultât Stiinte Cernăuti*, 1 (1927) 14.

Chapter 14

The second series of rare earths (the actinides)

14.1. Introduction

Although early in 1940, when neptunium and several elements of the 7th period were synthetized, only 3 actinides were known, viz., thorium, protactinium and uranium, the history of this series had begun long before. As far back as 1892, Bassett[1] considered the elements then known with a greater atomic weight than that of bismuth, i.e. thorium and uranium, to be analogous to cerium and praseodymium, respectively, with respect to their properties. He also reserved spaces for undiscovered actinides. Bassett was therefore the first to create a second extra-long period to include the actinides. Few investigators accepted Bassett's views until the electron structures were found to indicate a similarity between the actinides and lanthanides.

Indeed, long after Bohr formulated his atomic theory and it had been developed by him and other scientists, the majority of investigators remained unconvinced that thorium, protactinium and uranium were not primarily homologues of hafnium, tantalum and tungsten, even though Bohr himself saw them as members of a second series of rare earths and in 1926 Goldschmidt demonstrated their analogies on the basis of a contraction in the volume of the ions of these elements in the same way as those of the lanthanides.

The delay in accepting this second series of rare earths was mainly due to the energy states of the 5f-orbits, which differ only slightly from those of the 6d-orbits; furthermore, the highest valences of these elements suggested that they should be classified as sub-group elements.

14.2. Classification of the actinides before 1922

By 1922 Bohr had extended his atomic theory and constructed a periodic system of elements (Fig. 55, p. 156) which made it impossible to visualize the elements after actinium as sub-group elements. Yet, very gradually the new concept came to be accepted.

---

[1] Bassett shifted neodymium and praseodymium as the result of incorrect atomic weight determinations.

We have already mentioned Bassett *[1]* as the first investigator to regard thorium
and uranium as homologues of cerium and praseodymium, respectively; his system
(Fig. 52, p.150) also showed that he even expected other homologues of the lanthanides.
At that time, there was no real series of actinides—actinium was not discovered till
1899—but thorium and uranium and the 12 elements expected by Bassett must be
described as members of the 2nd series of rare earths. It is all the more to Bassett's
credit that he correctly classified both elements before any element had been discovered between bismuth and thorium. He reserved 5 places for the elements polonium, astatine, francium, radium and actinium. For protactinium a vacant space was

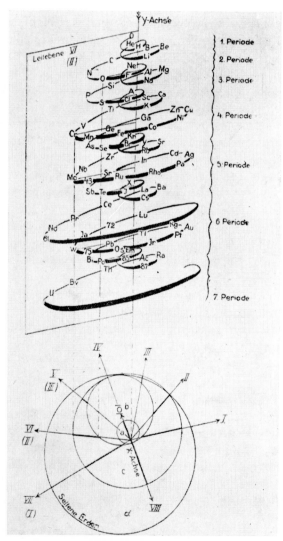

Fig. 134   Screw-shaped system of Schaltenbrand.

available between thorium and uranium. The final member of this series of 14 elements could have an atomic weight of 263. Werner *[2]*, who, as we have seen (*6.5* and *10.7*), set up an almost ideal system in 1905 (Fig. 54, p. 153), also assumed a second series of rare earths, but assigned incorrect places to actinium, which had been discovered a few years before, as well as to thorium and uranium. The place belonging to actinium as a homologue of lanthanum was assigned by Werner to a radioactive disintegration product, Laα (see *13.1*). Werner's system was therefore inferior to Bassett's.

When Bohr's atomic theory appeared in 1913, only a few scientists drew from it the conclusions relating to the elements after actinium before Bohr did so himself in 1922. Langmuir *[3]* presumed on the grounds of the electronic theory that this series of homologues of the rare earths must begin with thorium and end with element 104. But he refrained from making more concrete predictions since he could not see uranium as chemically analogous to neodymium, which followed from the assumption.

A system dated 1920, in which the elements concerned were correctly placed, was designed by Schaltenbrand *[4]*, who considered a screw-shaped system (Fig. 134) to be ideal. This system is exactly the same as Vogel's *[5]*. To make his meaning clearer, Schaltenbrand projected his system on a plane surface.

14.3. Computations on the starting point of the 2nd series of rare earths

The starting-point of the second series of rare earths is marked by the identification of the element in which a $5f^1$-electron structure[2] appears for the first time. However, as the energy levels of the $5f$-orbits show only small mutual differences and the difference between these levels and those of the $6d$-orbits is likewise small, many measurements and calculations were required before a final conclusion could be drawn. In 1926, Victor Moritz Goldschmidt *[6]* (1888–1947), professor of crystallography, mineralogy, and petrography in the University of Kristiania (Oslo), came to the conclusion, on the basis of volume contraction of the ions of the elements of this series, that actinium heads a series of elements analogous to that of the rare earths. In the same year Yoshikatsu Sugiura, who specialized in spectroscopy and quantum mechanics in the Institute for Physical Chemistry at Tokyo, and Harold Clayton Urey *[7]* (b. 1893), the future professor of chemistry at Columbia University, New York, discoverer of heavy hydrogen and in 1939 Nobel Prize winner, tried to shed some light on this question. They calculated quantum-mechanically the strength of the central fields of the atoms from the X-ray terms and concluded that the $5f$-orbit would not be occupied sooner than in element 95. Also in 1926, partly because of the different valences of thorium, protactinium and uranium, Swinne *[8]* was uncertain about the starting-point of the second series of the rare earth metals.

McLennan *[9]*, Rudy *[10]* and Perrin *[11]* were more positive on this point. On the

---

[2] See *12.2* for terminology.

basis of the available spectroscopic data they assumed thorium to be the first member of the new series. Seven years later, in 1933, Wu and Goudsmit [12] calculated, by means of the Schrödinger equation, the energy of the various ions with a radon configuration and concluded that it was highly probable that element 93 in its ground state has one or more 5f-electrons. Karapetoff [13] and Grosse [14] came to a similar conclusion, Grosse because of property analogies between the elements preceding element 93 and those of the sub-group elements of the 6th period. Karapetoff's argument was his belief that the 5f-energy state had not yet been demonstrated for any element. On the basis of the available data and suggestions given by Hahn, Meitner and Strassmann about the bombardment product of uranium, Quill [15] in 1938 considered the filling-up of the 5f-orbit to occur only in element 98 or element 99. Mrs. Goeppert-Mayer [16] in 1941 computed the $5f^1$-structure for element 91 or 92, i.e., for protactinium or uranium. In 1950 Cap [17] concluded from computations according to Slater, that the energy for the $5f^1$-electron state could belong to the newly-discovered element 94, i.e., plutonium.

From the foregoing it is evident that neither the old quantum theory nor modern quantum mechanics led to consistent results, even after a substantial number of elements of this series had been synthetized. In 1947 Meggers [18] believed, on spectroscopic grounds, that he could demonstrate that the thorium atom had a $6d^2$- rather than a $5f^1$-structure, the converse of which had not yet been conclusively proved (see Table 7). Glenn Theodore Seaborg [19] (b. 1912), head of the Department of Nuclear Chemistry at Berkeley, discoverer of several transuranium elements, and Nobel laureate (1951), thought in 1949 of the possibility of a regular completion of the 5f-orbit up to americium. This element could be assumed to have a $5f^7$-structure analogous to that of europium, which was the first to have a 7th electron in its 4f-orbit. This hypothesis was confirmed by Tomkins and Fred [20]. Seaborg considered the electronic state of thorium, protactinium and plutonium to be uncertain. A few years later, Dawson [21] thought, on the basis of magnetic data, that the 5f-orbit only started at plutonium but with a $5f^5$-structure. The most probable electronic arrangement is the one deduced by Katz and Seaborg [22] (see Table 7).

Not even the chemical properties of the transuranium elements indicate unmistakably that all actinides are homologues of the lanthanides. Only after plutonium does the most stable ionic state become 3^+. The most stable valences of thorium, protactinium and uranium are 4, 5 and 6, respectively, i.e., the same as for Hf, Ta and W, which are elements of the sub-group. For protactinium, tervalence does not occur. Bohr had soon recognized all these facts, and later Bedreag [23] also pointed to their importance. The observed valences and the atomic structures are only remotely interrelated. It is presumably because of the larger radius of its ions that U can become sexivalent whereas the related Nd can become tervalent. Compounds derived from bivalent as well as quadrivalent members of the actinides are a quite recent finding [24].

TABLE 7

PARTIAL ELECTRON STRUCTURE OF THE ACTINIDES PROPOSED BY SEVERAL INVESTIGATORS

number of the element	name of the element	filling up of the 5f-orbit				
		Meggers [18] 1947	Seaborg [19] 1949	Dawson [21] 1952	Katz and Seaborg [22] 1957	Seaborg [24] 1966
90	Th	0	(1)	0	0	0
91	Pa	2	(2)	0	2	2
92	U	3	3	0	3	3
93	Np	4	4	0	(4) (5)	4
94	Pu	5	(5)	5	6	6
95	Am	6	7	6	7	7
96	Cm	7	7	7	7	7
97	Bk			8	(8) (9)	(8) (9)
98	Cf			9	10	10
99	Es				11	11
100	Fm				12	12
101	Mv				13	13
102	No				14	14
103	Lw					14

( ) configuration uncertain

14.4. Systematization of the actinides between 1922 and 1940

Between 1922 and 1940, the electronic theory was extended and verified but no new elements were added to the series of actinides. Bohr [25, 26], who enlarged Thomsen's system, minimized all future controversy about the inception of the 5f-orbit by placing the elements starting with actinium as homologues of the elements starting with lanthanum and putting 14 assumed rare earths in the 7th period by considering this series beginning with or after uranium. He indicated this in his system (Fig. 55, p. 156) by a rectangle drawn with a dotted line. He also considered Ac, Th, Pa and U to be related to La (and/or Lu)[3], element 72, Ta and W, respectively. We have already mentioned that in 1926 Goldschmidt [6] was led, by the volume contraction of a few actinide ions analogous to that of the lanthanide ions, to conclude that a second series of rare earths had been demonstrated, although two years earlier [27] he had regarded the undiscovered elements with atomic numbers 94, 95 and 96 (which he called the neptunium group) as homologues of the platinum metals. This testifies to great intuition, because he knew only the ionic radii of Th^{4+} and U^{4+}. Ineffective attempts to demonstrate eka-osmium, eka-iridium and ekaplatinum in the ores of the platinum metals undoubtedly influenced Goldschmidt's reasoning. Yet some years later Goldschmidt [28] did not place these elements correctly. Curiously enough, he ignored

[3] See 10.11.

protactinium, which made uranium a homologue of praseodymium instead of that of neodymium. He called the new series of rare earths thorides.

Let us now review a few systems in which the elements thorium, protactinium and uranium were regarded as homologues of the rare earths cerium, praseodymium and neodymium. One of the most striking examples is that of Janet [29, 30, 31] (see also 6.11) who used a spiral system[4] (Fig. 70, p. 172) and classified Ac, Th, Pa and U as rare earths below the elements of the first series of these substances. Janet [33] in one of his systems depicted the elements on circular and helical figures, in which all the elements of the two series of rare earths are noted. An example of this design is given in Fig. 128 on p. 281.

A system (Fig. 73, p. 175) worthy of special attention consists of 4 cylinders with a common tangent [34]. The cylinders with the smallest diameter show the elements H, Li, Na, etc. at the back; He, Be, Mg, etc. are at the front.

Folded around the first cylinder are the groups

B,	Al,	Sc	...
C,	Si,	Ge	...
N,	P,	Va	... at the back,
O,	S,	Se	...
F,	Cl,	Br	...
Ne,	Ar,	Kr	...

while the front is covered by

In this way the rare earths La (55) - Ny[5] (70) come to lie on the 4th cylinder, as do the elements Ac through 102. The 3rd cylinder carries the elements 103 through 112, while the 2nd and 1st cylinders conclude with 113–118 and 119–120, respectively. In Fig. 74 (p.175) this system is shown as it were in portions, which makes it easy to study comprehensively. It is obvious at first sight that this system shows the lengths of its periods extremely well.

In 1931 Le Roy [35] supported the introduction of a second series of rare earths. Initially, he indicated the singular position of these two series of elements by placing them at the extreme left, and later at the extreme right of the system (Fig. 129, p. 281), directly preceded by the elements of the 1st and 2nd main group and the scandium group (see also Chapter 12). The smaller places occupied by the rare earths in this system are probably explained by technical difficulties in printing. This system would have been too long if these elements had been given the same amount of space as the other elements.

The spectroscopic measurements made in the 1930s, as a result of which the $5f^1$-structure was indicated not for thorium but protactinium or one of the next elements, modified the views held on the analogy of elements[6]. Ephraïm and Mezener [40] accordingly drew a parallel between La, Ce, Pr, and Th, Pa, U, in close agreement with Goldschmidt. The question of the actinium analogue was left open. But because thorium has properties in common with cerium, the conviction was growing (e.g., by German E. Villar [41, 42]) that the former element probably had a $5f^1$-structure. Just

---

[4] In his book on the law of Mendeleev, Kurbatov [32] favours, at least theoretically, a helical system. In his view, such a system would also make it possible to classify the planets and the animals of inferior order, for example, molluscs.

[5] Neoytterbium, the penultimate member of the series.

[6] But several investigators working on electronic structure did not attempt to classify these elements, e.g., Main Smith [36] and Haïssinsky [37]. Conversely, Bedreag [38, 39] included a number of undiscovered elements, i.e. the numbers 93 to 96, which he placed as homologues of Re to Pt.

before the transuranium elements were synthetized, and from experimental data collected by Hahn, Lise Meitner and Strassmann, he predicted 11 of these elements, to be obtained by bombarding uranium with neutrons. After these elements, more subgroup and main group elements would follow until the 7th period of the periodic system had been completed with the noble gas No. 118.

This prediction by Villar [43] had been anticipated, however, by Andrews [44] as early as 1928 (see 9.5). Swinne [45] also foresaw the analogy with the configuration of the rare earth elements. According to him, the 4th subshell of the O-shell had to be divided into 6 + 8 electron places (also compare p. 278, note 28). It is well worth noting that in 1959, and even in 1966, Villar [46, 47] maintained that lanthanum and actinium were not homologues of scandium and yttrium. He inclined, rather, to analogies between these elements and lutetium and lawrencium, notably on the basis of a formula he constructed [48] in 1938. This expresses the theoretical regularity in the build-up of the atomic shells, but has not been confirmed for the rare earths. The atoms of lanthanum and actinium do not possess 4f- or 5f-electrons respectively, in contrast to lutetium and lawrencium which do have them, as Villar recognized. Finke [49] had nevertheless already made the same proposal (Fig. 107, p. 207)[7] in 1949 and Dockx [50] in 1959. They thus preferred the analogy of the elements having a complete f-orbit and one d-electron with scandium and yttrium, to that of the latter with the elements having a complete s- and p-orbit and one d-electron, i.e., lanthanum and actinium.

14.5. Classification of the actinides after the discovery of the transuranium elements

The synthesis of the transuranium elements, starting in 1940, at first had little influence on the classification of the elements after actinium. In the first years after their discovery the properties of these elements were hardly known and their electronic structure had not yet been conclusively established. However, the main reason was that there was still no agreement on the electronic structure of the far better known elements thorium, protactinium and uranium. Many scientists preferred to regard the actinides provisionally as sub-group elements, for one reason because it was customary and also because it was not certain that the 5f¹-structure began with thorium [51, 52]. It could equally well start with uranium or even further on, as Starke [53] argued in 1943 and as Meggers [18] assumed in 1947 on the basis of spectroscopic data (see Table 7). Although thorium also had a valence of 4, protactinium of 5, and uranium of 6, the septivalence of neptunium had not been demonstrated.

Often, to be on the safe side, the actinides were classified in two ways at the same

---

[7] Although the new elements, including curium, are placed in his system, it is strange to find place 61 occupied by the symbol for the assumed element illinium instead of promethium, which had been synthetized in 1947.

time: as sub-group elements and as a second series of rare earths[8]. This was done by Wheeler *[58]* and Stedman *[59]* (Fig. 82, p. 185), for instance, as well as by Seaborg *[60]*, although the last made several positive contributions to the structural determinations. He considered a $5f^1$- and a $5f^6$-structure for Np and Pu, respectively *[61]*, and predicted great stability for element 96 with a $5f^7$-structure which, like its homologue gadolinium, could be assumed to possess a half-filled f-orbit *[62]*, as was confirmed after the discovery of curium. This element has only a valence of 3.

Seaborg's first attempts to synthetize americium and curium failed, merely because their properties were thought to be analogous to those of plutonium *[63]*. Once it was assumed that these elements belonged to a separate series, i.e., the actinides, they were isolated immediately, in 1944 and 1945.

			Ce	Pr	Nd	Il
Sm	Eu	Gd	Tb	Dy	Ho	Er
Tm	Yb	Cp				
			Th	Pa	U	Np
Pu	Am	Cm				

Fig. 135   Rare earth metals in the system of Schenk.

Among the authors who stressed the analogy in electron structure and chemical properties between the lanthanides and actinides were Szabó *[64]*, Simmons *[65]*[9], Villar *[66]* (1947), Achimov (Fig. 58, p. 159), Vallet (Fig. 108, p. 207), Wringley *et al.* *[67]* (a system in the shape of a *flight of steps* (Fig. 109, p. 208) in which each subperiod occupies a step of its own[10]) and Scheele (Fig. 60, p. 160). Schenk *[68]* represented the stability of the tervalent lanthanide and actinide ions by an empty, a half-filled and a totally filled f-level, each with the same s^2d^1-configuration, placing the elements below each other. The elements concerned are La, Gd, Cp(= Lu), Ac and Cm (Fig. 135). A somewhat similar system (Fig. 80, p. 183) was published by Clark *[69]*.

[8] Bedreag *[54, 55, 56]* evidently found classification difficult, judging by the different suggestions he put forward, according to one of which elements 93 and 94 should form a quite separate group between the chromium and manganese groups on the basis of spectroscopic data and the valences of the elements after actinium. Soon after this, Bedreag *[57]* reconsidered this hypothesis when he became aware of the 1933 computations of Wu and Goudsmit *[12]*. This does not mean that he now accepted a series of actinides. According to Bedreag *[23]*, the series of elements in which the $5f$-orbit is built up should run from plutonium to element 106. Nevertheless, he preferred to regard the known elements of this series, i.e., plutonium through curium, as each other's homologues. In fact, he ended the whole series with element 100, although he did not explain why he did so.

[9] The classification in Simmons' system of the two series of rare earths apart from the other elements might have been caused by technical difficulties in printing, as the small reference arrows also suggest.

[10] That the *steps* of the two series of elements with an f-configuration are not of equal height, is due to the principle of arrangement according to the main quantum number.

References

1 H. BASSETT, *Chem. News*, 65 (1892) 3, 19.

2 A. WERNER, *Ber.*, 38 (1905) 914, 2022.

3 I. LANGMUIR, *J. Am. Chem. Soc.*, 41 (1919) 868.

4 G. SCHALTENBRAND, *Z. Anorg. Allgem. Chem.*, 112 (1920) 221.

5 R. VOGEL, *Z. Anorg. Allgem. Chem.*, 102 (1918) 117.

6 V. M. GOLDSCHMIDT, *Geochemische Verteilungsgesetze der Elemente*, Kristiania 1926, Vol. VII p. 56.

7 Y. SUGUIRA and H. C. UREY, *Kgl. Danske Videnskab. Selskab*, 7 (1926) No. 13.

8 R. SWINNE, *Z. Tech. Physik*, 7 (1926) 166, 205.

9 J. C. McLENNAN, A. B. McLAY and H. GRAYSON SMITH, *Proc. Roy. Soc. (London)*, A 112 (1926) 76.

10 R. RUDY, *Rev. Gén. Chim.*, 38 (1927) 661.

11 J. PERRIN, *Grains de Matière et de Lumière*, Paris 1935, Vol. II, Part II p. 30.

12 F. Y. WU and S. GOUDSMIT, *Phys. Rev.*, 43 (1933) 496.

13 V. KARAPETOFF, *J. Franklin Inst.*, 210 (1930) 609.

14 A. V. GROSSE, *J. Am. Chem. Soc.*, 57 (1935) 440.

15 L. L. QUILL, *Chem. Revs.*, 23 (1938) 87.

16 M. GOEPPERT-MAYER, *Phys. Rev.*, 60 (1941) 184.

17 F. CAP, *Experientia*, 6 (1950) 291.

18 W. F. MEGGERS, *Science*, 105 (1947) 514.

19 G. T. SEABORG, *Nucleonics*, 5 (1949) No. 5, 16.

20 F. S. TOMKINS and M. FRED, *J. Optical Soc. Am.*, 39 (1949) 357.

21 J. K. DAWSON, *Nucleonics*, 10 (1952) No. 8, 39.

22 J. J. KATZ and G. T. SEABORG, *The Chemistry of Actinide Elements*, London 1957, p. 464.

23 C. G. BEDREAG, *Bull. Inst. Polytech. Jassy*, 3 (1948) 317.

24 G. T. SEABORG, *Transurane, Synthetische Elemente*, Stuttgart 1966, p. 97; *Man-made Transuranium Elements*, New Jersey 1963.

25 N. BOHR, *Drei Aufsätze über Spektren und Atombau*, Brunswick 1922, pp. 131 ff.

26 N. BOHR, *The Theory of Spectra and Atomic Constitution*, three Essays, Cambridge 1922, p. 70.

27 V. M. GOLDSCHMIDT, *Geochemische Verteilungsgesetze der Elemente*, Kristiania 1924, Vol. II p. 23.

28 V. M. GOLDSCHMIDT, *Fortschr. Mineral. Krist. Petrog.*, 15 (1931) 73.

29 C. JANET, *Chem. News*, 138 (1929) 372.

30 C. JANET, *La Structure du Noyau de l'Atome considérée dans la Classification périodique*, Beauvais 1927.

31 C. JANET, *Essais de Classification hélicoïdale des Eléments chimiques*, Beauvais 1928.

32 V. KURBATOV, *Zakon D. I. Mendeleeva (The Law of Mendeleev)*, Leningrad 1925, p. 398.

33 C. JANET, *Chem. News*, 138 (1929) 388.

34 C. JANET, *La Classification hélicoïdale des Eléments chimiques*, Beauvais 1928.

35 R. H. LE ROY, *J. Chem. Educ.*, 8 (1931) 2052.

36 J. D. MAIN SMITH, *J. Chem. Soc.*, (1927) 2029.

37 M. HAÏSSINSKY, *Génie Civil*, 96 (1930) 210.

38 C. G. BEDREAG, *Compt. Rend.*, 197 (1933) 838.

39 C. G. BEDREAG, *Bul. Facultât. Stiinte Cernăuti*, 6 (1932) 197.

40 F. EPHRAÏM and M. MEZENER, *Helv. Chim. Acta*, 16 (1933) 1257.

41 G. E. VILLAR, *Ann. Acad. Brasil. Sci.*, 12 (1940) 51.

42 G. E. VILLAR, *J. Chem. Educ.*, 19 (1942) 329.

43 G. E. VILLAR, *J. Chem. Educ.*, 19 (1942) 286.

44 W. S. ANDREWS, *Sci. Monthly*, 27 (1928) 535.

45 W. SWINNE, *Wiss. Veröffentl. Siemens-Konzern*, 10 (4) (1931) 137.

46 G. E. VILLAR, *J. Inorg. Nucl. Chem.*, 28 (1966) 25.

47 G. E. VILLAR, *Bol. Fac. Ing. Agrimensura Montevideo*, 7 (1959) 35; *C.A.*, 54 (1960) 14834.

48 G. E. VILLAR, *Bol. Fac. Ing. Agrimensura Montevideo*, 3 (1938) 233; *Anales Asoc. Quím. Argentina*, 26 (1938) 126.

49 W. FINKE, *Z. Physik*, 126 (1949) 106.

50 S. DOCKX, *Théorie fondamentale du Système périodique des Eléments*, Brussels 1959, p. 82.
51 L. TALPAIN, *J. Phys. Radium* [8], 6 (1945) 176.
52 Y. TA, *Compt. Rend.*, 221 (1945) 441.
53 K. STARKE, *Z. Anorg. Allgem. Chem.*, 251 (1943) 251.
54 C. G. BEDREAG, *Compt. Rend.*, 215 (1942) 537.
55 C. G. BEDREAG, *Ann. Sci. Univ. Jassy*, 281 (1942) 139, 143.
56 C. G. BEDREAG, *Naturwissenschaften*, 31 (1943) 490.
57 C. G. BEDREAG, *Bull. Sect. Sci. Acad. Roumaine*, 25 (1943) 410.
58 T. S. WHEELER, *Chem. & Ind.*, 25 (1947) 639.
59 D. F. STEDMAN, *Can. J. Research*, 25B (1947) 199.
60 G. T. SEABORG, *Chem. Eng.*, 23 (1945) 2190; 24 (1946) 1192.
61 G. T. SEABORG, *Nucleonics*, 5 (1949) No. 5, 16.
62 G. T. SEABORG, *Science*, 104 (1946) 379.
63 G. T. SEABORG, *Transurane, Synthetische Elemente*, Stuttgart 1966, p. 25; *Man-made Transuranium Elements*, New Jersey 1963.
64 Z. SZABÓ, *Phys. Rev.*, 76 (1949) 147.
65 L. M. SIMMONS, *J. Chem. Educ.*, 24 (1947) 588.
66 G. E. VILLAR, *Bol. Soc. Quím. Peru*, 13 (1947) 73.
67 A. N. WRINGLEY, W. C. MAST and T. P. McCUTCHEON, *J. Chem. Educ.*, 26 (1949) 248.
68 P. W. SCHENK, *Österr. Chemiker-Ztg.*, 50 (1949) 52.
69 J. D. CLARK, *Science*, 111 (1950) 661.

Chapter 15

Limits of the periodic system

15.1. Introduction

We have already seen *(7.10)* that more than one investigator was led to predict the existence of elements with an atomic weight below that of hydrogen. The examination of the spectra of the sun and other stars, which attracted so much attention at the turn of century, seemed to support this view strongly. Indeed, before the atomic theory was advanced and experimentally verified, there was no reason to consider hydrogen as absolutely the lightest element. It was therefore an open question whether the scale of atomic weights had a lower limit and what element was the first in the periodic system. Although there were many hypotheses about the existence of elements lighter than hydrogen, no conclusion about a limit could be drawn. Only the neutron was ever considered, but that came after more was known about the atomic nucleus. Some investigators wanted to include the electron in the periodic system, although it was known that this elementary particle was not an element.

The hypotheses about an upper limit of the periodic system were even more spectacular. These had a more satisfactory scientific basis, certainly after the 1920s. In connection with the absence of elements heavier than uranium, two main questions were asked:

(1) Is the absence of these elements connected with the instability of heavier atomic nuclei?

(2) Could heavier elements occur in other parts of the universe?

An answer to these questions could be obtained in various ways: from qualitative calculations alone (by working from a hypothesis), or by taking quantitative calculations as the basis and thus entering the field of theoretical physics. Nevertheless, more than one such prediction was verified experimentally.

It is a remarkable fact that Newlands *[1]* was the only discoverer of the periodic system to consider other possibilities beyond the accepted limits of atomic weight. He saw the necessity for a lower limit because of his introduction of natural and ordinal numbers, which strongly accentuated the relative character of the atomic weights (see *16.3*).

References p. 335

15.2. Atomic weights of less than one[1]

Reynolds *[2]* was one of the first investigators to put forward hypotheses relating to elements with a lower atomic weight than that of hydrogen. When he constructed his graphical representation of the elements in 1886 (Fig. 111, p. 224; *6.13*), he concluded that there must be another period of elements having hydrogen as its 7th member. It seemed quite conceivable that the zigzagging series of elements could originate at the O-axis in the graph, and this would imply missing elements with valences of 2, 3, 4, -3, -2, -1. There were insufficient grounds for prediction regarding the first period, even though Mendeleev had not decided whether or not there were 6 elements between hydrogen and lithium (see *7.11*). Two years later, Samuel Haughton *[3]* referred to Reynolds' conclusion in suggesting that these elements with atomic weights between 0 and 1 might have something to do with heat, light or electricity, which was a peculiar thought at a time when it had been accepted for certain that the latter were weightless. Haughton apparently did not realize that classification in the periodic system required analogies with properties of elements already placed, unless a new group was considered. These "super-light" elements cannot be classified on the basis of their properties, certainly not of their atomic weight.

It is extremely interesting that Mendeleev should have put forward a hypothesis on this point. In 1904, towards the end of his life, Mendeleev began to consider the more spectacular aspects of chemistry. In a series of articles entitled *An attempt towards a chemical conception of ether [4]*, in which he mentioned the classification of the noble gases in the zero group as new evidence of the validity of his system, he assumed the existence of a few more elements. Above helium in the zero row of the periodic system he saw a place for an undiscovered element x, and above neon in the first row a place for an element y next to hydrogen. Mendeleev also thought there might be an element between hydrogen and helium and a halogen with an atomic weight of 3, because there were only 4 halogens but 5 alkali metals. On the analogy of the ratios between other atomic weights, the ratio of He : y should be about 10 : 1. The atomic weight of y would then be no more than about 0.4. Here Mendeleev thought of coronium (*7.10*), which many investigators considered to be a new element occurring on the sun. In 1911, on the basis of spectroscopical data, Nicholson gave 0.5131 as its atomic weight.

The element x, according to Mendeleev, could be the world ether, with an atomic weight of no more than 0.17. Mendeleev provisionally suggested that this entity be called newtonium. If its atomic weight proved to be far lower, then this element could only exist as a gas free in the atmosphere and would therefore be the world ether, which must have an atomic weight between 9.6×10^{-7} and 5.3×10^{-11} owing to its velocity in space.

The idea of including "elements" with low atomic weights in the system was actually originated by Mendeleev's countryman Nikolai Nikolavitsch Beketov *[5]* (1827–1911), who in this connection also thought of ether and of the electron, which he regarded as a structural component.

It is strange that Mendeleev, who had always been a fervent empiricist, should have become so much less critical towards the end of his life. Seen in this light, the unsuccessful attempts by the investigators to be discussed below (who were men of far less reputation than Mendeleev) to determine the limits of the periodic system are

[1] See also *7.10*.

easier to understand. In 1906, Sima Losanitsch [6] (1847–1935), professor of organic chemistry in Belgrade, classified coronium and newtonium in his system (Fig. 136), without any comment, as first member of the zero and first periods, respectively. This created an appreciable number of vacant spaces.

periods	O	I	2	3	4	5	6	7											
small 0	Nw	–	–	–	–	–	–	–											
1	Ch	H	–	–	–	–	–	–											
2	He	Li	Be	B	C	N	O	F											
large 3	X	Cs	Ba	La															
4	–	–	–	Yb 172	–	Ta 182	W 183	–	(Os 190	Ir 192	Pt) 193	Au 196	Hg 199	Tl 203	Pb 205	Bi 207	–	–	
5	–	–	Ra 223	Em	Th 231	Ac	U 237	–	(–	–	–)	–	Hg^r	Cn 254	Pb^r	Po	Te^r	–	

Fig. 136   A detail of the system of Losanitsch (1906).

In 1910 Morosoff [7] assumed protohydrogen as an element with an atomic weight of 1.000, i.e., lower than that of hydrogen (1.008), as the basic structural component of all elements. The element that Nicholson [8] called protohydrogen in 1914 was assigned an atomic weight of 0.0818 on the basis of spectroscopic data.

During the first and second decades of the present century, the classification of coronium, ether, the electron, etc., seemed imperative to an appreciable number of investigators. In 1911 Emerson [9] placed coronium in the series of the noble gases and ether (see 9.5), which he identified with the electron[2], as a homologue of hydrogen (Fig. 63, p. 165). We have seen that Loring [12] assumed, in the space next to hydrogen, a noble gas satellite (St) with an atomic weight of 0.27. In 1923, on very slender grounds (Fig. 137), Loring [13, 14] assumed the existence of an element with the

Active elements preceding	He=	2	At. No.	He=	2	2 −	2= 0
	Ne=	2 + 7= 9		Ne=	10	10 −	9= 1
	Ar=	9 + 7= 16		Ar=	18	18 −	16= 2
	Kr=	16 + 17= 33		Kr=	36	36 −	33= 3
	Xe=	33 + 17= 50		Xe=	54	54 −	50= 4
	Rn=	50 + 31= 81		Rn=	86	86 −	81 = 5

Fig. 137   A system devised by Loring to demonstrate the existence of the element number 0.

atomic number 0, but could not make any predictions about its properties. In his opinion the electron should be excluded, because of its negative charge [15]. Ether and coronium he retained, however, as possibilities [16]. In 1942, building on Darrow's work [17], Loring [18, 19] set up more realistic relationships (Fig. 138), but these actually indicate only long-known regularities in the differences between the atomic numbers of the noble gases. Loring also placed the neutron[3], which had been discovered 10 years before, in this series, at place 0. He ignored the fact that the neutron had nearly the same atomic weight as hydrogen and therefore could be seen as a limit

[2] The electron, with an atomic number of 0, was assumed by Rydberg [10, 11] as a homologue of helium.
[3] Vosdvishenshi [20] suggested the name *mendelevium* instead.

of the periodic system unless coronium, for instance, was rejected. At the same time, Loring considered the possibility of placing the positron before hydrogen. He also put forward (as late as 1942) the remarkable idea of placing the α-particle as first member of the group of noble gases on the basis of the above-mentioned numerical relationships. This procedure doubled all the differences in atomic number between the noble gases in the series: 1^2, 2^2, etc., and Loring also had to assume a negative atomic number. In classifying the neutron he had been preceded much earlier by Antonoff [21], who soon after its discovery assigned it to place 0 in the periodic system, where he also put the combination of the electron and the positron, the neutrino.

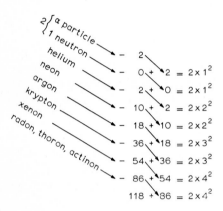

Fig. 138   Atomic number relations of Loring (1942).

In 1926, Von Antropoff [22, 23, 24] also pointed to the vacant space next to hydrogen with the order number 0. He reserved this place for an element yet to be discovered and meanwhile called this neutronium. E. Meyer [25] thought Von Antropoff's argument was unacceptable; he visualized this place as occupied by the proton: since its shell has zero electrons, it would fit very well next to the hydrogen atom, which has one electron. Von Antropoff [26] rightly objected to this hypothesis, because the question of whether or not the place at 0 is occupied by an element neutronium is unimportant in the absence of any experimental data. He also argued that the nuclear charge determines incorporation into the periodic system; consequently, an element with nuclear charge 0 would have to occupy the place 0, which excludes the proton. According to Meyer's reasoning, for example, Na would occur in two places, as ion and as atom. When Meyer [27] countered by suggesting that both conceptions be maintained, he exposed his ignorance of the fundamental difference between an element and an ion, or at least his inability to employ the two conceptions[4] separately, even as late as 1926.

---

[4] Meyer's deduction of the other limit of the periodic system is also meaningless. He argued that uranium should be the last element because of its remarkably regular octahedral structure. He did not disclose the basis of his assumption.

In a system published by Achimov [28] in 1946, the electron (e) and the neutron (n) occur above hydrogen and helium, respectively. The need to insert this basic matter in the periodic system was evidently very persistent.

It is difficult to avoid the conclusion that most of the above suggestions had almost no scientific basis.

15.3. Early attempts to define the upper limits of the periodic system

In 1884 Mills [29] (1840–1921) from Glasgow was the first to attempt a determination of the last of the series of elements. What he in effect did was to seek a formula for the atomic weights that would show the atomic weight of uranium to be the highest possible value. His success is hardly surprising, because a formula containing sufficient numbers chosen at random always leads to the desired object, even when the number of different known and unknown quantities is small[5].

The relationships between the atomic weights of the elements (y) and the numbers of parts of the periods of the periodic system he chose ($p = 1 \to 16$) are:

$$y = p \cdot 15 - 15 \, (0.9375)^x \text{ (where } x \text{ is a whole number)}^{[6]}.$$

For the series including uranium, $p = 16$. With the highest value for x, viz. ∞, the atomic weight $y = 240$, which led Mills to assume that uranium was indeed the final element of the periodic system, the more so as the sum of all terms $15(0.9375)^x$, x being $0, 1, 2, \ldots \infty$ is also exactly[7] 240. Mills cannot be said to have been very scientific in his reasoning; his formula was unfounded and the correct result mere chance, but his work deserves mention because it represents the first attempts to solve the problem.

When radioactivity was discovered around the turn of the century, several scientists saw a possible connection between this phenomenon and the limits of the periodic system. In 1903 Beketov [5] took up this point, but not in any great depth.

Losanitsch [6], who, like Beketov, took very small particles as the basis of the

---

[5] In his formula the number 15 occurs three times; and 0.9375 is 15:16. The highest value for p is 16.
[6] For the reader's guidance, we compare below the determined and calculated values for the atomic weights of the part of a period, i.e., the series for which $p = 3$ holds. The formula is then $y = 45 - 15 \, (0.9375)^x$.

	x	y	present atomic weight
P	1	30.96	30.98
S	2	31.98	32.07
Cl	7	35.37	35.46
K	14	39.02	39.10
Ca	17	39.90	40.08
Sc	42	43.98	44.96

$$[7] \; \sum_{0}^{\infty} 15 \cdot \left(\frac{15}{16}\right)^x = \frac{15}{1 - \frac{15}{16}} = 15 \times 16 = 240.$$

structure of the periodic system (he saw hydrogen as containing 1000 of these particles or corpuscles), made each series of homologous elements end with a radioactive element (Fig. 136, p. 327). For these end elements he chose the elements announced as novel[8] shortly after 1900. He could do so because little was known about them, which also holds for his introduction of elements with an atomic weight greater than that of uranium. The first element of this series mentioned by name was radio-lead[9]. The assumed element carolinium[10] had an atomic weight of 254. Losanitsch [40] thought he had demonstrated dvimercury in mercury ores as a radioactive homologue of mercury. That he did not classify polonium as ekatellurium must have been partly due to the fact that this element, which had been discovered in 1898, had as yet no definite atomic weight. Although Losanitsch did not mention this explicitly or indicate it in his system, he must have considered dvi-iodine to be the final element of the periodic system.

Loring [14], although many years later, also related the end of the series of elements to the instability of the elements. When, in 1923, he placed the still undiscovered elements in a series with scandium as the first element, the differences in atomic weight formed the next series (Van den Broek series):

atomic number   difference

$$
\begin{array}{ccl}
93 \\
& 2 & = 2 \times (1) \\
91 \\
& 6 & = 2 \times (1 + 2) \\
85 \\
& 10 & = 2 \times (1 + 2^2) \\
75 \\
& 14 & = 2 \times (3 + 2^2) \\
61 \\
& 18 & = 2 \times (5 + 2^2) \\
43 \\
& 22 & = 2 \times (7 + 2^2). \\
21 \\
\end{array}
$$

[8] Losanitsch classified the element emanium (Em), announced in 1904 by Giesel [30] to be a homologue of lanthanum, as a quinquevalent homologue of tantalum. Debierne [31] could only see this element as the actinium already discovered by him in 1899.

[9] Hofmann and Strauss [32, 33], who in 1901 isolated radio-lead from thorium and uranium minerals, first determined its atomic weight as > 260. Shortly afterwards [34], they came to the conclusion that the atomic weight was below that of ordinary lead. In seeking a place for radio-lead between the other elements they noticed the place of ekamanganese, and the space between tin and lead, but this latter place only occurs when the rare earths are considered as homologues of the remaining elements. According to Debierne [35], radio-lead and radio-tellurium (demonstrated by Marckwald [36] in bismuth minerals) were identical with the newly-discovered polonium.

[10] While working with thorium salts, Brauner [37] believed in 1901 that he had found a new element with an atomic weight of 280.7. Baskerville [38], who made similar studies, gave the name carolinium to this element and, considering it as quadrivalent, determined its atomic weight at a value between 260 and 280. Three years later, Baskerville [39] determined this atomic weight at 255.6 and designated the residual pure thorium as berzelium after its discoverer.

As the smallest difference in the series, i.e., 2, was reached with element 93, Loring considered the existence of elements heavier than uranium to be impossible. This conclusion had little foundation, and furthermore element 91 had already been discovered as protactinium in 1917; element 75 was found to be rhenium in 1925, and element 87 (francium) had not yet been discovered in 1923.

15.4. Deduction of the atomic weight of the heaviest element by means of physical theories

In 1923 Rosseland [41] approached the problem in a more promising way. He started with the formula for the smallest distance (d) between the atomic nucleus and the nearest electron orbits, i.e., with the azimuthal quantum number $k = 1$. The formula is by approximation:

$$d = \frac{d_0}{2N} (1 - \alpha^2 N^2),$$

where d_0 = radius of the hydrogen atom,
  N = atomic number, nuclear charge,
  $\alpha = \dfrac{2\pi e^2}{hc} = 7.2 \times 10^{-3}$ (fine-structure constant),
  h = constant of Planck,
  c = velocity of light, and
  e = charge of an electron.

For the uranium atom this distance amounts to $d = 15 \times 10^{-12}$. Calculation of the radius of the uranium nucleus from the energy of the α-particles emitted by the nucleus, and assumption that this energy must be ascribed to the electrostatic repulsion of the nucleus, gave Rosseland the same value. Because the electron orbits would come even closer to the nucleus and these nuclei would acquire larger dimensions in elements with a higher nuclear charge, Rosseland came to the surprising conclusion that the occurrence of many more elements after uranium was unlikely[11].

In the same year, Bohr [44] stated that each electron orbit with the quantum number n_k touches the atomic nucleus, the equation being

$$\frac{N}{k} = \frac{hc}{2\pi e^2} = 137.$$

---

[11] Several investigators tried to demonstrate on similar grounds that the existence of elements with a higher atomic weight than 92 was unlikely. According to Kossel [42], great magnetic forces might induce electrons of these large atoms to fall on the nucleus. In Noddack's opinion [43], it was likely that the penetration-orbits of the electrons approached the nucleus very closely, which was in agreement with Kossel but was not convincing, at least with respect to the limit at element 92.

If in this formula the smallest integer, i.e., $k = 1$, is taken as the minor ($=$ azimuthal) quantum number, it follows that $N = 137$ for the nuclear charge. If so simple a calculation were valid, the limit of the system would lie at a nuclear charge of 137. Bohr approached the problem from another point of view and also reached a different result from Rosseland's, even though both investigators reasoned from the fine-structure constant.

According to Sommerfeld [45], the same result as Bohr's could be obtained by considering the formula for the energy of the fine structure as follows:

$$1 + \frac{W}{m_0 c^2} = \sqrt{1 - \frac{\alpha^2 Z^2}{k^2}} \text{ (for a circular orbit), where}$$

$$W = E_{\text{kin}} + E_{\text{pot}}; \quad Z = \text{atomic number.}$$

To retain the energy and the corresponding quantum orbit, the second member must not be imaginary. If this expression is taken as zero[12], $Z = 137$ applies to $k = 1$ and $Z = 68.5$ to $k = \frac{1}{2}$. This only holds if the theory of relativity is left out of consideration, so the last word on the subject had not yet been spoken.

According to the geochemist Goldschmidt [47], the system of elements seemingly ends with element 92, where the lithophile elements cease to exist. But he thought these should be followed by siderophile elements (see also p. 223), which, after the element with a nuclear charge of 119, are followed in turn by the former. Goldschmidt arrived at this assumption by extrapolation of the curve of the atomic volumes, also employing analogies occurring in the external electron structure of the atoms.

Flint and Richardson [48, 49] tried to apply the theory of relativity as well as Heisenberg's uncertainty principle to this problem. For an electron moving with a velocity v about the nucleus, the time τ, considered in the rest frame of the electron, is expressed in ordinary time by the relation

$$\tau = t\sqrt{1 - v^2/c^2} \tag{1}$$

where $c =$ velocity of light. Heisenberg's uncertainty principle can be expressed as follows:

$$\Delta E \times \Delta t > h,$$

where $E =$ energy and $h =$ Planck's constant. By substituting in this relation the time

[12] Walter Gordon [46] (1893–1939), scientific collaborator at the State Institute of Physics in Berlin and afterwards professor of physics in the University of Hamburg, who developed the more advanced quantum theory of Dirac, remarked in his study on the energy levels of the hydrogen atom

that $\sqrt{1 - \alpha^2 Z^2}$ should be $> \frac{1}{2}$. It follows for the nuclear charge that $Z < \dfrac{\sqrt{3}}{2\alpha} = 118.7$, and this

condition is fulfilled.

of revolution of an electron $t = 2\pi r/v$, where $r =$ radius of the electron orbit, as well as the energy $E = m_0c^2$, where $m_0 =$ rest mass, this relation becomes:

$$\sqrt{1 - v^2/c^2} \times 2\pi r/v > h/m_0c^2 \qquad (2)$$

It follows from the expressions of the quantized angular moment $mrv = h/2\pi$ and the equality of the centrifugal force to the coulomb attraction between electron and nucleus that

$$\frac{mv^2}{r} = \frac{Ne^2}{r^2} \qquad (3)$$

$$v = \frac{2\pi Ne^2}{h} \qquad (4)$$

and $\qquad r = \dfrac{h}{2\pi mv} \qquad (5)$

in which $\quad m = \dfrac{m_0}{\sqrt{1 - v^2/c^2}} \qquad (6)$

Substitution of (5) and (6) in (2) gives:

$$1 - v^2/c^2 > v^2/c^2 \quad \text{or} \quad v < c/\sqrt{2} \qquad (7)$$

To find the value for N, v must be substituted in (4):

$$N < \frac{1}{2\sqrt{2}} \times \frac{hc}{e^2}$$

When taking this into account Flint and Richardson came to the conclusion that N must be smaller than 98, a far smaller value than the one obtained without the introduction of the theory of relativity.

Six years later, Glaser and Sitte [50] applied both the quantum theory of Dirac and the classical Lorentz formula to this problem and after calculations which cannot be considered within the scope of this work, came to an N-value of 90.5 ± 0.5. That this value is smaller than even the atomic number of the heaviest known element, was ascribed by the quantum theorists to the approximations they had applied.

The *universe theory* of Eddington [51] also influenced the calculation of the atomic number of the heaviest possible element. In 1932 Narliker [52] applied this theory. His aim was to harmonize Bohr's value of 137 with the Eddington theory. Stated briefly, his results showed that both the electrical charge of protons and electrons and the number of these particles forming a universe, defined as the small isolated world of elementary particles, amounts to $2 \times 3 \times 2^4 = 96$. According to Narliker, the maximum number of quantum states for a nucleus with corresponding satellite electrons, i.e., 137, should be equal to the number of degrees of freedom in a system of two charges. Eddington showed further that 136 of these degrees of freedom were relativity rotations. When it had also been shown that the distribution of charge is invariant for 91 of these rotations, it was only a small step for Narliker to conclude that the highest possible number of electrons in a system is 91 + 1. Until 1935 no objection, at least of a practical nature, was raised against the result of this calculation.

After Fermi and Hahn and his co-workers concluded that they had demonstrated the existence of heavier elements, arguments were advanced to support the view that additional structural components of the atom should be included in the theoretical deduction of the limit of the periodic system. The importance of this pronouncement by Diersche [53] was enhanced by the fact that the new elements were obtained by bombardment with neutrons. Although the new elements proved to be radioactive, so that Feather [54] did not expect a very high upper limit of stability, one transuranium element after another was synthetized [55], with the result that several theories and hypotheses about the upper limit of the periodic system collapsed. We may mention that of Masriera [56] who, like Narliker, extended Eddington's theory and obtained an atomic number of 96 as the result of his hypothesis. This meant only that Masriera had to assume that the number of particles in a *universe*, deduced by Eddington, i.e., $N_1 = 2 \times 3 \times 2^4 = 96$, is equal to the largest number of protons in one nucleus. In addition to accepting this hypothesis, Masriera had to assume that neutrons were not independent particles. In spite of the author's defense of the hypothesis [57, 58], its stand against criticism was comparatively short-lived; only four years after its formulation in 1946, the elements 97 (berkelium) and 98 (californium) were synthetized [59].

A more recent development in the computation of the upper limit of the periodic system is attributable to Wheeler [60] and Werner [61]. Using the available data on the stability of the radioactive nuclei, they came to the conclusion that the limit did not lie at a nuclear charge of 137 and that an atom with an atomic number of 147 and an atomic weight of 500 could still be stable. The practical possibility of producing such heavy elements is doubted, among others by Seaborg [62], who has synthetized many transuranium elements.

The question of nuclear stability was semi-empirically discussed by Von Weizsäcker [63] on the basis of the competition between attractive nuclear forces and repulsive Coulomb forces. As the parameters in this formula had been determined fairly accurately, it was shown that reliable predictions about nuclear stability can be made.

Further, the wealth of data on isotope production, e.g., by using high-current particle accelerators or nuclear detonations, has provided a more thorough understanding of nuclear structure and nuclear stability. Since then, it has been assumed that at least some stable nuclides having $Z > 100$ can be synthesized (and, in fact, have been isolated), thereby invalidating most of the earlier theories. In retrospect it may be seen that earlier speculations, e.g., those involving the number of atomic electrons, left the nuclear force entirely out of consideration.

It may also be mentioned that there is a difference in point of view between chemists and physicists as to the definition of nuclear stability. To a physicist, for instance, a nuclide such as $^{232}_{90}\text{Th}$ (having $1.5 \times 10^{-3}\,\%$ weight abundance in the earth's crust) is not essentially more stable (half-life 1.4×10^{10} years) than, say, the isotope $^{256}_{102}$ No (having a half-life of 2.7 sec).

15.5. Conclusion

Since atomic weights are relative numbers, the *lower limit* need not be unity. Deviations of the atomic weights from Prout's hypothesis (whole numbers) cast some doubt on the assumption that there are no elements with atomic weights smaller than unity.

Newlands, the only discoverer of the periodic system to discuss its lower and upper limits, assumed that the relative significance of the atomic weights justified the introduction of "ordinal numbers". His conclusion that there should be a simple relation between order numbers and atomic weights seemed quite logical in those days, although we now know that such correlations are limited to the lighter elements.

Before the atomic number was definitely substituted for the atomic weight as a consequence of Bohr's atomic theory, there could not be any certainty that hydrogen really is the lightest element. In fact, erroneous interpretations of solar and other astral spectra led several investigators—including Mendeleev—to assume the existence of elements with atomic weights below unity, the so-called proto-elements.

Even after it became known that the number of protons indicated the order number of the elements in the periodic system, some scientists continued to create places ahead of hydrogen, e.g., for the neutron, the electron, and even the α-particle or the hydrogen ion.

The introduction of the atomic number had less effect on speculations about the *upper limit* of the system. Before some information became available about atomic and nuclear structures, it was assumed that several disintegration products were trans-uranium elements. After this (1913), the upper limit of the periodic system was considered more seriously, and the application of quantum mechanics to the problem of the stability of the atom was emphasized. The complexity of matter made it difficult to arrive at definitive conclusions, however. New elements are still being synthetized fairly regularly, but since element 99 (einsteinium) they have all been short-lived isotopes.

References

1 J. A. R. NEWLANDS, *Chem. News*, 37 (1878) 255.
2 J. E. REYNOLDS, *ibid.*, 54 (1886) 1.
3 S. HAUGHTON, *ibid.*, 58 (1888) 93, 102.
4 D. I. MENDELEJEF, *Prometheus*, 15 (1904) 97, 121, 129, 145; *An Attempt towards a chemical Conception of Ether*, London 1904.
5 N. N. BEKETOV, *Zhur. Russ. Fiz. Khim. Obshchestva*, 35 (1903) 189.
6 S. M. LOSANITSCH, *Die Grenzen des periodischen Systems der chemischen Elemente*, Belgrade 1906.
7 N. A. MOROSOFF, *Die Evolution der Materie auf den Himmelkörpern, eine theoretische Ableitung des periodischen Systems*, Dresden 1910.

8 J. W. Nicholson, *Compt. Rend.*, 158 (1914) 1322.

9 B. K. Emerson, *Am. Chem. J.*, 45 (1911) 160.

10 J. R. Rydberg, *Acta Regia Soc. Physiograph. Lundensis*, 24 (1913) No. 18.

11 J. R. Rydberg, *Lunds Univ. Årsskr.*[2], 9 (1923) No. 18.

12 F. H. Loring, *Chem. News*, 100 (1909) 281.

13 F. H. Loring, *Chem. News*, 126 (1923) 307.

14 F. H. Loring, *The Chemical Elements*, London 1923.

15 F. H. Loring, *Chem. News*, 126 (1923) 325, 371.

16 F. H. Loring, *Chem. News*, 127 (1923) 225.

17 K. K. Darrow, *The Renaissance of Physics*, New York 1937, p. 208.

18 F. H. Loring, *Chem. Products*, 6 (1943) 51.

19 F. H. Loring, *Chem. Products*, 5 (1942) 69.

20 H. S. Vosdvishenshi, *Zhur. Obshcheĭ Khim.*, 19 (1949) 1653.

21 G. Antonoff, *J. Soc. Chem. Ind.*, 53 (1934) 728.

22 A. von Antropoff, *Z. Angew. Chem.*, 39 (1926) 722.

23 A. von Antropoff, *Z. Elektrochem.*, 32 (1926) 423.

24 A. von Antropoff and M. von Stackelberg, *Atlas der physikalischen und anorganischen Chemie* Berlin 1929.

25 E. Meyer, *Z. Elektrochem.*, 33 (1927) 189.

26 A. von Antropoff, *Z. Elektrochem.*, 33 (1927) 475.

27 E. Meyer, *Z. Elektrochem.*, 33 (1927) 476.

28 E. I. Achimov, *Zhur. Obshcheĭ Khim.*, 17 (1947) 1241.

29 E. J. Mills, *Phil. Mag.* [5], 18 (1884) 393.

30 F. Giesel, *Ber.*, 37 (1904) 1696, 3963; 38 (1905) 775.

31 A. Debierne, *Compt. Rend.*, 139 (1904) 538.

32 K. A. Hofmann and E. Strauss, *Ber.*, 33 (1900) 3126; 34 (1901) 3033.

33 K. A. Hofmann and E. Strauss, *Ber.*, 34 (1901) 407.

34 K. A. Hofmann and E. Strauss, *Ber.*, 34 (1901) 907.

35 A. Debierne, *Compt. Rend.*, 139 (1904) 281.

36 W. Marckwald, *Ber.*, 35 (1902) 4239; 36 (1903) 2662.

37 B. Brauner, *Proc. Chem. Soc. (London)*, 17 (1901) 67.

38 C. Baskerville, *J. Am. Chem. Soc.*, 23 (1901) 761.

39 C. Baskerville, *J. Am. Chem. Soc.*, 26 (1904) 922.

40 S. M. Losanitsch, *Ber.*, 37 (1904) 2904.

41 S. Rosseland, *Nature*, 111 (1923) 357.

42 W. Kossel, *Naturwissenschaften*, 16 (1928) 298.

43 I. Noddack, *Angew. Chem.*, 47 (1934) 301.

44 N. Bohr, *Ann. Physik* [4], 71 (1923) 266.

45 A. Sommerfeld, *Atombau und Spektrallinien*, 4th ed., Brunswick 1924, p. 465.

46 W. Gordon, *Z. Physik*, 48 (1928) 11.

47 V. M. Goldschmidt, *Geochemische Verteilungsgesetze der Elemente*, Kristiania 1924, Vol. II p. 23.

48 H. I. Flint and O. W. Richardson, *Proc. Roy. Soc. (London)*, A 117 (1928) 637; *Nature*, 129 (1932) 746.

49 See e.g. H. I. Flint, *Nature*, 129 (1932) 746.

50 W. Glaser and K. Sitte, *Z. Physik*, 87 (1934) 674.

51 See e.g. A. Eddington, *Relativity Theory of Protons and Electrons*, Cambridge 1936, p. 136.

52 V. V. Narliker, *Nature*, 129 (1932) 402.

53 M. Diersche, *Chem. Ztg.*, 59 (1935) 833.

54 N. Feather, *Proc. Roy. Soc. Edinburgh*, A62 (1945) 211.

55 See e.g. J. W. van Spronsen, *Chem. Weekblad*, 63 (1967) 145; G. T. Seaborg, *Actinides Reviews* 1 (1967) 3.

56 D. M. Masriera, *Mem. Acad. Ciènc. Arts Barcelona*, 28 (1946) No. 4, 579.

57 D. M. Masriera, *Ibérica* [2], 2 (1946) 210.

58 D. M. Masriera, *Phys. Rev.*, 71 (1947) 458.

59 J. W. van Spronsen, *Chem. Weekblad*, 46 (1950) 667.

60 J. A. WHEELER, *Niels Bohr and the Development of Physics*, London 1955, Chap. 9; *Proceedings of the International Conference on the Peaceful Uses of Atomic Energy, Geneva August 1955*, New York 1966, p. 155.

61 F. G. WERNER and J. A. WHEELER, *Phys. Rev.*, 109 (1958) 126.

62 G. T. SEABORG, *Transurane*; *Synthetische Elemente*, Stuttgart 1966; *Man-made Transuranium Elements*, New Jersey 1963.

63 C. F. VON WEIZSÄCKER, *Z. Physik*, 96 (1935) 431.

Chapter 16

Priority aspects and further development of the periodic system by its discoverers [1]

16.1. Introduction

In order to establish priorities, one needs both a satisfactory definition of the subject matter involved and reliable information as to who was the first to describe this information in the terms of that definition. Obviously, in our case the latter is much more difficult than the former. As we have emphasized earlier, scientific communication before 1870 was much slower and less complete than it is today.

Thus, the simplest criterion—also used, for example, in patent law—would be the date of publication or that of submission to a recognized scientific journal; in addition, priority could be based on a lecture attended by qualified scientists. However, this modern criterion does not seem to us to be quite applicable to the period of discovery of the periodic system. A striking example of the difficulties involved is formed by Couper's work on the theory of atom binding. He requested that his paper should be read by one of the members of the French Academy of Sciences, but this presentation was postponed [2, 3, 4]. When, somewhat later, Dumas declared his willingness to do so, it was already too late: similar theories had been published by Kekulé four weeks before Couper's work appeared in the *Comptes Rendus*. Another example is the failure of the French Academy of Sciences to publish the illustrations accompanying Béguyer de Chancourtois' paper on his periodic system—apparently because of typographical difficulties. To some extent, Newlands was similarly thwarted. The papers he read before the Chemical Society in 1866 and 1873 were not inserted in the Society's periodical. Fortunately, they appeared in the *Chemical News*.

To investigate priority in the discovery of the periodic system of elements, we shall start from the following definition: The periodic system of elements is a sequence of all the (known) elements arranged according to increasing atomic weight in which the elements with analogous properties are arranged in the same group or column.

In principle, the same discovery can be made independently by several scientists. They are all to be regarded as discoverers, even when their results were arrived at years apart. That the discoverer should be aware immediately of all the consequences of his discovery seems to us to demand too much. We disagree with the view of Mendeleev, who refused to recognize Meyer as an independent discoverer because he had not predicted any elements; Meyer's system did have vacant spaces.

Study of the sources leads to the conclusion that six scientists had about the same stature if slight differences in the dates of publication are ignored. Thus, we are inclined to disagree with conclusions strongly affected by national pride[1] favouring Mendeleev (Russia), Meyer (Germany), Newlands (England), or to some extent De Chancourtois (France) as the main discoverer! In fact, we would add Odling as well as Hinrichs to the list of co-discoverers, though the latter two did not even claim priority for themselves. Hinrichs, for example, merely republished his system in some of his books.

Between the discoveries of De Chancourtois in 1862 and those of Mendeleev in 1869, seven years elapsed in which one more element was discovered and atomic weights were more accurately determined. It would therefore be wrong to demand that all systems be identical: relative equality suffices. Although Mendeleev is generally acknowledged as the chief discoverer, a plea may certainly be entered for the Frenchman who was the first to construct a system, even though it was not irreproachable in all respects. It is simply that De Chancourtois' work was virtually unknown for a long time.

Newlands regarded himself, even after he had become aware of the work of his rivals, as the first and, properly speaking, also the most important discoverer. When Meyer designed his first system, he knew no more about his predecessors than did Mendeleev, Odling and Hinrichs. Odling did not take part in that conflict, but others took up arms in his behalf. Although he lived to a great age (1829–1921), Odling published very little after 1870. He himself thought that most of the credit should go to Newlands.

Before proceeding to discuss the claims of priority by or for the discoverers of the periodic system, we must first pay some attention to the claim of priority entered by Pettenkofer.

16.2. Pettenkofer's claim

Pettenkofer is the only predecessor[2] who felt justified in claiming priority, which he did a few years after he had discovered the relationships between the atomic weights of

---

[1] An example of an opinion motivated by chauvinism is found in Arija's booklet [5] on the history of the atomic theory: "Die Urkunde zur Entdeckungsgeschichte des Perioden Gesetzes beweist unbestreitbar die Priorität Mendelejews. Behauptungen über *Mitautorschaft oder über einen unabhängigen Weg* auf dem ausländischen Gelehrte zur Entdeckung des Perioden Gesetzes gekommen seien, sind legendenhaft, sind reine Erfindungen."

[2] Gladstone's share in the work preparatory to the discovery of the periodic system became the subject of discussion as late as 1895. In a dispute with Wilde [6, 7, 8], who spoke rather disparagingly of the insertion of the noble gases in the periodic system, Gladstone [9–12] referred to his results of 1853 (see 4.8). All this means is that Wilde thought more of his own share in the development of the periodic system in 1878 (see 6.20) than of the contribution made, no less than 25 years before, by Gladstone. Wilde could not be convinced that this great difference in time made it impossible to compare the two contributions.

the elements (see *4.5*). He based his claim *[13]* in 1858 on three things. First, on Dumas' second publication of 1857 in the same field, evidently because for Pettenkofer Dumas' first publication did not attack his priority (see *4.6* and *4.11*). Secondly, on the fact that Pettenkofer's lecture of 1850 was in fact published in the *Gelehrte Anzeigen der Akademie der Wissenschaften zu München*, even though it was not printed in a scientific (chemical) periodical[3], and lastly, on Pettenkofer's belief that nobody had taken any notice of his publication. This last is not quite true, because Liebig and Kopp *[14]* referred to the article in their *Jahresbericht*. Indeed, both investigators compared the groups of analogous elements with the radical series in organic chemistry, and in this Pettenkofer was ahead of Dumas. Pettenkofer's belief in the validity of the arithmetical progressions was so great that he presumed to correct equivalent weights. The value determined for the atomic weight of lithium in 1850 was 6.51, whereas the relationship with the atomic weights of the other alkali metals, viz., sodium and potassium (7–23–39), gave a value of 7, which was experimentally confirmed in 1858. The fact that the difference in atomic weights in the range Cr–Mo–W was given as 22 by Dumas and 20 by him, was considered by Pettenkofer to have little bearing on the theoretical principle.

Pettenkofer took this opportunity to point out that he certainly would have done further research after 1850 if he had been granted financial support for such work.

16.3. Newlands' claim

Newlands *[15, 16]* stated his claim to priority in the discovery of the *law of octaves* even before this discovery had been disputed. He put forward his claim in 1866 during a lecture before the Chemical Society in London (see *5.3*). After reading another paper before the same Society in 1873, in which he claimed priority over Lothar Meyer, Newlands *[17]* asked why his previous paper read to the Chemical Society had not been published in that Society's Journal. Odling, the President, replied that the Society never published purely theoretical work. This may partly explain why Newlands launched a priority campaign, as it were, which culminated in the publication of a booklet in 1884 *[18]*. He stated his views very forcibly, mainly in his own country *[19–23]*, but also in Germany *[24]*.

R. Gerstl, the London correspondent of the *Deutsche Chemische Gesellschaft*, aided him by drawing attention to the fact that Newlands had reserved places in his system for elements yet to be discovered *[25]*. But the difference between Newlands and Mendeleev is that the former made few concrete statements about the vacant places, whereas the latter made detailed predictions. Newlands' only prediction was the correct atomic weight of germanium, for which he of course claimed full priority; a claim which cannot be disputed. He did not know of the work of his predecessor, De Chan-

---

[3] The complete paper of 1850 was finally included in the *Annalen der Chemie und Pharmazie [13]*.

courtois, any more than his fellow-discoverers has been aware of it. It is noteworthy that Newlands as well as De Chancourtois, Meyer and Mendeleev, all mentioned Dumas as their predecessor, thus showing that they could properly evaluate the work of the French scientist.

In the competition for priority between Mendeleev and Meyer, the former *[26]* casually denied that Newlands had set up a system analogous to his own in 1864. Mendeleev reproached Newlands for having based his system on a dubious philosophical hypothesis, where he himself always attempted to proceed from the experimental facts. Although we prefer Mendeleev's system, he had no right to reject Newlands' precedence completely[4]. Mendeleev compared his and Newlands' contributions with the difference between the contributions of Mariotte and Lavoisier to the law of conservation of mass. Although Mariotte had stated, a century before Lavoisier, "La nature ne fait rien de rien et la matière ne se perd point", the latter must nonetheless be regarded as the discoverer of the law. It is surprising to note that Mendeleev conceded his obligation to Lenssen and Dumas (see *5.8*) but not to Newlands.

In 1884, at the termination of the bitter battle for priority, Newlands published a brochure containing all his articles on his discovery, supplemented by some quotations as evidence that he had played the leading role in the discovery. He intended comparison of his share in the discovery of the periodic system with that of the atomic theory originated by Dalton—although the latter was later somewhat modified by other investigators—to confirm his claim and make it clear that this was the central idea which Mendeleev cast into a more complete form.

It is interesting to cite some of the quotations given in Newlands' brochure. Mendeleev *[26, 29]* wrote: "It is possible that Newlands has prior to me enunciated something similar to the periodic law, but even this cannot be said of H. L. Meyer." The following passage occurs in Odling's paper *[30]* presented to the British Association for the Advancement of Science: "Mr. Newlands was the first chemist to arrange the elements in such a seriation that new ones might be predicted to exist where certain gaps are observed in the seriation of atomic weights." Lastly, from a publication by Carnelley *[31]*: "It was not, however, till within the last fifteen years that these relations were traced in a systematic manner; and it is to Newlands and especially to Mendeleev, that we owe a new field of research and a new powerful method of attacking chemical problems. The importance of the work of Newlands and Mendeleev cannot be easily overrated. The principle proposed independently by each of them will serve in the future, and has done to some extent already, to indicate those directions in which research is most needed, and in which there is most promise of interesting results. The application of the principle will also enable us to make predictions of phenomena still unknown, and will at the same time prevent many fruitless researches. It is and will be, in fact, for some time to come, the finger-post of chemical science."

---

[4] In referring to Newlands' law of octaves in his *Faraday Lecture [27]* of 1889, Mendeleev mistook the columns of analogous elements for octaves; this error was pointed out to him by Newlands himself *[28]*.

Not yet satisfied with the results of his claim, Newlands also brought out a German version of the brochure [32]. In his last German article [24] he made clear once more that Meyer and Mendeleev had indeed made their contributions to the further development, but "Es sei nicht mehr als billig und liegt im Interesse aller wahren Forschung, sei dieselbe praktisch oder theoretisch, dass dem Urheber einer Entdeckung das Verdienst seiner Arbeit zukomme."

Newlands became so preoccupied with his claims to priority that he did little work on the further development of the periodic system and contributed scarcely any fundamental improvements to it. The nature of his independent professional position in applied chemistry may have had something to do with this [15]. In 1872 Newlands [19] pointed out that, of the elements most widely distributed in nature, two occur in each of the principal groups of the periodic system:

1 H and Cl		5 C and Si
2 Na and K		6 N and P
3 Mg and Ca		7 O and S
4 Al and Fe		

Newlands was the first to realize the importance of an order number for the classification of the elements (see 5.3). He believed in a natural order of the elements. To approach these natural numbers as closely as possible, Newlands [20] divided the atomic weights by 2.3, thus obtaining a new series of atomic weights with Na = 10 as standard. According to Newlands [23], this standard could be varied (e.g. Cl = 15 or C = 5). His elucidation of this point was based on some modifications of the periodic system (Fig. 25, p. 109). This procedure, of course, led to a large number of vacant spaces in his system, and even a whole empty period. The order number of uranium moved up to 101. At the same time, the law of octaves was nullified. With Cl = 15 (strictly speaking 14.98), the periods have 10 elements. The last mention of Newlands [28] dates from 1890 and refers to the occasion on which he tried to correct Mendeleev on a point concerning his own system at the time of Mendeleev's Faraday Lecture (see p. 348).

16.4. Priority conflict between Mendeleev and Meyer

The longest battle over priority was fought between Mendeleev and Meyer. Their difficulties arose largely from the fact that Meyer published his first periodic system after Mendeleev's views had been officially stated. When Meyer published his article in 1870 (see 5.7), he knew of these views only from the brief report in the *Zeitschrift für Chemie*. But before Meyer presented his ideas in a periodical, he had already worked them out in 1868 into a system intended for the new edition of his *Moderne Theorien der Chemie*. Mendeleev, of course, was not aware of this when Meyer's publication reached him. In the fight for priority Meyer did not refer to this later system at all. After he had given the manuscript to Remelé in July 1868, he did not recall it until May 1893, when he read a paper on the periodic system before the *Deutsche Chemische Gesellschaft [33]*. We cannot altogether accept the contention that in 1864 Meyer had classified only elements with analogous properties and had not considered their interrelationships. The elements—even though only a fraction of those already known were included—occur in the order of increasing atomic weight, but Meyer put the main emphasis on valence. Both investigators had to make alterations in their subsequent systems, e.g. in the atomic weights of In, Ce and U. Mendeleev indeed had, as we have seen (5.8), a better idea of the consequences of his discovery than Meyer, although both had immediately accepted the periodic system as the basis of inorganic chemistry. The textbooks of both provide indisputable evidence of this.

In his first two publications of 1869, Mendeleev already related the atomic volume to the atomic weight, whereas Meyer did not expound his views on this point until 1870. Mendeleev *[70]* expressed his claim to priority as follows: "Obgleich ein Feind aller Prioritätsfragen, habe ich mich doch entschlossen, die niedergeschriebenen Bemerkungen zu machen, um so mehr, als mir die H.H. Gerstl, Meyer und theilweise Hr. Blomstrand die Priorität meines Systems streitig machen, gegen einander aber mit solchen Ansprüchen nicht auftreten, obschon solche der Zeit des Erscheinens obenerwähnter Abhandlungen nach eher gerechtfertigt wären. —Schon die Afzählung so verschiedener Ansprüche beweist an und für sich zur Genüge, dass meine Schlussfolgerungen den Aufgaben, welche sich obenerwähnte eminente Chemiker gestellt haben, entsprechen, ohne zugleich nur Wiederholungen ihrer Aussagen zu sein; ich glaube auch voraussetzen zu dürfen, dass nach genauer Bekanntschaft mit den von mir erhaltenen Ergebnissen man meinen Ideen die Selbständigkeit nicht absprechen wird."

The priority affair proper did not start until ten years later, Meyer having already in 1876 referred to the problem in a footnote of the third edition of his textbook *[34]*. The conflict broke out because of a letter Wurtz *[35]* wrote to the *Deutsche Chemische Gesellschaft* in 1880. In this letter Wurtz stated that the German translator of his book *La Théorie atomique [36]* had given too much credit to Meyer. According to Wurtz, there was no need to do so, since Meyer *[37]* himself ascribed the fundamental idea to Mendeleev. Meyer, however, had not expressed himself in the terms stated by Wurtz and the interpretation was also incorrect, because the fundamental idea was already known and had originated from Meyer himself. Meyer put the matter literally: "Die nachstehende Tabelle ist im Wesentlichen identisch mit der von Mendelejef gegebenen." In the second edition of his textbook (1872) Meyer credited Mendeleev only with the arrangement of *all* the elements, adding that the system of the Russian resembled his own system closely. Wurtz' conclusion, however, was quite different: Meyer had added important things, but the fundamental idea came from Mendeleev.

The only discrepancy we have been able to find between the original edition and Wurtz' book *[36]* is that on page 172 the German translator *[38]* has added the name of Lothar Meyer, which he placed before that of the Russian in the line on page 137 of the French edition, reading "M. Mendelejef a démontré que leurs variations sont une fonction périodique de leur poids atomique." What is of much greater importance to us is the fact that Wurtz mentioned De Chancourtois. Although Wurtz only noted that Mendeleev's work was analogous to that of De Chancourtois and made no reference to the literature, this statement is remarkable because it was the first to be made on this point.

In 1880 this rectification led to detailed claims of priority by both scientists. Meyer [39] returned to the history of the system, because he did not want "aus allzu grosser Bescheidenheit den Antheil, den ich an derselben genommen, der Vergessenheit anheim fallen zu lassen."

Meyer [39] defended himself against Mendeleev's contention [40, 44] that his 1864 system was a simple composition of groups of analogous elements. It included, according to the author himself, a tendency both towards an arrangement of the elements in the order of their atomic weight and the attainment of a periodicity in properties. The group Cu, Ag, Au was placed by Meyer at the end of his arrangement (Fig. 36, p. 126) because of the different valences of the elements of this group; he believed that they did not belong together. He could not classify the elements Zn, Cd and Hg in any other group of bivalent elements. In his opinion, the incorrectly determined atomic weights of Mo, Nb, V and Ta were responsible for the fact that these elements could not be given their correct places. After they had been properly determined, Meyer was able to include all of them in one system, but, before he could make this public, the report of Mendeleev's lecture had appeared. As regards the prediction of elements, Meyer gave priority to Mendeleev only if Newlands' claim to priority was not justified, which Meyer could not yet confirm.

In his turn, Meyer criticized Mendeleev's system (Fig. 39, p. 129) because it contained not one but three series of elements that were not of equal magnitude. This discontinuity was caused by seven elements whose atomic weights had, as we now know, been incorrectly determined. These were Er, Yt, In, Ce, La, Di and Th. If Mendeleev had wanted to create a simple system, he would have changed those atomic weights. On the contrary, Meyer claimed, he had made a one-series system (Fig. 38, p. 129) but these seven elements had not been included here either, with the exception of indium. From his graphic representation of the atomic volumes (Fig. 40, p. 130), Meyer drew the conclusion that the first long period began only with the third period. Mendeleev's division, however, began after the very first row of elements, which certainly did cause faulty placing of some rather poorly investigated metals. To explain the limitations of the earlier publication Meyer said: "Ich wäre in meiner Arbeit gern auf die Verschiedenheiten unserer Tafeln näher eingegangen; aber bei dem damals beschränkten und fest begrenzten Raume der Annalen durfte ich die Freundlichkeit der Redaktion, die mir ganz ausnahmsweise die Veröffentlichung einer keine neuen experimentellen Daten enthaltenden Abhandlung verstattete, nicht missbrauchen und musste mich der äussersten Kürze befleissigen. Ich sagte daher, meine Tafel sei *im Wesentlichen* (d.h. in der Anordnung nach der Grösse der Atomgewichte) *identisch mit der von Mendele-*

jeff gegebenen. Dies war vielleicht etwas zuviel Höflichkeit aber jedenfalls besser, als hätte ich mir zuviel Verdienst zugeschrieben."

Meyer pointed out that Mendeleev did not mention the improvements Meyer had added to his system. He concluded that, even without his own modest share in the development of the periodic system, i.e., the setting up of a simple series according to increasing atomic weight, as well as the discovery of the periodicity, Mendeleev's merit still remained very great. He hoped that the whole matter could now be considered settled, and concluded: "Es ist nicht leicht, gegen jemanden, der einem die eigenen Lieblingsgedanken unerwartet durchkreuzt, völlig objektiv gerecht zu bleiben."

To this defence by Meyer, which was really also an attack, Mendeleev *[26]* replied by sending his first original Russian publication of March 1869 to the editors of the *Berichte der deutschen chemischen Gesellschaft* (see *5.7* and *5.8*). Furthermore, Mendeleev reprinted the official report of the meeting of scientists held on 23rd August 1869, at which he had spoken on the atomic volume of the elements. Both papers had already appeared in December 1869, before Meyer's publication. Mendeleev assumed that Meyer had read only a report of these publications and not the original texts. Mendeleev then indicated, by means of many quotations from his first work, that he had copied nothing from Meyer, including the divisions in the system, which, he understood, Meyer stated he had. We have seen, however, that this was not what Meyer had meant. Mendeleev remarked that the expression *periodicity of the properties* had been created by him. Meyer repeated only, according to Mendeleev, what had already been found by Mendeleev himself. The discovery of gallium in 1875 elicited from him the words: "Ich gestehe, dass ich einem so glänzenden Beweis des periodischen Gesetzes, wie diese Entdeckung des Hrn Lecoq de Boisbaudran, bei Lebzeiten nicht erwartet habe."

In March 1869, Mendeleev believed that U (116?) was a homologue of B and Al. A year later he assigned to indium the place he had earlier given to uranium. Meyer did so too, at about the same time. Nor did Meyer precede Mendeleev in the correction of the atomic weights of Ce, U and Y. When Meyer proposed doubling the atomic weight of uranium, Mendeleev had already altered this magnitude.

Mendeleev concluded that, if Meyer had discovered a periodic system in 1864, he could not have failed to see that the ratio between B = 11 and Al = 27 was the same as between C = 12 and Si = 28. The criticism was indeed justified that in 1864 Meyer had classified only according to valence and not to atomic weight and had also made use of wrong valences, e.g., quadrivalence or sexivalence for aluminium.

In his defence, Mendeleev summarized his work published in 1869 in ten points. He claimed that his system not only:

(1) *die chemische Aehnlichkeit der Elemente ausdrückt, sondern*
(2) *auch der Eintheilung der Elemente in Metalle und Metalloïde entspricht;*
(3) *ihre Werthigkeit unterscheidet;*
(4) *ähnliche Elemente verschiedener Gruppen zusammenstellt (z.B. B, C, Si, Al, Ti);*

(5) *die der Homologie ähnliche Uebereinstimmung der Elemente, auf welche viele Chemiker hinwiesen, erklärt;*
(6) *Wasserstoff als ein typisches Element ausscheidet, was auch die gegenwärtige Wissenschaft anerkennt;*
(7) *die verbreitetsten und in der Natur sich gegenseitig begleitenden Elemente neben einanderstellt;*
(8) *auf die Mangelhaftigkeit Prout's Hypothese und*
(9) *auf die Beziehungen zwischen den Elementen gemäss ihrer gegenseitigen Verwandschaft hinweist. Ausserdem weist*
(10) *ein Vergleich der specifischen Gewichte und specifischen Volumina der verschiedenen Reihen angehörigen Elemente bis zu einem gewissen Grade auf die Naturgemässheit des Systems auch in dieser Beziehung hin.*

Point 10 was intended to show that Mendeleev had been earlier than Meyer in drawing a conclusion as to the specific volume. He wrote on this point: "Als mir (im Anfang 1870) aus Moskau die Correctur meiner Abhandlung *Über Atomvolum der Elemente* zugeschickt wurde, setzte ich am Schlusse derselben folgende Anmerkung, aus welcher zu ersehen ist, wie wenig ich geneigt bin, Prioritätsfragen selbst anzuregen. Seite 71 (Anmerkung): *Das hier erörterte habe ich auf der Versammlung im August 1869 mitgetheilt. 1870 erschien in Liebig's Annalen (nachdem diese Abhandlung zum Drucken abgeschickt war) ein denselben Gegenstand behandelnder Aufsatz des Hrn L. Meyer. Die Schlussfolgerungen des Hrn Meyer gründen sich auf die Zulassung des von mir gegebenen Systems der Elemente und stimmen mit den von mir hinsichtlich der Atomvolumen gezogenen überein. Die Schlussfolgerungen haben durch die der Abhandlung beigegebene graphische Darstellung an Klarheit gewonnen.* Mit dem Niederschreiben dieser Nachschrift will ich nicht die Frage bezüglich der wissenschaftlichen Priorität anregen (meiner Ansicht nach haben diese Fragen oft gar kein wissenschaftliches Interesse), sondern nur die Aufmerksamkeit auf die, dieser Abhandlung des Hrn Meyer beigegebene Tafel, als auf ein Mittel, das bei der Aufkärung der complicirten Beziehungen, auf welche in der vorhergehende Zeilen hingewiesen wurde, behülflich sein könnte, lenken."

Mendeleev claimed further:

(1) *To have given expression in August, 1869, to all the ideas which at the time of writing formed the basis of the periodic system;*
(2) *that Meyer had not been prior to him and had added nothing new;*
(3) *that Meyer was the first German to set up the outward form of the periodic system, but*
(4) *that Meyer did not comprehend the deeper meaning. He had predicted no atomic weights and did not alter any.*

It is clear from this contest for priority that the two discoveries were made at about the same time, and that one of the discoverers had indeed been the first to publish, but might not have been the first to conceive the idea. It is in fact of little use to discuss this point; they must both be given the credit due to them.

Mendeleev was of the opinion that, if anyone had introduced a new factor, it had been Carnelley [41] with his discovery of the periodicity in magnetic properties. He did acknowledge a debt to Lenssen and Dumas, as he had pointed out before.

Mendeleev entered into this detailed discussion only because Meyer had written that Mendeleev had set up the system "ohne ihn zu nennen" and because Meyer personally sent him a reprint. He ended with the words, recognizably inspired by the Scriptures: "Auf einen Brief hätte ich mit einem Brief geantwortet, auf die Abhandlung antworte ich mit einer Abhandlung, auf Tafeln mit Tafeln, auf 1870 mit 1869, auf Dezember (Meyer's publication was dated December 1869) mit März und August, weil ich die von einem so berühmten Gelehrten, wie L. Meyer, gemachten Ansprüche für nichts anderes als einen Irrtum halten kann."

This reply was quite contrary to Meyer's expectations [42]. He believed he had given an exposition of the facts with complete detachment. He had not known of the abstract of Mendeleev's publication and claimed credit only for ideas not originated by Mendeleev. Meyer thought it was asking too much ".....dass wir deutschen Chemiker, ausser den in germanischen und romanischen auch noch die in slavischen Sprachen erscheinenden Abhandlungen lesen und die deutschen Berichte über ihren Inhalt auf ihre Genauheit prüfen sollen."

Meyer persisted in the opinion that Mendeleev should have mentioned him. He himself mentioned Mendeleev in all the editions of his textbook. In 1872 he credited Mendeleev with the honour of having been the first investigator to classify all the elements. In his view, the whole conflict would never have arisen if Wurtz had not allowed his letter to be printed [33]. These publications became known in England through abstracts [29, 43, 45]. In France, attention was once again drawn to Mendeleev by Krakau, the correspondent of the Russian Chemical Society, who gave an abstract of Mendeleev's lectures [46]. These lectures dealt with the element scandium, just discovered by Nilson and already predicted by Mendeleev. The system found in his French [47] and English [48, 49] publications of 1879 was reproduced there.

The foregoing considerations indicate that Mendeleev did not tend to overestimate the value of the contributions of other scientists. He justified this particularly by saying, as he also stated later [50], that they predicted no elements and therefore gave no evidence of having comprehended the broad basis of the new natural law. Even Meyer [37] said: ".........es würde voreilig sein, auf unsichere Anhaltspunkte hin eine Änderung der bisher angenommenen Atomgewichte vorzunehmen." In this Meyer had justice on his side, because, although some rearrangements of the elements had been made correctly, others had not.

16.5. Further development of the periodic system by Meyer and Mendeleev

Lothar Meyer took a large share in the development of the periodic law, partly because he had to prepare new editions of his *Die modernen Theorien der Chemie [51]* in which the periodic system was always kept up to date. There was also his professorship of chemistry at the University of Tübingen, where sixty graduate students took a doctor's degree under him. When the periodic system presented difficulties, Meyer attempted to solve them, as in the incorporation of beryllium (see *12.6*) and of some rare earths (see *10.4*).

After the years in which the conflict of priority reached its culmination, Meyer's publications on the periodic system decreased in number. Only just before his death did he publish again *[52]*. In 1893 he emphasized the value of the periodic system of elements for students of inorganic chemistry and gave many examples *[33]*. His colleague Seubert stressed Meyer's part in the development of the periodic system by republishing the first papers of Meyer and Mendeleev in 1895, the year of Meyer's death *[53]*.

Mendeleev's share in the development of the periodic system was very important. His special attention to the application of the system to inorganic chemistry in the several editions of his textbook *[54]*, which was also translated into English *[55]*, German *[56]* and French *[57]*, established the book as a permanent contribution to fundamental chemistry. Mendeleev owes his authority in large measure to the success of his predictions (see *7.6*).

His intense interest in the periodic system persisted throughout his life, starting with his contributions to the discovery and continued in his many discussions on priority matters, his attempts to classify the noble gases (1895; see *10.7*), and his highly influential *Faraday Lecture* of 1889. This *Lecture [27]*, which Mendeleev was prevented from delivering himself by the illness of one of his children on the appointed day, May 23, 1889, was read by Henry Edward Armstrong, one of the editors of the *Journal of the Chemical Society*. Soon after, on June 4, Mendeleev himself read a paper *[58, 59]* in which he made his first mention of Béguyer de Chancourtois. He also discussed the parts played by Rydberg *[60]*, Flavitsky, Crookes, Haughton, Mills and Tchitcherine in the development of the periodic system (mostly in the field of the relationships between the atomic weights and other properties of the elements), which he stated again in the English edition of his textbook *[61]* published two years later. Mendeleev evinced the greatest esteem for the work of Rydberg, Carnelley, Flavitsky (see p. 250) and Tchitcherine *[62]*. The latter assumed the existence of a primary substance, but Mendeleev was highly sceptical of what he called the metaphysical speculations of the ancient Greeks. He considered the ideas of the Pythagoreans to be utopian and denied that the periodic system of elements was historically connected with these "relics of the torments of classical thought" *[27, 63]*. He therefore objected vehemently to Berthelot *[64]* when the latter mentioned the periodic system, Prout's hypothesis, alchemy and Greek speculations in the same breath *[65, 66]*. In the 1905 translation of his textbook he said "......it appears to me that the whole question of primary matter belongs to the province of fancy and not of the science", in spite of which he shortly afterwards formulated his ether theory *[67]* (see also p. 228 and *15.2*) Mendeleev also gave particular attention to all kinds of numerical relationships and regularities in connection with the magnetic properties of the elements. Before that time Mendeleev had already spoken highly of Carnelley's work (see p. 347).

Nevertheless, Mendeleev remained critical of graphical representations attempting to illustrate a relation between the atomic weights of the elements and several other properties. He also rejected a continuous function indicating this relation. His principal objection was that the number of groups of homologous elements was not limited and there must be continuous transitions in the properties of the elements. This also accounts for the fact that Mendeleev never appreciated De Chancourtois' work (see *16.7*). Mendeleev apparently did not consider the possibility of a new group of elements in his periodic system, as was later required for the noble gases.

Mendeleev's last mention of the periodic system dates from 1904, a few years before his death. In contrast to his earlier predictions, he then risked a few far-reaching and, unfortunately, not too well-grounded statements about the world ether, induced by his aversion to the concept of subatomic particles and, also, by his preference for unity and simplicity *[68]* (see *7.10, 15.2*).

16.6. Priority concerning Odling

Although Odling himself did not claim priority[5], he was nevertheless drawn into the conflict. It was again Gerstl *[69]*, the London correspondent of the *Deutsche Chemische Gesellschaft* who took up the cause of an Englishman. In 1871 he pointed to Odling's system as given by Watts in his *Dictionary of Chemistry* (see *5.5*), which Gerstl thought bore close resemblance to that of Mendeleev.

Mendeleev *[70]* was unable to obtain a copy of this system or the one Odling had published four years earlier, although he was aware of their existence and knew that Odling had, as he put it, accepted several systems one after another. Mendeleev's first acquaintance with Odling dates from April 1869, i.e. after Mendeleev had read his first paper *[71]* on his system to the Russian Chemical Society in March of 1869. At the April meeting Savshchenkov, the Russian translator of Odling's *A Course of Practical Chemistry* (1867 ed.), called his attention to Odling's system, but added the comment that the composer of the system had not grasped its meaning, upon which Mendeleev remarked that if Odling had seen its theoretical implications he would no doubt have discussed them.

Odling's work of 1857 was better known to Mendeleev, although he classed it with

---

[5] Odling *[30]* never thought of his own part as important. As late as 1877, he still gave Newlands far too much credit.

the contributions of Gladstone, Cooke, Kremers, Lenssen, Pettenkofer, Dumas and other investigators of the eighteen-fifties. Mendeleev apparently considered Odling's subsequent contributions to be extensions to the series of analogous elements, e.g., the groups B, Si, Ti; Be, Y, Th; Al, Zr, Ce, U; and Hg, Pb, Ag, and saw no periodic system in them, in spite of the fact that Odling qualified his system as natural, as Mendeleev had also done. If Odling had discovered a kind of system, then, according to Mendeleev, he had abandoned it by the time he wrote *A Manual of Chemistry [72]*. We have already concluded above (p. 115) that Odling's 1868 system was regressive rather than a step forward as compared with that of 1864.

Mendeleev at that time knew of Odling's *A Course of Practical Chemistry [73]*, in which the elements were arranged according to their atomic weight but without elucidation, and his *Outlines of Chemistry [74]*. Mendeleev's criticism concentrated on the elements S, Fe, Mn, Ca and Hg, which Odling regarded as belonging to diads, i.e., the bivalent elements. Mendeleev was of the opinion that what he called Gerstl's misapprehension could be reduced to ignorance of his Russian publication and of the similar designation of the two systems. Mendeleev certainly lost sight of the fact that Odling had published his first, and also best, system five years before Mendeleev's appeared.

16.7. The claims to priority of Béguyer de Chancourtois [75, 76]

There is not the smallest doubt as to the priority of De Chancourtois. Although he himself never felt any need to take part in the fierce fight for priority, there can be no question that his share in the discovery of the periodic system of elements was not only an important one but even constituted the very first contribution. De Chancourtois never again referred to his discovery, at least in public, although he was certainly entitled to point to his work, the more so because in the first three decades after its publication hardly anyone[6] knew of the existence of the *Vis tellurique*. Only Charles Joseph Sainte-Claire Deville [78], de Chancourtois' colleague at the Collège de France, pointed out in 1862 that in 1855 he had set up a system of simple substances that must be regarded as an anticipation of the *Vis tellurique [79]*. Since Deville's system was not based on the equivalent weights, his claim for priority cannot be honoured. De Chancourtois was mentioned only incidentally by Wurtz [36] as late as 1880, when the fight for priority arose between Meyer and Mendeleev. As a fellow-countryman, Wurtz might certainly have been expected to claim priority. De Chancourtois was also mentioned by Berthelot [80] in 1885, in an unexpected place, namely in his work on alchemy, and by Mendeleev [27] in his *Faraday Lecture* given in 1889, three years after the death of De Chancourtois. In the English version of his textbook [61] published in 1891, he also devoted a few lines to De Chancourtois' system. Mendeleev regarded

---

[6] Like most communications with a chemical purport, De Chancourtois' work was included in the *Jahresbericht* published by Kopp and Will [77].

this system, like those of Huth and Baumhauer, as a spiral system with many draw-backs, including the unlimited number of groups of elements. Since Mendeleev did not recognize Newlands and Meyer as his fellow-discoverers, he did not credit De Chancourtois with the discovery of the periodic system because he, too, failed to predict elements and did not recognize the system as natural [81]. This last contention was certainly not justified.

Mendeleev's lecture in London explains the fact that the first claim for priority came from England, having been made by Hartog[7] [82] in 1889. Hartog may also have had a partiality for French chemists because he had studied in France. From 1882 to 1884 he was at the Sorbonne, where Wurtz was one of his teachers. From 1885 to 1889 he was a student of Berthelot's at the Collège de France. As a matter of fact, these scientists both knew De Chancourtois' work, as we have seen.

As Frenchmen, the chemist Lecoq de Boisbaudran and the mineralogist and geologist De Lapparent[8] [83, 84] claimed priority for De Chancourtois. This scientist had indeed tried, at the time of his discovery, to make this public as best he could. He was really compelled to do so, because the *Comptes Rendus* had not included the diagram accompanying his report to the Académie Française (see 5.2). De Chancourtois therefore published his system—in a rather inconvenient form—in a pamphlet, but this publication was not widely circulated.

Only in 1891, two years after reference had been made in an English periodical to De Chancourtois' priority and nearly 30 years after its publication, was De Chancourtois' work properly publicized in France itself. Not only Hartog but also De Boisbaudran and De Lapparent published a simplified *Vis tellurique* to illustrate De Chancourtois' systematization. These last scientists first presented their claims after they had become acquainted with Newlands' attempts to obtain recognition in 1884. Lecoq de Boisbaudran had become aware of the periodic system as early as 1875, through his discovery of the element gallium, which had been predicted by Mendeleev, but at that time he was not acquainted with the *Vis tellurique*. Crookes [85], then editor of the *Chemical News* to which the two French scientists had sent in their claim for priority for De Chancourtois, thought the latter's system of little importance and gave them his opinion accordingly. He also believed that they should not have been so concerned about the general ignorance of De Chancourtois' work, since they themselves had only just discovered the *Vis tellurique*.

That De Chancourtois himself continued to take great interest in the field, is shown by the fact that just before his death in 1886 he made inquiries about the great new developments in the systematization of chemical elements. We know from Fuchs [86], who wrote his obituary, that De Chancourtois also wanted to classify in his sytem the

---

[7] Philip Joseph Hartog (1864–1947) became senior lecturer in chemistry at Owens College of the Victoria University of Manchester in 1892.
[8] Albert Auguste de Lapparent (1839–1908) was appointed professor of mineralogy and geology at the Catholic Institution at Paris in 1875.

substances found in the most widely distributed mineral veins, according to their atomic and equivalent weight.

It should be observed here once again that we must regard De Chancourtois' *Vis tellurique* as the first periodic system of elements. A glance at his work shows us that the homologous elements are not always in a perfectly vertical line, one below the other. To understand the deviations we must realize that this system is a graphic representation, although based on the same principle as that used by such authors as Newlands, Meyer and Mendeleev. The form these investigators gave their systems made it possible to find the elements in a vertical line one below the other.

16.8. Conclusion

In the foregoing we have attempted to show that six discoverers independently constructed a periodic system of elements during the period from 1862 to 1869. We cannot agree with those who recognize only one or two discoverers. However tempting it may be to defend the priority rights of a fellow-countryman, the rights of others must be seen objectively. We would refer in the first place to the extensive claims for priority made by Kedrov in favour of Mendeleev. We need not list all similar claims, but one case is worth mentioning. Crookes *[85]*, who himself had contributed to the further development of the periodic system, said about Newlands and De Chancourtois (in those days Hinrichs and Odling were not known at all), "We may be permitted to doubt whether they can be fairly considered as the germ of the Periodic Law." Seubert *[87]* went even further: "einen wirklichen Anteil an der Begründung des Periodischen Systems haben sie nicht gehabt."

When Rudorf *[88]*, one of the historians in this field, remarks that Meyer and Mendeleev could have set up their system even without the work of Newlands and De Chancourtois, we fully agree. They did not even know of either of them. But this does not detract from the value of the discovery by Newlands and De Chancourtois.

Although recognition has been given to the work of Meyer[9] and Mendeleev[10], and

---

[9] E.g., in 1930, at the centenary celebrations of Meyer's birth *[89]*. On the facade of the house at Varel (near Oldenburg) where he was born, a plaque was placed bearing the inscription:

> "Am 19 August 1830 wurde
> Lothar Meyer
> Professor der Chemie
> in diesem Hause geboren.
> Er ist der Begründer des periodischen
> Systems der Elemente."

[10] E.g., in 1934, at the centenary celebrations of Mendeleev's birth, also held in England *[90]* and France *[91]*. A monument in honour of Mendeleev's discovery was unveiled in the Meteorological Bureau of the U.S.S.R. Leningrad. This memorial (Fig. 139), which included the representation of a large periodic system of elements, was placed next to the Main Chamber of the Department of Weights and Measures *[92]*, of which Mendeleev was director from 1893 until his death.

Fig. 139   Monument erected near the Main Chamber of Weights and Measures at Leningrad in 1934

official honours were bestowed on Meyer, Mendeleev and Newlands[11], hardly any attention was paid to De Chancourtois' contribution, even in his native country. Not until a century after the discovery of his *Vis tellurique* was he commemorated as one of the discoverers of the periodic system by the present writer *[75]*, also in France *[76]*.

References

1 J. W. VAN SPRONSEN, *Chem. Weekblad*, 47 (1951) 807.

2 H. C. BROWN, *J. Chem. Educ.*, 36 (1959) 104.

3 A. S. COUPER, *Compt. Rend.*, 46 (1858) 1157; *Phil. Mag.* [4], 16 (1858) 104; see also *Alembic Club Reprints* No. 21, Edinburgh 1933; 2nd ed. 1953.

4 J. W. VAN SPRONSEN, *Chem. Weekblad*, 59 (1963) 665.

5 S. ARIJA, *Erforschung der Atome und Mendelejews periodisches System der chemischen Elemente* Markneukirchen, no date.

6 H. WILDE, *Chem. News*, 72 (1895) 291.

7 H. WILDE, *Chem. News*, 73 (1896) 35.

8 H. WILDE, *Manchester Lit. Phil. Soc.* [4], 9 (1895) 67.

9 J. H. GLADSTONE, *Chem. News*, 72 (1895) 223.

10 J. H. GLADSTONE, *Nature*, 52 (1895) 537.

11 J. H. GLADSTONE, *Chem. News*, 72 (1895) 267.

[11] Although Meyer and Mendeleev received the Davy Medal from the Royal Society in 1882, Newlands did not receive this award until 1887, at the proposal of Frankland.

12 J. H. GLADSTONE, *Chem. News*, 72 (1895) 305.

13 M. PETTENKOFER, *Ann.*, 105 (1858) 187.

14 J. LIEBIG and H. KOPP, *Jahresber. Fortschr. Chem.*, 4 (1850) 292.

15 See also J. W. VAN SPRONSEN, *Chymia*, 11 (1966) 125.

16 J. A. R. NEWLANDS, *Chem. News*, 13 (1866) 113.

17 J. A. R. NEWLANDS, *Chem. News*, 27 (1873) 318.

18 J. A. R. NEWLANDS, *On the Discovery of the periodic Law and on Relations among Atomic Weights*, London 1884.

19 J. A. R. NEWLANDS, *Chem. News*, 26 (1872) 19.

20 J. A. R. NEWLANDS, *ibid.*, 25 (1872) 252.

21 J. A. R. NEWLANDS, *ibid.*, 32 (1875) 21.

22 J. A. R. NEWLANDS, *ibid.*, 38 (1878) 106.

23 J. A. R. NEWLANDS, *ibid.*, 37 (1878) 255.

24 J. A. R. NEWLANDS, *Ber.*, 17 (1884) 1145.

25 J. A. R. NEWLANDS, *ibid.*, 11 (1878) 516.

26 D. MENDELEJEFF, *ibid.*, 13 (1880) 1796.

27 D. MENDELEJEF, *J. Chem. Soc.*, 55 (1889) 634.

28 J. A. R. NEWLANDS, *Chem. News*, 11 (1890) 136.

29 D. I. MENDELEJEF, *Chem. News*, 43 (1881) 15.

30 W. ODLING, *Pharm. J.* [3], 8 (1877/78) 144.

31 T. CARNELLEY, *Phil. Mag.* [5], 8 (1879) 305.

32 J. A. R. NEWLANDS, *Zur Geschichte des periodischen Gesetzes*, London 1884.

33 L. MEYER, *Ber.*, 26 (1893) 1230.

34 L. MEYER, *Die modernen Theorien der Chemie*, 3rd ed., Breslau (Wroclaw) 1879, p. 290.

35 A. WURTZ, *Ber.*, 13 (1880) 6.

36 A. WURTZ, *La Théorie atomique*, Paris 1879.

37 L. MEYER, *Ann.*, Suppl. VII (1870) 354.

38 A. WURTZ, *Die atomistische Theorie*, Leipzig 1879.

39 L. MEYER, *Ber.*, 13 (1880) 259.

40 D. MENDELEEV, *Proceedings 2nd Meeting of Scientists* (Russ.), 23 Aug. 1869, p. 62.

41 T. CARNELLEY, *Ber.*, 12 (1879) 1958.

42 L. MEYER, *Ber.*, 13 (1880) 2043.

43 L. MEYER, *Chem. News*, 41 (1880) 203.

44 B. N. MENSCHUTKIN, *Nature*, 33 (1934) 946.

45 L. MEYER, *Chem. News*, 43 (1881) 15.

46 D. MENDELEJEF, *Bull. Soc. Chim. France* [2], 38 (1882) 139.

47 D. MENDELEJEF, *Mon. Sci. Docteur Quesneville* [3], 9 (1879) 691.

48 D. MENDELEJEF, *Chem. News*, 40 (1879) 231, 243, 255, 279.

49 D. MENDELEJEF, *Chem. News*, 41 (1880) 2, 27, 39, 49, 61, 71, 83, 93, 106, 113, 125.

50 D. MENDELÉEFF, *Principes de Chimie*, Paris 1899, Vol. II, pp. 436 ff.

51 L. MEYER, *Die modernen Theorien der Chemie*, 2nd ed., Breslau (Wroclaw) 1872; 3rd ed. Breslau (Wroclaw) 1876; 4th ed. Breslau (Wroclaw) 1883; 5th ed. Breslau (Wroclaw) 1884.

52 See e.g. L. MEYER, *Grundzüge der theoretischen Chemie*, Leipzig 1890; 2nd ed., Leipzig 1893; *Die Atome und ihre Eigenschaften*, 6th ed., Breslau (Wroclaw) 1896.

53 *Das natürliche System der chemischen Elemente, Abhandlungen von Lothar Meyer und D. Mendelejeff*, ed. by Karl Seubert, Ostwald's Klassiker No. 68, Leipzig 1895.

54 D. I. MENDELEEV, *Osnovy Khimii*, St Petersburg, 1st ed., Vol. I, 1869, Vol. II, 1871; 2nd ed., Vol. I, 1872, Vol. II, 1873; 3rd ed., 1877; 4th ed., Vol. I, 1881, Vol. II, 1889; 5th ed., 1889; 6th ed., Vol. I, 1894, Vol. II, 1895; 7th ed., 1903; 8th ed., Vol. I, 1905, Vol. II, 1906; 9th ed., (posthumous) Vol. I, 1927, Vol. II, 1928.

55 D. MENDELÉEFF, *Principles of Chemistry*, London 1891; 2nd ed., 1897; 3rd ed., 1905.

56 D. MENDELEJEFF, *Grundlagen der Chemie*, St. Petersburg–Leipzig 1892.

57 D. MENDELÉEFF, *Principes de Chimie*, Paris 1899.

58 N. N., *The Chemist and Druggist*, 34 (1889) 786.

59 D. MENDELEEV, *Zhur. Russ. Fiz. Khim. Obshchestva*, 21 (1889) 233.

60 J. R. RYDBERG, *Bihang Kongl. Svenska Vetenskaps Akad. Handlingar*, 10 (1885) No. 2; see also *Kongl. Svenska Vetenskapsakad. Handl.*, 23 (1890) No. 11, e.g. pp. 9, 140.

61 D. MENDELÉEFF, *Principles of Chemistry*, London 1891, Vol. I, pp. 19 ff.

62 B. TCHITCHÉRINE, *Bull. Soc. Imp. Naturalistes Moscou* [2], 4 (1890) 42.

63 D. MENDELÉEFF, *Principles of Chemistry*, London 1891, Vol. II p. 144.

64 M. BERTHELOT, *Les Origines de l'Alchimie*, Paris 1885, p. 312.

65 See also F. A. PANETH, *Travaux du Congrès Jubilaire Mendéléev*, Leningrad 1936; *Scientia* 58 (1935) 219, 272; H. DINGLE and G. R. MARTIN, *Chemistry and Beyond, A Selection from the Writings of the late Professor F. A. Paneth*, New York–Sydney 1964, p. 53.

66 For a biography of Mendeleev see e.g. P. WALDEN, *Ber.*, 41 (1908) 4719.

67 D. MENDELÉEFF, *Principles of Chemistry*, 3rd ed., London 1905, Vol. II pp. 33, 460.

68 R. KARGON, *J. Chem. Educ.*, 42 (1965) 388.

69 R. GERSTL, *Ber.*, 4 (1871) 132, 484, 533.

70 D. MENDELEJEFF, *Ber.*, 4 (1871) 348.

71 D. MENDELEEV, *Zhur. Russ. Khim. Obshchestva*, 1 (1869) 60.

72 W. ODLING, *A Manual of Chemistry, descriptive and theoretical*, Vol. I, London 1861; Germ. transl., Erlangen 1863; French transl., Paris 1868; Russ. transl., St. Petersburg 1863.

73 W. ODLING, *A Course of Practical Chemistry*, London 1868, p. 226; *Course de Chimie pratique*, Paris 1869, p. 4.

74 W. ODLING, *Outlines of Chemistry*, London 1870.

75 J. W. VAN SPRONSEN, *Chem. Weekblad*, 58 (1962) 576.

76 J. W. VAN SPRONSEN, *L'Histoire de la Découverte du Système périodique des Eléments chimiques et l'Apport de Béguyer de Chancourtois*, Paris 1965.

77 H. KOPP and H. WILL, *Jahresber. Fortschr. Chem.*, 15 (1862) 6; 16 (1863) 14.

78 C. J. SAINTE-CLAIRE DEVILLE, *Compt. Rend.*, 54 (1862) 782.

79 C. J. SAINTE-CLAIRE DEVILLE, *Compt. Rend.*, 40 (1855) 177.

80 M. BERTHELOT, *Les Origines de l'Alchimie*, Paris 1885, p. 302.

81 D. MENDELÉEFF, *Principles of Chemistry*, 3rd ed. London 1905, Vol. II pp. 18, 22, 28.

82 P. J. HARTOG, *Nature*, 41 (1889) 186.

83 P. E. LECOQ DE BOISBAUDRAN and A. DE LAPPARENT, *Compt. Rend.*, 112 (1891) 77.

84 P. E. LECOQ DE BOISBAUDRAN and A. DE LAPPARENT, *Chem. News*, 63 (1891) 51.

85 W. CROOKES, *Chem. News*, 63 (1891) 51.

86 E. FUCHS, *Ann. Mines* [8], *Mémoires*, 9 (1887) 505; *Notice nécrologique sur M. A. E. Béguyer de Chancourtois*, Paris 1887.

87 *Das natürliche System der chemischen Elemente, Abhandlungen von Lothar Meyer und D. Mendelejeff*, ed. by K. Seubert, Ostwald's Klassiker No. 68, Leipzig 1895, p. 122.

88 G. RUDORF, *Das periodische System, seine Geschichte und Bedeutung für die chemische Systematik* Hamburg–Leipzig 1904, p. 30.

89 See e.g. F. PANETH, *Naturwissenschaften*, 18 (1930) 964.

90 LORD RUTHERFORD, *J. Chem. Soc.*, (1934) 635.

91 G. URBAIN, *Rev. Sci.*, 72 (1934) 657.

92 D. Q. POSIN, *Mendeleyev, The Story of a great Scientist*, New York 1948.

Name Index

Subject Index